Standards of
Mouse Model Phenotyping

Edited by
Martin Hrabé de Angelis,
Pierre Chambon,
and Steve Brown

Standards of Mouse Model Phenotyping

Edited by
Martin Hrabé de Angelis, Pierre Chambon,
and Steve Brown

WILEY-VCH Verlag GmbH & Co. KGaA

The Editors

Prof. Dr. Martin Hrabé de Angelis
GSF National Research Center
for Environment and Health
Institute of Experimental Genetics
Ingolstädter Landstraße 1
85764 Neuherberg
Germany

Prof. Dr. Pierre Chambon
Université Louis Pasteur
Institute of Genetics
67404 Illkirch Cedex
France

Prof. Dr. Steve Brown
Medical Research Council
MRC Mammalian Genetics Unit
Harwell, Oxfordshire OX11 ORD
UK

Library of Congress Card No.: applied for
British Library Cataloguing-in-Publication Data
A catalogue record for this book is available from the British Library.
Bibliographic information published by Die Deutsche Bibliothek
Die Deutsche Bibliothek lists this publication in the Deutsche Nationalbibliografie; detailed bibliographic data is available in the Internet at <http://dnb.ddb.de>.

Typesetting: primustype Hurler GmbH, Notzingen
Printing: betz-Druck GmbH, Darmstadt
Binding: J. Schäffer GmbH, Grünstadt
Cover: Grafik-Design Schulz, Fußgönheim

Printed in the Federal Republic of Germany

Printed on acid-free paper

ISBN-13: 978-3-527-31031-9
ISBN-10: 3-527-31031-2

Foreword

Mutations with an effect on coat color or behavior were recorded by mouse fanciers well before the science of Genetics was established. They were curiosities, occurring by chance at very low frequency, and their main advantage was to make the mouse an even more interesting pet animal with many phenotypic variations that could be produced in different combinations by breeding.

Over the last century, mouse geneticists, especially the most sagacious, collated and studied a great variety of mutations, many of them exhibiting neuromuscular, eye or skeletal defects and abnormal fur or coat colors. These mutations were discovered either spontaneously, in the nucleus of inbred strains, or as side-products of the many experiments that were undertaken to assess the genetic risks associated with the use of nuclear radiations. All have been extremely helpful, contributing for example, to the development of the genetic map. A few of these mutants have also been used as animal models for human diseases while others, such as the nude or SCID mice, which can both permanently accept a variety of grafts including xenografts, were and still are used as tools for research in immunology or oncology.

More recently, by taking advantage of the exceptional mutagenic activity of ethyl-nitroso-urea, programs aimed at the mass-production of new mutations have been undertaken in several laboratories worldwide. With such on-going programs, mutations are no longer rare events occurring spontaneously, but random hits still occur in the mouse genome. Their phenotypes can be studied in great detail but the characterization of the molecular defect will necessarily follow (forward genetics).

With the development of highly efficient techniques of genetic engineering in pluripotent embryonal stem cells (ES cells) and the availability of a nearly complete sequence of the mouse genome, the situation has changed dramatically. Here again, large projects involving a network of laboratories have been undertaken with the aim of producing a very large number of knockout mutations, ideally one in every gene of the mouse genome, and it is likely that these projects will reach their goal within the next five years. Geneticists will then have at their disposal a collection of ES cell libraries, with up to 20,000 genes potentially inactivated. Any genetic defect can then be accurately identified but the main problem will then be to describe the phenotype of the mutant mice as precisely and comprehensively as possible.

The reason for developing such projects is quite clear: producing many mutations and phenotyping them very precisely is the best and most logical way of as-

Standards of Mouse Model Phenotyping. Edited by Martin Hrabé de Angelis, Pierre Chambon, and Steve Brown

ISBN: 3-527-31031-2

sessing the function of the genes in the mouse genome. Indeed, when a gene becomes non-functional after a mutation has occurred, careful comparison of the mutant and normal phenotype in addition to taking into account the molecular defect generated by the mutational event, is an excellent (not to say the best) method of assessing the function(s) of the gene in question. In short, the production of new mutations and the precise phenotyping of the mutant genotype are the two sides of the same coin.

Geneticists have worked out many strategies for the efficient production of new mutations in the mouse genome using either chemical mutagenesis or gene trapping or gene targeting but until recently phenotyping was not their main concern and as a result has received less attention. In other words, whilst it was technically possible to inactivate almost any gene in the mouse genome, there was limited scientific interest in the subsequent analysis of the resulting phenotype as the relevance of any data thus obtained was thought to be questionable. Indeed, many knockouts produced over the last 10 years in genes which were thought to be extremely important, were reported to be phenotypically "normal" to the great surprise of their creators!

There is a wide range of difficulties associated with the process of phenotyping. Although it is easy to detect and describe a cerebellar defect in the mouse or a disorder which leads to the animal losing its fur after a few days, it is more difficult to detect an inner ear defect with a relatively late onset or a very subtle degenerative disorder of the retina and it is virtually impossible to detect the phenotype of certain mutations in genes involved in the innate mechanisms of defense unless a specific challenge test is carried out to reveal the mutation. The situation is even more complex when modifier genes in the genetic background interact with the pathology of the mutant allele.

This book, edited by Professor Martin M. Hrabe de Angelis in cooperation with Steve Brown and Pierre Chambon, is original and the timing of its publication is opportune. It consists of 13 chapters, all written by expert scientists who are members of the EUMORPHIA consortium and work in different research institutes across Europe. This volume describes in detail a series of screens known as EMPReSS (European Mouse Phenotyping Resource for Standardized Screens) that encompass more than 150 standard operating procedures (SOPs) covering all the main body systems of the mouse.

This book will definitely be of major interest to those creating or using a variety of mutant mice and the authors must be warmly thanked for this initiative.

April 2006 Jean Louis Guénet

Table of Contents

Standards of Mouse Model Phenotyping. Edited by Martin Hrabé de Angelis, Pierre Chambon, and Steve Brown

ISBN: 3-527-31031-2

Preface

The speed with which information regarding mammalian genomes has accumulated over the last few years is remarkable. Yet, despite this wealth of information, its immediate use in the diagnosis and therapy of human diseases is limited since only a small fraction of mutations causing congenital malformations or other human diseases has been identified.

Animal models are essential to the understanding of the genetics and pathogenesis of human diseases. The mouse is intensively used as a model system because of its similarity to humans in genome organization, development, biochemical pathways, and physiology. Mouse models have been the key to unraveling several fundamental scientific findings which are important for understanding the molecular mechanisms underlying human diseases in addition to the development and testing of drugs and therapies. Specific advantages of the mouse as a model system include:

- The genome is 90% identical to the human genome.
- It is possible to alter the genome in the mouse using gene-driven and phenotype-driven approaches and to produce models of human diseases, including genetic diseases.
- Alteration of the mouse genome may also produce changes in the normal functioning of organs, systems, and behavior, giving insight into the mechanisms behind their normal function and possible treatments for malfunction.
- The mouse model is used for drug screening and testing of therapies, including gene delivery and gene therapy.

The bottleneck in the process of establishing suitable mouse models is quite often appropriate phenotyping. From my own experience as a postdoctoral fellow, phenotypic analysis of "my" mouse mutants were focused on very specific organ systems and their function.

This strategy was successful and unraveled several important functional aspects of genes but at the same time I was not able to detect additional phenotypic alterations in the very same mouse model. These additional alterations were caused by the pleiotropic effect of the gene of interest. I simply missed additional alterations because I did not look for them or because of the lack of equipment and experience in specific methods used in other areas of research.

Triggered by this experience and the expertise in phenotype-driven forward genetics screens the idea of the German Mouse Clinic was born. The German

Standards of Mouse Model Phenotyping. Edited by Martin Hrabě de Angelis, Pierre Chambon, and Steve Brown

ISBN: 3-527-31031-2

Mouse Clinic is a unique platform for the comprehensive standardized phenotyping of mouse lines. Fourteen laboratories specializing in different areas of research, work under one roof and measure over 240 parameters in every mouse line and as a result many new findings have emerged. For almost all lines, including well-known mutant mouse lines, new phenotypes have been identified. This confirms the power and feasibility of standardized comprehensive phenotyping.

The German Mouse Clinic works in close collaboration with pan-European projects such as EUMORPHIA and EUMODIC. Together with my colleagues Steve Brown and Pierre Chambon we were able to bring together experts in the field of mouse functional genomics to assemble a book that presents a wide set of standardized phenotyping assays in 13 research fields.

This book should be seen as a starting point rather than as an end-product since mouse phenotyping will be developed further over the coming years and additional chapters and research fields such as "genome–environment interaction" might be added in future editions.

Implementation of the "German Mouse Clinic" led to a unique platform for comprehensive phenotyping. Baselines for more than 240 parameters have been established, and "Proof of Principle" has been shown in several mouse lines; for example, through the GMC an additional severe metabolic disorder was demonstrated in the mouse line ABE17 which was previously known only as a neurological model for prion disease. Comprehensive phenotyping was essential in the discovery of this additional feature, which will impact upon the interpretation of affected pathways. Japan, the USA, and other countries in Europe are implementing organizations similar to the GMC, underlining the need for these enterprises. The realization of the GMC was only possible with substantial financial support from the NGFN and the GSF. The GMC has already produced important scientific results through the isolation and characterization of various mouse models. In the lung function screen, we have built a unique database of reference values regarding the phenotypic variance of respiratory function in inbred mouse strains. We have been able to detect strong inter-strain variance (e. g. a factor of 3 for lung compliance), and a high genetic-to-total variance suggests a significant genetic contribution to phenotypic variability. Mouse strains with an obviously unfavorable lung function, which should be prone to lung diseases, may serve as ideal animal models.

However, the phenotypic analysis of mouse mutants is often focused on the individual research interests of the particular laboratory or limited to specific tests because of the lack of equipment and experience of specific methods in other research areas.

In order to take better advantage of the existing mutant mouse lines and to provide the scientific community with a platform for systematic, standardized, and comprehensive phenotyping of mouse models, we have established the German Mouse Clinic (GMC) at the GSF in Munich. We have brought together experts from different institutions (Universities of Bonn, Marburg, Munich (TU and LMU), GBF) to work side by side in one building. Within NGFN 1, the coordinating team of the GMC and the GMC staff built up the German Mouse Clinic in a concerted effort (the set-up of the laboratories, establishment of a unique and comprehensive primary screen, standardization of methods, etc). Because the GMC is unique in its concept and organization, it sets standards for SOPs and comprehensive analysis of mouse models. The phenotyping platform covers the research areas of dysmorphology, behavior, neurology, ophthalmology, clinical chemistry, immunology, allergy, nociception, molecular phenotyping, lung function, energy metabolism, and pathology and is well equipped with the newest technologies (e. g. microcomputer tomography, blood auto-analyzer). We offer phenotypic analysis on the basis of scientific collaboration and have the facilities to house guest scientists in dedicated guest laboratories.

Neuherberg, April 2006 Martin Hrabé de Angelis

List of Contributors

Koichiro Abe
GSF – National Research Centre
for Environment and Health
Institute of Experimental Genetics
German Mouse Clinic
Ingolstädter Landstr. 1
85764 Neuherberg
Germany

Bernhard Aigner
Ludwig-Maximilians-University
Institute of Molecular Animal Breeding
and Biotechnology
Feodor-Lynen-Str. 25
81377 Munich
Germany

Antonio Aguilar
Technical University Munich
ZAUM – Center for Allergy
and Environment
Biedersteiner Str. 29
80802 Munich
Germany
and
GSF – National Research Center
for Environment and Health
Division of Environmental
Dermatology
and Allergy GSF/TUM
Ingolstädter Landstr. 1
85764 Neuherberg
Germany

Francesca Alessandrini
Technical University Munich
ZAUM – Center for Allergy
and Environment
Biedersteiner Str. 29
80802 Munich
Germany
and
GSF – National Research Center
for Environment and Health
Division of Environmental
Dermatology
and Allergy GSF/TUM
Ingolstädter Landstr. 1
85764 Neuherberg
Germany

Carmen A. Argmann
Institut de Génétique et de Biologie
Moléculaire et Cellulaire
CNRS/INSERM/Université
Louis Pasteur
67404 Illkirch
France

Standards of Mouse Model Phenotyping. Edited by Martin Hrabé de Angelis, Pierre Chambon, and Steve Brown

ISBN: 3-527-31031-2

Johan Auwerx
Institut Clinique de la Souris
BP10142
67404 Illkirch Cedex
France
and
Institut de Génétique et de Biologie
Moléculaire et Cellulaire
CNRS/INSERM/Université
Louis Pasteur
67404 Illkirch
France

Rudi Balling
German Research Centre
for Biotechnology
Mascheroder Weg 1
38124 Braunschweig
Germany

Johannes Beckers
GSF – National Research Centre
for Environment and Health
Institute of Experimental Genetics
German Mouse Clinic
Ingolstädter Landstr. 1
85764 Neuherberg
Germany

Heidrun Behrendt
Technical University Munich
ZAUM – Center for Allergy
and Environment
Biedersteiner Str. 29
80802 Munich
Germany
and
GSF – National Research Center
for Environment and Health
Division of Environmental
Dermatology
and Allergy GSF/TUM
Ingolstädter Landstr. 1
85764 Neuherberg
Germany

Gonzalo Blanco
MRC Mammalian Genetics Unit
Harwell
Didcot
Oxfordshire OX11 0RD
UK

Steve Brown
MRC Mammalian Genetics Unit
Harwell
Oxfordshire OX11 0RD
UK

Dirk H. Busch
GSF – National Research Centre
for Environment and Health
Institute of Experimental Genetics
German Mouse Clinic
Ingolstädter Landstr. 1
85764 Neuherberg
Germany
and
Technical University Munich
Institute for Medical Microbiology,
Immunology and Hygiene
Frankfurter Str. 107
81675 Munich
Germany

Pierre Chambon
Institut Clinique de la Souris
BP10142
67404 Illkirch Cedex
France

Marie-France Champy
Institut Clinique de la Souris
BP10142
67404 Illkirch Cedex
France

André Constantinesco
Service de Biophysique et Médecine Nucléaire
CHU de Hautepierre
1 av. Molière
67098 Strasbourg
France

Claudia Dalke
GSF – National Research Centre for Environment and Health
Institute of Experimental Genetics
German Mouse Clinic
Ingolstädter Landstr. 1
85764 Neuherberg
Germany

Jack Favor
GSF – National Research Centre for Environment and Health
Institute of Experimental Genetics
German Mouse Clinic
Ingolstädter Landstr. 1
85764 Neuherberg
Germany

Tobias J. Franz
GSF – National Research Centre for Environment and Health
Institute of Experimental Genetics
German Mouse Clinic
Ingolstädter Landstr. 1
85764 Neuherberg
Germany

Helmut Fuchs
GSF – National Research Centre for Environment and Health
Institute of Experimental Genetics
German Mouse Clinic
Ingolstädter Landstr. 1
85764 Neuherberg
Germany

Hilary Gates
MRC Mammalian Genetics Unit
Harwell
Oxfordshire OX11 0RD
UK

Georgios Gkoutos
MRC Mammalian Genetics Unit
Harwell
Oxfordshire OX11 0RD
UK

Jochen Graw
GSF – National Research Centre for Environment and Health
Institute of Experimental Genetics
German Mouse Clinic
Ingolstädter Landstr. 1
85764 Neuherberg
Germany

Eain Green
MRC Mammalian Genetics Unit
Harwell
Oxfordshire OX11 0RD
UK

Jan Gutermuth
Technical University Munich
ZAUM – Center for Allergy and Environment
Biedersteiner Str. 29
80802 Munich
Germany
and
GSF – National Research Center for Environment and Health
Division of Environmental Dermatology and Allergy GSF/TUM
Ingolstädter Landstr. 1
85764 Neuherberg
Germany

Martin Hrabé de Angelis
GSF – National Research Centre
for Environment and Health
Institute of Experimental Genetics
German Mouse Clinic
Ingolstädter Landstr. 1
85764 Neuherberg
Germany
and
Technical University Munich
Experimental Genetics
Am Hochanger 8
85350 Freising
Germany

Thilo Jakob
Technical University Munich
ZAUM – Center for Allergy
and Environment
Biedersteiner Str. 29
80802 Munich
Germany
and
GSF – National Research Center
for Environment and Health
Division of Environmental
Dermatology
and Allergy GSF/TUM
Ingolstädter Landstr. 1
85764 Neuherberg
Germany

Anahita Javaheri
GSF – National Research Centre
for Environment and Health
Institute of Experimental Genetics
German Mouse Clinic
Ingolstädter Landstr. 1
85764 Neuherberg
Germany
and
Technical University Munich
ZAUM – Center for Allergy
and Environment
Biedersteiner Str. 29
80802 Munich
Germany
and
GSF – National Research Center
for Environment and Health
Division of Environmental
Dermatology
and Allergy GSF/TUM
Ingolstädter Landstr. 1
85764 Neuherberg
Germany

Svetoslav Kalaydjiev
GSF – National Research Centre
for Environment and Health
Institute of Experimental Genetics
German Mouse Clinic
Ingolstädter Landstr. 1
85764 Neuherberg
Germany

Martina Klempt
Ludwig-Maximilians-University
Institute of Molecular Animal Breeding
and Biotechnology
Feodor-Lynen-Str. 25
81377 Munich
Germany

Gabriele Köllisch
Technical University Munich
ZAUM – Center for Allergy
and Environment
Biedersteiner Str. 29
80802 Munich
Germany
and
GSF – National Research Center
for Environment and Health
Division of Environmental
Dermatology
and Allergy GSF/TUM
Ingolstädter Landstr. 1
85764 Neuherberg
Germany

Heena Lad
MRC Mammalian Genetics Unit
Harwell
Oxfordshire OX11 0RD
UK

Andreas Lengeling
Junior Research Group Infection
Genetics
German Research Centre
for Biotechnology
Mascheroder Weg 1
38124 Braunschweig
Germany

Thomas Lisse
GSF – National Research Centre
for Environment and Health
Institute of Experimental Genetics
German Mouse Clinic
Ingolstädter Landstr. 1
85764 Neuherberg
Germany

Martin Mempel
Technical University Munich
ZAUM – Center for Allergy
and Environment
Biedersteiner Str. 29
80802 Munich
Germany
and
GSF – National Research Center
for Environment and Health
Division of Environmental
Dermatology
and Allergy GSF/TUM
Ingolstädter Landstr. 1
85764 Neuherberg
Germany

Laurent Monassier
Mouse Clinical Institute
CNRS
INSERM
Université L. Pasteur de Strasbourg
67404 Illkirch Cedex
France

Werner Müller
German Research Centre
for Biotechnology
Department of Experimental
Immunology
Mascheroder Weg 1
38124 Braunschweig
Germany

Angelika Neuhäuser-Klaus
GSF – National Research Center
for Environment and Health
Institute of Human Genetics
Ingolstädter Landstr. 1
85764 Neuherberg
Germany

Patrick M. Nolan
MRC Mammalian Genetics Unit
Harwell
Didcot
Oxfordshire OX11 0RD
UK

Markus Ollert
Technical University Munich
ZAUM – Center for Allergy
and Environment
Biedersteiner Str. 29
80802 Munich
Germany
and
GSF – National Research Center
for Environment and Health
Division of Environmental
Dermatology
and Allergy GSF/TUM
Ingolstädter Landstr. 1
85764 Neuherberg
Germany

Oliver Puk
GSF – National Research Center
for Environment and Health
Institute of Developmental Genetics
Ingolstädter Landstr. 1
85764 Neuherberg
Germany

Ildikó Rácz
University of Bonn
Laboratory of Molecular Neurobiology
Life & Brain Center
Sigmund-Freud-Str. 25
53125 Bonn
Germany

Birgit Rathkolb
GSF – National Research Centre
for Environment and Health
Institute of Experimental Genetics
German Mouse Clinic
Ingolstädter Landstr. 1
85764 Neuherberg
Germany
and
Ludwig-Maximilians-University
Institute of Molecular Animal Breeding
and Biotechnology
Feodor-Lynen-Str. 25
81377 Munich
Germany

Johannes Ring
Technical University Munich
ZAUM – Center for Allergy
and Environment
Biedersteiner Str. 29
80802 Munich
Germany
and
GSF – National Research Center
for Environment and Health
Division of Environmental
Dermatology
and Allergy GSF/TUM
Ingolstädter Landstr. 1
85764 Neuherberg
Germany

Karen P. Steel
Wellcome Trust Sanger Institute
Wellcome Trust Genome Campus
Hinxton
Cambridge CB10 1SA
UK

Valter Tucci
MRC Mammalian Genetics Unit
Harwell
Didcot
Oxfordshire OX11 0RD
UK

Eckhard Wolf
Ludwig-Maximilians-University
Institute of Molecular Animal Breeding
and Biotechnology
Feodor-Lynen-Str. 25
81377 Munich
Germany

Andreas Zimmer
University of Bonn
Laboratory of Molecular Neurobiology
Life & Brain Center
Sigmund-Freud-Str. 25
53125 Bonn
Germany

1
Characterizing Hearing in Mice

Karen P. Steel

1.1
Introduction

Hearing impairment is very common in humans. One child in 1000 is born with a significant hearing impairment, and another one in 1000 develops progressive hearing loss during the first few years of life [1]. Age-related progressive hearing loss affects large numbers of people, and by the age of 70 years, more than half of the UK population has a 25-dB or greater hearing impairment, sufficient to benefit from wearing a hearing aid [2]. Hearing impairment often causes serious communication problems in sufferers, with much resulting social and economic isolation from the rest of the community.

Deafness is a very heterogeneous disorder, with a wide range of causes. This makes it difficult to study directly in humans. Many different genes are known to be involved in deafness. For example, for non-syndromic human deafness, over 80 loci have been defined and 30 genes identified [3], and Online Mendelian Inheritance in Man lists over 400 distinct syndromes including deafness as a feature. In most clinical collections reported, *GJB2* mutations are a major contributor (for example, associated with 33 % of severe or profound familial childhood deafness in the UK, [4]), but the vast majority of other cases, including most later-onset progressive hearing loss cases, have no molecular diagnosis. There are probably several hundred genes involved in deafness in humans, any one of which can be mutated and cause deafness in an individual. Mouse mutations are available for a relatively small proportion of these genes. Around 200 mouse mutants with some sort of auditory system defect have been described [5, 6] but despite the rapid progress in identifying genes underlying deafness over the past few years, many deafness genes have not yet been identified in mouse or human. More mouse mutants with hearing or balance defects will give us access to more of the molecules critical for normal hearing, as well as more candidate genes for deafness in humans.

In addition to single-gene causes of deafness, minor variations in multiple different genes (genetic background) can also interact to make a person more or less likely to develop hearing loss as they get older, and twin, sib and family studies have demonstrated a range of heritabilities up to around 0.5 for age-related hearing loss [7–9]. Noise, drugs and infections can all contribute to hearing impairment, but

Standards of Mouse Model Phenotyping. Edited by Martin Hrabé de Angelis, Pierre Chambon, and Steve Brown

ISBN: 3-527-31031-2

these will interact with the particular gene variants carried by an individual to influence the degree of damage. For example, the A1555G mutation of the human mitochondrial genome makes carriers highly susceptible to ototoxic drug-induced deafness [10], and there are several mouse mutations that predispose the carriers to noise-induced hearing loss [11–14]. Genetics is therefore an important factor in hearing impairment.

Mice are excellent models for human deafness. The structure and function of the auditory system is very similar in the two species. The range of pathological features observed in deaf mice appears to be very similar to the pathology in human deafness, although it is inevitably much more difficult to investigate the development of the pathology in humans than it is in an animal model like the mouse. Not surprisingly, the same genes appear to underlie deafness in the two species. There are many examples where the mouse deafness gene has been identified by positional cloning and this has led very rapidly to the finding of mutations in the orthologous human gene in people with inherited deafness. Similarly, genes found to be involved in human deafness often give essentially the same phenotype when mutated in the mouse. Comparisons of mouse and human genes involved in deafness are given in a useful website edited by Zheng and Johnson [6].

Sensory deficits are often difficult to detect in a mutant mouse, yet are of obvious importance in human disease as well as influencing behavioral phenotypes of newly-created mutant mice. Complete deafness (for example deafness, *Tmc1*^{dn}), rapidly progressive blindness (for example retinal degeneration, *Pde6b*rd) or specific anosmias can go undetected for generations because of the lack of overt signs that are obvious to those handling the mice. Many standard inbred strains carry mutations causing sensory defects, complicating assessment of new mutations created on these backgrounds. For example, *Pde6b*rd is carried in C3H strains, C57BL mice show a specific inability to detect the smell of isovaleric acid, and many inbred strains such as C57BL and DBA carry mutations contributing to progressive hearing loss [15–18]. In this chapter, I focus on ways of characterizing the hearing ability of mouse mutants, including simple screening methods. This is not intended to be a comprehensive catalog of all the ways that the auditory system could be studied, but simply highlights the major approaches that might be considered.

1.2 Behavioral Tests of Hearing

Although there have been a few reports of conditioned behavioral tests for hearing, mice are very difficult to train, and tests like these reflect other features in addition to sensory function. However, a simple test for hearing is to elicit a Preyer reflex. This is a flick backwards of the pinna upon hearing a sharp sound, and in young mice with very good hearing, this is sometimes part of a startle response in which the whole body of the mouse jumps. The mice often stay still for a second after the first exposure, but with repeated exposures, they usually stop responding. The Preyer reflex is a suprathreshold response, not an indication of normal thresholds, so it can be used to pick out non-responding mice that have a severe or profound

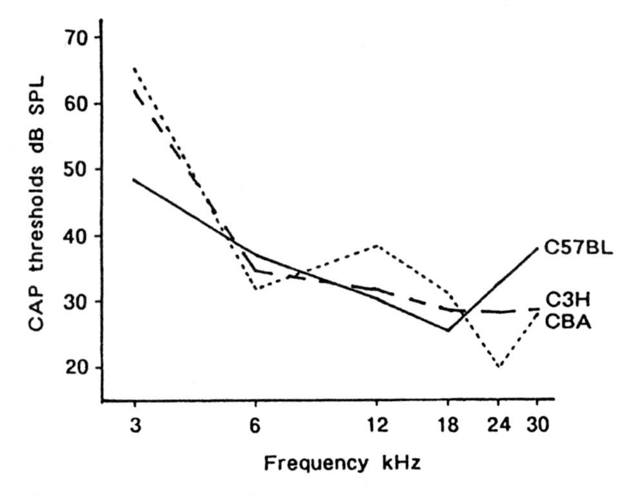

Fig. 1.1 Mean thresholds for a CAP response measured with a round-window electrode for three different inbred strains of mouse: C57BL/6J ($n = 8$; age 20–30 days), C3H/He ($n = 5$; age 30–58 days), and CBA/Ca ($n = 16$; age 43–95 days). The lowest thresholds are in response to stimuli around 18 to 24 kHz.

hearing impairment, but will not detect mild or moderate hearing impairments.

Startle responses can also be measured using platforms that detect the movement of a mouse in response to sound or other stimuli within a sound-attenuating chamber, as used in measurement of pre-pulse inhibition. This approach permits standardization of the response measurement, rather than depending upon the tester observing a response, but is not as widely used in large-scale screens as the simple observation method.

What is a suitable stimulus to use to elicit a Preyer reflex? Mice are sensitive to a broad range of frequencies, and respond best to frequencies that are much higher than the human frequency range. Fig. 1.1 illustrates this range. Their most acute hearing is around 18 to 24 kHz. A sharp metallic click made with large forceps is often used, as it will contain plenty of high frequency components, but the output is not calibrated or accurately reproducible. We use a custom-built clickbox which generates a brief 20-kHz soundburst at an intensity of 90 dB SPL when held 30 cm directly above the mouse. This screen has been particularly useful in detecting new deaf mutants in the large-scale ENU (*N*-ethyl-*N*-nitrosourea) mutagenesis programs run at Harwell, UK and Munich, Germany, as well as other screens elsewhere [19, 20]. The Preyer reflex test using the clickbox is part of the SHIRPA protocol [21].

1.3
Physiological Tests of Hearing

The most frequently used physiological tests of auditory function are auditory brainstem response measurements (ABR) and round-window response recordings. ABRs represent responses from neurons from the cochlear nerve onwards through the central auditory pathways. Round-window responses are a well-established method for assessing cochlear activity, and give response thresholds close to those of single units [22]. Both require the mouse to be anesthetized, but ABR recording can be carried out with recovery so that repeated measures over time can be taken. Round-window recordings are generally carried out in terminal experiments in fully anesthetized mice, because more extensive surgery is required. Care needs to be taken to keep the mouse warm and to monitor its condition during the procedures, and as for any surgical procedure, appropriate training and licensing by the regulatory authorities are required. Response recordings equivalent to ABRs and round-window responses are used in humans. In the clinic, human ABRs can usually be measured without anesthetic, while electrocochleography is carried out under general anesthetic with an electrode placed near rather than on the round window using a less invasive approach than is possible in mice.

For ABR response recording, differential pin electrodes are placed subcutaneously at the midline on the top of the head (the vertex) and just behind the ear pinna of the side to be recorded (for example [23]). The mouse is then presented with appropriate stimuli (see below) and the responses recorded as differences between the signals from the two electrodes. Many responses (100 to several thousand) are collected and computer-averaged, because each individual response will be lost in the background noise generated by the body of the mouse. Averaging the sweeps will enhance the signal-to-noise ratio, because the response will occur at the same time within the sweep after the onset of the sound, while the background noise will be random and will tend to flatten, or average out, as more and more sweeps are included. ABRs include several waves, and while the first negative deflection is believed to represent cochlear nerve activity and the later waves reflect function within the central auditory pathways, there is no consensus about the exact origins of the later waves. Studies in the cat suggested origins for ABR waves [24], and a comparison with the waveforms recorded from the mouse may reveal parallels.

For round-window recordings, a fine Teflon-coated silver wire electrode is placed on the round window, which is exposed via a small hole made in the bone of the middle ear wall just caudal of the pinna and tympanic membrane (eardrum) [22, 25–27]. The reference electrode for differential recording is placed on the tissue just dorsal of the ear. Fig. 1.2 gives a diagrammatic representation of the arrangement. Three distinct response types can be recorded using a round-window electrode. Cochlear microphonics (CM) are an ac (alternating current) response mimicking the waveform of the stimulus and are generated primarily by nearby outer hair cells [28, 29]. The compound action potential (CAP) is a sharp single or double negative deflection occurring just after the onset of the stimulus, and reflects summed synchronous activity of cochlear nerve neurons. Summating potentials (SP) are offsets

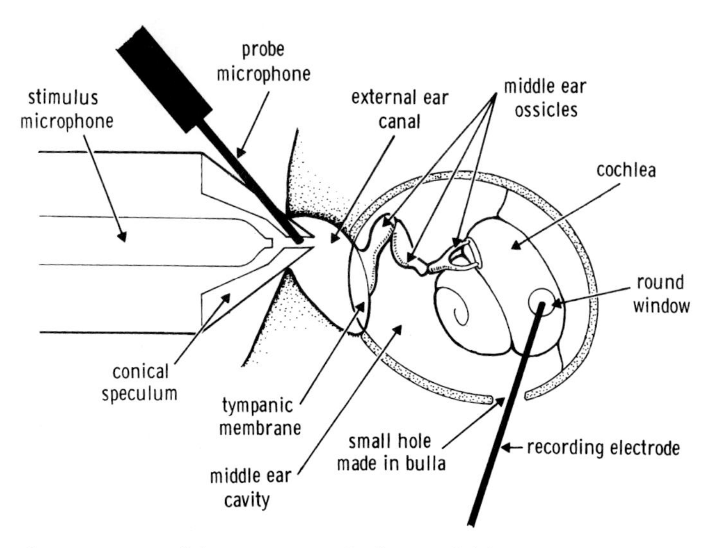

Fig. 1.2 Diagram of the arrangement for the sound delivery system and round-window recording electrode. The middle ear cavity is shown cut away for illustration only; normally only a small hole is made to avoid unnecessary disturbance of the middle ear ossicles. Reprinted from [25] with kind permission. Copyright © 1983 American Psychological Association.

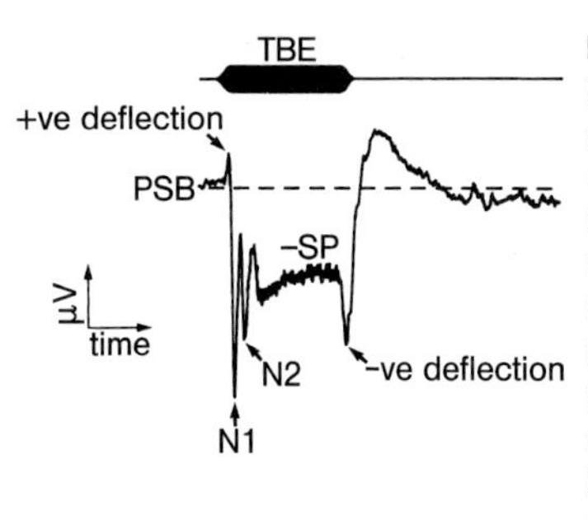

Fig. 1.3 Top, toneburst envelope (TBE), showing duration (15 ms) and gradual (1 ms) rise and fall of the amplitude. Bottom, typical waveform recorded from a normal mouse to a 3-kHz, 100-dB SPL, showing a negative SP (–SP, downward offset for the duration of the toneburst) and a CAP with two negative dips (N1 and N2) at the start of the response. Higher frequency stimuli tend to give a positive SP rather than the negative SP shown here. PSB, prestimulus baseline. Adapted from [32] with kind permission of Elsevier Science. Copyright © 1992.

in the response sustained for the duration of the stimulus which can be positive or negative in polarity, and are thought to represent the intracellular dc responses of sensory hair cells [30–32]. Fig. 1.3 shows an example of a waveform showing a CAP and SP response. The additional information provided about activity of different cell types within the cochlea make round-window recordings rather than ABR the preferred approach in my own laboratory.

Just as for the Preyer reflex, some thought needs to be given to appropriate stimuli for eliciting an ABR or round-window response. Stimuli can be presented free-field or through a closed sound system. The advantage of a free-field system is its ease of use, while the advantage of a closed sound system is that higher sound intensities can be obtained and the level can be calibrated within the closed sound field, effectively specifying the intensity at the ear drum. The calibrating system used in our laboratory consists of a probe tube inserted into the closed cavity leading to a measuring microphone, as illustrated in Fig. 1.2. We have found the high sound levels obtained in a closed sound system (up to 130 dB SPL or more at some frequencies) useful in investigating responses in severely hearing-impaired mice. With any delivery method, it should be remembered that high frequency sounds are highly directional, so any small change in the relative position of sound source and eardrum can change the stimulus intensity at the eardrum.

Either broadband clicks or frequency-specific tonebursts are commonly used as stimuli. Clicks will stimulate a large proportion of cochlear hair cells, so may give a lower threshold for a response, but tonebursts will give information that is more specific to the place along the length of the cochlear duct that responds optimally to the test frequency: high frequencies at the base and low frequencies at the apex of the cochlear duct. We use a toneburst that is 15 ms long and shaped with a 1 ms rise and fall time at the start and finish of the burst, because an instantaneous rise and fall will produce spectral splatter in the output of the microphone used to produce the sound, and degrade the frequency-specificity of the stimulus.

Responses can be analyzed in several ways. As a minimum, thresholds are usually obtained by recording at moderate intensities that would normally elicit a response, and then stepping down in 5- or 10-dB steps until the response is lost. Then, intensity is increased in smaller steps, and stepped up and down around the minimum intensity level at which a response is just detected by visual inspection of the waveform, defined as the threshold. Step size is normally reduced near threshold, and we routinely use 2- or 3-dB steps. These thresholds are then plotted with respect to frequency, to give the mouse equivalent of an audiogram, as in Fig. 1.1. Suprathreshold features can also be analyzed, such as input–output functions plotting stimulus intensity against response amplitude. Latencies to different components of the waveform can be measured. The shape of the waveform itself can be examined, and has proved helpful in studying several hearing-impaired mouse mutants [33, 34].

Further measurements of cochlear function can be used as appropriate. Another response which is often used for assessing cochlear function in the mouse is the distortion product otoacoustic emission, which probably reflects outer hair cell activity at frequency-specific sites along the cochlear duct [27, 35–37]. This can be a useful screening tool, because although the mouse needs to be anesthetized, it is relatively non-invasive and can be used with recovery.

If gross recordings suggest the mutant might have a particularly interesting hair cell defect, then single cell recording can be carried out *in vitro* in acute or cultured organ of Corti samples [38–40]. A wide variety of basolateral hair cell currents can be measured this way, and transduction currents can be measured by deflecting hair bundles by fluid jets or using direct mechanical deflection. The integrity of transduction channels can be assessed using the styryl dye FM1–43, which is useful in giving a broader view of many hair cells than single hair cell response measurements can provide, but does not give any information about properties of the basolateral membranes of hair cells [40, 41].

Single unit recordings in the cochlear nucleus and higher auditory nuclei are relatively straightforward to carry out, but it is difficult to record reliably from cochlear nerve neurons and there are very few reports of successful intracellular recording from an intact cochlea *in vivo*. However, the recent report from Taberner and Liberman [42] suggests that most features of mouse auditory neurons are similar to those of other mammals.

If a stria vascularis defect is suspected (as for example in a mouse with a white spotting phenotype or signs of autoimmune problems), strial function can be assessed by recording the endocochlear potential, a high resting potential of around + 100 mV maintained within the scala media of the cochlear duct by the stria vascularis. This recording is made with a micropipette electrode filled with a suitable fluid (such as 150 mM KCl) inserted through the lateral wall of the basal turn of the cochlear duct in anesthetized mice, with a silver/silver chloride pellet electrode acting as the reference just under the dorsal head skin [26, 43, 44].

1.4 Anatomy of the Ear

The anatomy of the ear can be assessed in a variety of ways. Special techniques are adopted because the middle ear and inner ear contain extensive air or fluid-filled chambers but are surrounded by very dense bone which means that routine histology is often disappointing. However, it is straightforward to dissect the middle ear and inspect it under a dissecting microscope for signs of abnormality, such as excess calcification, abnormal vascular membranes lining the cavity, or presence of fluid or inflammatory material, all indicators of current or earlier middle ear disease such as otitis media. It is also possible to dissect out ossicles for later examination (see Fig. 1.4 for an example) [45–47].

The gross structure of the inner ear can be examined in isolated specimens dissected from the head and processed to clear the soft tissues to get a better view of the overall bony structure. Processing with potassium hydroxide and clearing with glycerol is a simple approach that we have used successfully to investigate mouse mutant inner ears for evidence of malformations [26]. We have detected a wide variety of malformations using this technique, ranging from narrow or truncated semicircular canals, shortened cochlear ducts, to barely-recognizable vesicles with no canals or coiled cochlear duct. The otoconia of the utriculus and sacculus can be viewed in these preparations, and gross defects such as giant otoconia, a lack of otoconia, or reduced size of the patch of otoconia can be detected. If a malformation of the inner ear is evident from glycerol-cleared preparations, we then proceed to use paint-filling to study the early development of the defect. This requires removal and fixation of embryonic or early postnatal specimens, usually as half-heads, followed by clearing in methyl salicylate and injection with a solution of white paint [48]. Some examples of such preparations are shown in Fig. 1.5. A recently-developed approach to studying the three-dimensional structure of the inner ear is optical tomography [49]. This requires labeling of the structures and clearing before the sample is imaged in fine virtual slices, which are then reconstructed and viewed from any desired angle.

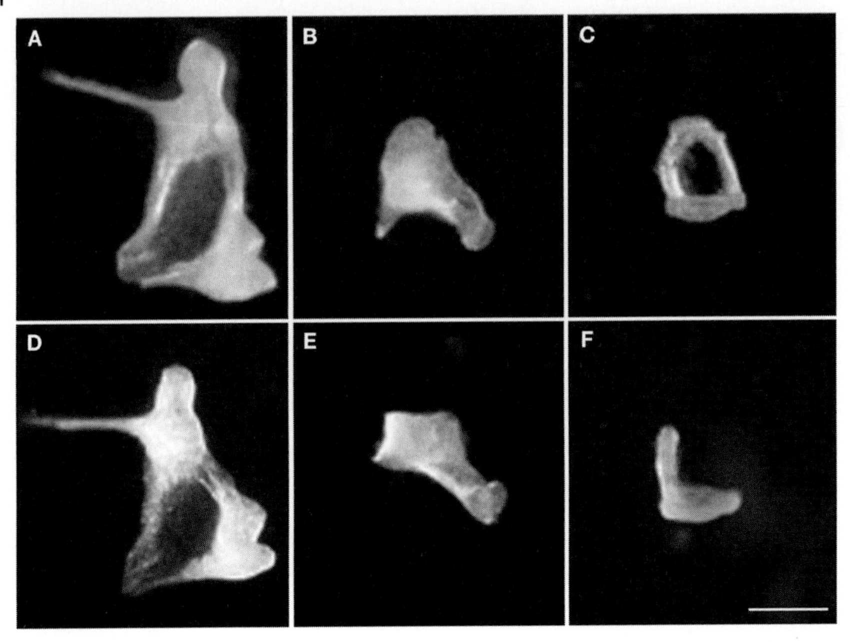

Fig. 1.4 Ossicle defects seen in the hearing-impaired mouse mutant hushpuppy. Normal malleus, incus and stapes are shown in A–C respectively. Mutants have a normal malleus (D) but have various incus abnormalities such as small bodies and reduced long and short processes (E). Mutants also show a characteristic stapes defect, with a reduced or absent posterior crus (F). Scale bar represents 500 μm. Images reproduced from [55] with kind permission of Lippincott, Williams & Wilkins.

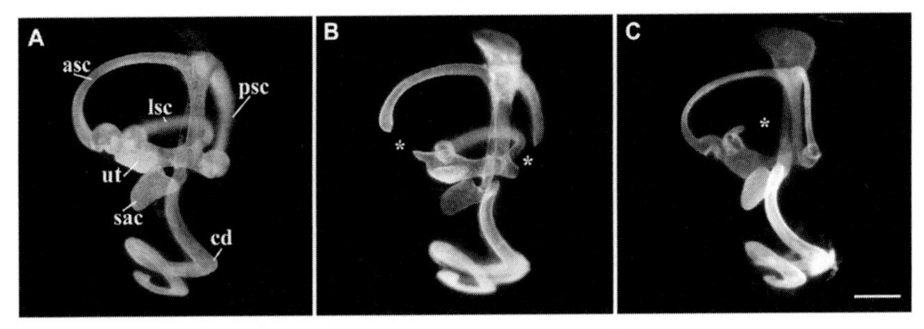

Fig. 1.5 Semicircular defects seen in mouse mutants, shown in paint-filled E16.5 embryos. (A) Normal structure of the inner ear labyrinth. (B) Semicircular canal truncations (marked with *) of the anterior and posterior semicircular canals in the headturner mutant. (C) Lateral semicircular canal truncation in the tornado mouse mutant. asc, anterior semicircular canal; cd, cochlear duct; lsc, lateral semicircular canal; psc, posterior semicircular canal; sac, saccule; ut, utricle. Scale bar 500 μm. Images reproduced from [56](Copyright National Academy of Sciences USA, Fig. 1.5 A and 1.5B) and [57] with kind permission of Springer-Verlag GmbH (Fig. 1.5C).

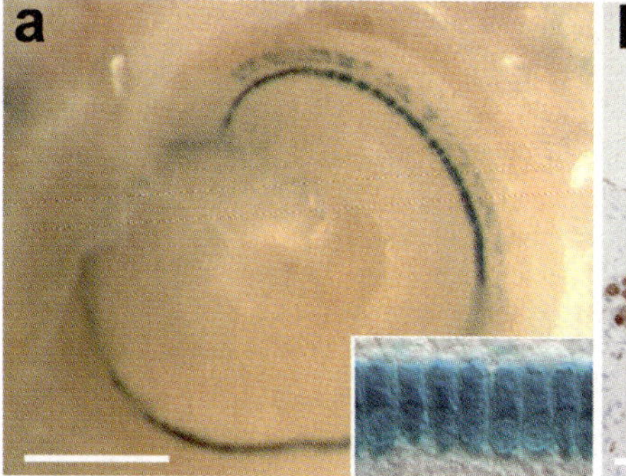

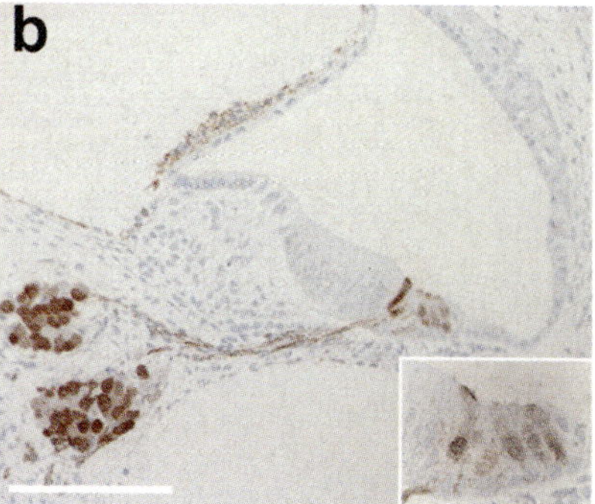

Fig. 1.6 (a) The apical turn of an adult transgenic mouse cochlea dissected to expose the organ of Corti, showing blue LacZ labeling in inner and outer hair cells. Inset shows high magnification images of LacZ-labeled inner hair cells of the same mouse. Scale bar = 500 μm. (b) Light microscope image of a section through the plastic-embedded cochlea of a normal postnatal day 1 mouse immunolabeled with anti-calretinin and counterstained with hematoxylin; hair cells, neurons and ganglion cells show positive immunoreactivity with the antibody used (DAB, brown stain). Inset shows hair cell region at higher magnification. Scale bar = 500 μm.

Surface preparations are often used to examine parts of the cochlear duct that can be removed and laid flat on a microscope slide and examined by a suitable technique for resolving images in thick specimens such as differential interference contrast microscopy or confocal microscopy. The organ of Corti is often examined in this way [25, 50], and the stria vascularis can also be examined in surface preparation [51]. An example of a surface preparation of the organ of Corti is shown in Fig. 1.6 a, using labeling of LacZ expressed specifically in the sensory hair cells. The inner ear can be embedded in a resin, then the block cut up into smaller pieces for examination and further sectioning, the block surface technique [52].

Serial sections of the inner ear can be prepared, but paraffin-wax embedding is only useful up to about 4 days after birth, after which the calcification of surrounding bone makes the whole structure too fragile to be supported by wax, giving poor section quality even after decalcification. Better preservation of the structure can be obtained using a harder embedding medium, such as a plastic resin. Plastic-embedded samples can also be cut as thinner sections (1 or 2 μm thick) if required. An example of a plastic-embedded section is given in Fig. 1.6 b, showing reasonable preservation of cochlear structures. If a suitable resin such as araldite is used, sectioning for light microscopy can be followed by thin sectioning for electron microscopy [53]. Much of the classical work in describing the inner ear of deaf mouse mutants was published by M. S. Deol, who used a combined celloidin/paraffin-wax embedding procedure, enabling him to produce good quality serial sections through the entire inner ear, but this approach is not easy and is not in common use today. Interpreting sections of the inner ear requires some care, and three-dimensional reconstruction of serial sections (either in one's head or by computer) is usually necessary to identify with confidence the various chambers and sensory epithelia within the labyrinth.

Fig. 1.7 Low magnification scanning electron micrograph of the cochlea of a normal mouse at postnatal day 1, dissected to reveal the surface of the organ of Corti. This approach allows measurement of the length and width of the organ of Corti, and will reveal if turns are missing. Top left shows the apex, and the middle right shows the hook region at the extreme base of the cochlea. The mouse cochlea has just over two turns. Scale bar = 1 mm.

Scanning electron microscopy is a technique that is used routinely to examine surfaces of the inner ear, particularly the upper surface of the organ of Corti in the cochlea. We generally use a modified version of the OTOTO technique first introduced by Hunter-Duvar [54] for inner ear studies: osmium tetroxide, followed by thiocarbohydrazide, osmium tetroxide, thiocarbohydrazide, osmium tetroxide. Thiocarbohydrazide serves as the cross-linking agent and allows more osmium (the fixative and staining agent) to be taken up by the sample, which allows samples to be examined in the microscope with minimal or no further metal coating, giving better preservation of very fine details of surface ultrastructure. Scanning electron microscopy allows inspection only of surfaces and not internal ultrastructure, but it is an excellent tool with which to examine the hair bundle (an array of modified microvilli called stereocilia) on the upper surface of sensory hair cells and the organization of the organ of Corti, as well as other surfaces such as the lining of the middle ear or vestibular sensory patches. We have discovered mutants with too many or too few hair cells, hair cells in abnormal positions, disorganized hair bundles, short or long stereocilia, thin stereocilia, rotated hair bundles, fused stereocilia, progressively degenerating hair cells, and no development of hair cells at all, all using routine scanning electron microscopy. Some examples of images of normal cochlear specimens at different magnifications are shown in Figs 1.7–1.9, and two examples of abnormal samples are shown in Fig. 1.8 b and 1.8 c, illustrating the utility of this approach.

Protocols for many of the techniques outlined above are given in the relevant publications or at the EUMORPHIA website, and each laboratory can choose and adapt the approaches according to the type of mutant under investigation and the facilities available. In my own studies of deaf mutant mice over the past few years, we have evolved a first-pass phenotyping approach that has been successful in detecting a first indication of the structural correlates of hearing or balance disorders in over 90% of the mutants we have studied. Our first-pass phenotyping includes: behavioral observations to detect the Preyer reflex and signs of balance defects; dissection of the middle ear to detect malformed ossicles or signs of inflammation; glycerol clearing of the inner ear to discover any gross malformation; and scanning

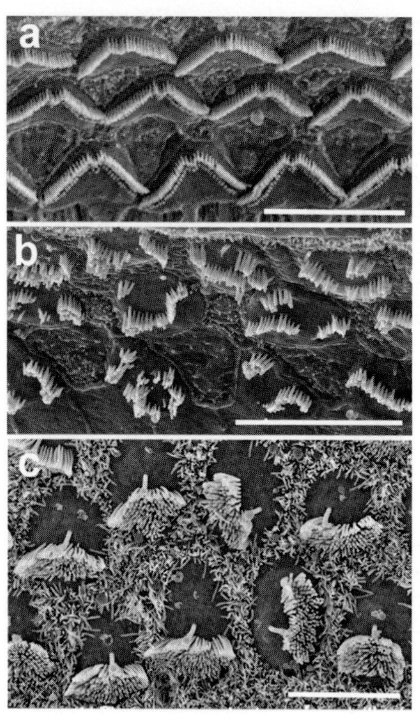

Fig. 1.8 Scanning electron micrographs of the surface of the basal turn of the organ of Corti. Scale bars = 10 μm. (a) A normal specimen at 21 days old, showing three rows of outer hair cells with their stereocilia forming hair bundles with a distinctive W- or V-shape. (b) Image of the basal turn of the organ of Corti of a mutant mouse at 21 days, showing severely disorganized stereocilia bundles on the top of outer hair cells. (c) Image of the surface of the middle turn of the organ of Corti of a mutant mouse at day 1 showing how some hair cells appear to be abnormally rotated.

electron microscopy of the organ of Corti to look for signs of abnormal development of degeneration. These approaches are generally used with mutants and littermate controls at around 3–4 weeks of age, although some mutants show progressive loss of the Preyer reflex so these are studied at an older age point, after the onset of deafness. Following these first-pass phenotyping tests, further work depends upon the nature of the defect, but in general the mutants are studied at younger stages to pinpoint the earliest sign of the defect. For example, if there is an inner ear malformation, a range of prenatal stages will be analyzed by paint-filling to define the first sign of abnormal development. If the hair bundles appear abnormal, these will be studied at younger stages and at higher resolution to detect the earliest defect. Physiological tests of cochlear function are carried out, as the Preyer reflex does not reveal thresholds for hearing. In the case of developmental defects of the inner ear, expression studies are often the next step, using either *in situ* hybridization or immunocytochemistry to investigate the expression of genes known to have important roles in early development. Other more detailed analyses can be carried out as appropriate.

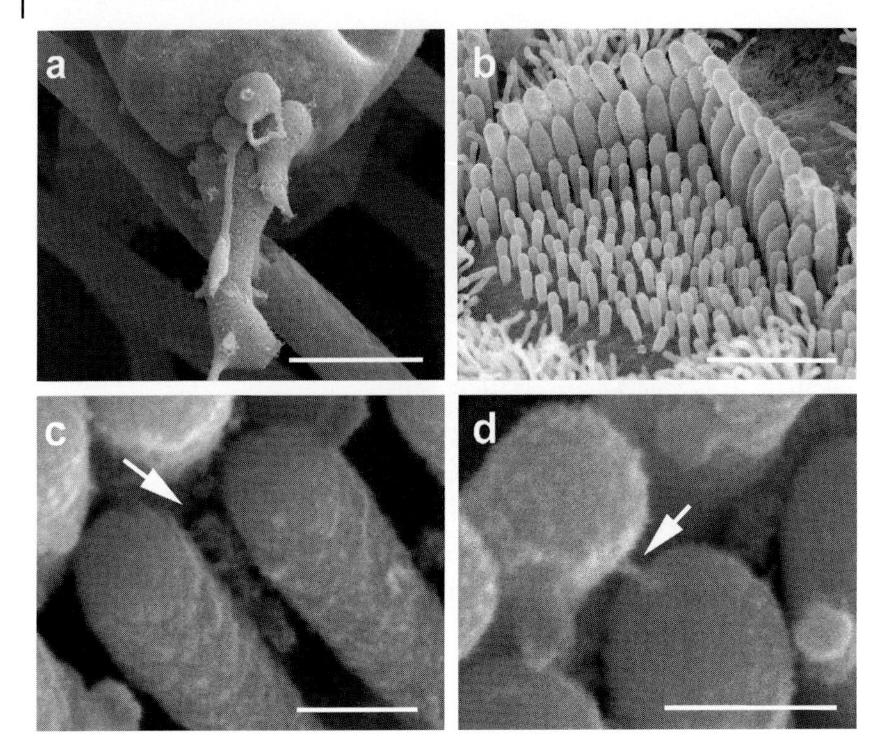

Fig. 1.9 Higher magnification scanning electron microscope images from the organ of Corti. (a) Image showing synapses on the bottom of an outer hair cell accidentally exposed during sample preparation. Scale bar = 2 μm. (b) High magnification image showing inner hair cell stereocilia bundle in the middle turn of the cochlea of a normal mouse at postnatal day 1, showing detail of the length and width of the stereocilia, the numbers of stereocilia, and the shape of the tips (note some are pointed in the middle row). Scale bar = 2 μm. (c) Ultra-high magnification image showing horizontal top connectors (arrow) between stereocilia of outer hair cells from middle turn of the cochlea of a 21-day-old normal mouse. Scale bar = 200 nm. (d) Ultra-high magnification image showing a tip link (arrow) connecting two neighboring stereocilia of an outer hair cell of the middle turn of the cochlea of a 21-day-old normal mouse. Scale bar = 200 nm. These images were taken from OTOTO-processed samples without any sputter coating using a field emission microscope (Hitachi FE 4800) at 5 kV.

1.5 Conclusions

We have developed a first-pass phenotyping approach to study many mouse mutants over the past few years, which is successful in establishing a structural correlate for hearing or balance defects in most of the new mutants studied. This gives a starting point for further more detailed investigations. In the case of mouse mutants from phenotype-driven screens, such as the ENU mutagenesis programs that have successfully generated large numbers of new mutants with hearing or balance defects, we have found it important to carry out mapping of the mutations alongside phenotypic characterization. Often it has been necessary to have a good insight into the phenotype in order to assess whether a mouse is affected or not, as variable expression of the phenotype is common especially in the dominantly-inherited traits we have studied. Furthermore, insight into the phenotype can give vital clues to the type of gene involved, aiding candidate gene selection.

Although we have discovered many different genes involved in deafness in mouse and humans, it is clear that many more remain to be found. This is especially true for late-onset, progressive hearing loss, as we know very little about the molecular basis of this to date, yet it represents the vast bulk of cases of hearing loss in the human population. Mouse mutants will continue to play an important role in discovering new genes underlying deafness, which in turn will be excellent candidates for investigation in hearing-impaired humans.

Acknowledgements

I thank Agnieszka Rzadzinska, Amy Kiernan, John Ambrose and Sarah Spiden for help with the figures, and the Wellcome Trust, MRC, EC and Deafness Research UK for financial support of the work leading to this review.

References

1 Fortnum HM, Summerfield AQ, Marshall DH, Davis AC, et al. *Brit Med J* **2001**; *323*: 1–6.
2 Davis AC. *Hearing in Adults*, Whurr: London, **1995**.
3 Van Camp G, Smith RJH. *Hereditary Hearing Loss Homepage* **2005**, http://webhost.ua.ac.be/hhh/
4 Hutchin T, Coy NN, Conlon H, Telford E, et al. *Assoc Res Otolaryngol Abstr* **2003**; *26*: 144.
5 Steel KP. *Deaf Mouse Mutants.* http://www.sanger.ac.uk/deafmousemutants, **2001**
6 Zheng QY, Johnson K. *Hereditary Hearing Impairment in Mice*, **2005**, http://www.jax.org/hmr/models.html
7 Gates GA, Couropmitree NN, Myers RH. *Arch Otolaryngol Head Neck Surg* **1999**; *25*: 654–659.
8 Karlsson KK, Harris JR, Svartengren M. *Ear Hear* **1997**; *18*: 114–120.
9 DeStefano AL, Gates GA, Heard-Costa N, Myers RH, et al. *Arch Otolaryngol Head Neck Surg* **2003**; *129*: 285–289.
10 Estivill X, Govea N, Barcelo E, Badenas C, et al. *Am J Hum Genet* **1998**; *62*: 27–35.
11 Ohlemiller KK, McFadden SL, Ding DL, Flood DG, et al. *Audiol Neurootol* **1999**; *4*: 237–246.
12 Kozel PJ, Davis RR, Krieg EF, Shull GE, et al. *Hear Res* **2002**; *164*: 231–239.
13 Holme RH, Steel KP. *J Assoc Res Otolaryngol* **2004**; *5*: 66–79.
14 Schick B, Praetorius M, Eigenthaler M, Jung V et al. *Cell Tissue Res* **2004**; *318*: 493–502.
15 Lyon MF, Rastan S, Brown SDM. *Genetic Variants and Strains of the Laboratory Mouse*, Oxford University Press: Oxford, **1996**.
16 Griff IC, Reed RR. *Cell* **1995**; *83*: 407–414.
17 Noben-Trauth K, Zheng QY, Johnson KR. *Nature Genet* **2003**; *35*: 21–23.
18 Johnson KR, Zheng QY, Weston MD, Ptacek LJ, Noben-Trauth K. *Genomics* **2005**; *85*: 582–590.
19 Hrabé de Angelis MH, Flaswinkel H, Fuchs H, Rathkolb B, Soewarto D, Marschall S, Heffner S, Pargent W, Wuensch K, Jung M, Reis A, Richter T, Alessandrini F, Jakob T, Fuchs E, Kolb H, Kremmer E, Schaeble K, Rollinski B, Roscher A, Peters C, Meitinger T, Strom T, Steckler T, Holsboer F, Klopstock T, Gekeler F, Schinde-

wolf C, Jung T, Avraham K, Behrendt H, Ring J, Zimmer A, Schughart K, Pfeffer K, Wolf E, Balling R. *Nature Genet* **2000**; *25*: 444–447.

20 Nolan PM, Peters J, Strivens M, Rogers D, Hagan J, Spurr N, Gray IC, Vizor L, Brooker D, Whitehill E, Washbourne R, Hough T, Greenaway S, Hewitt M, Liu X, McCormack S, Pickford K, Selley R, Wells C, Tymowska-Lalanne Z, Roby P, Glenister P, Thornton C, Thaung C, Stevenson JA, Arkell R, Mburu P, Hardisty R, Kiernan A, Erven A, Steel KP, Voegeling S, Guenet JL, Nickols C, Sadri R, Nasse M, Isaacs A, Davies K, Browne M, Fisher EM, Martin J, Rastan S, Brown SD, Hunter J. *Nature Genet* **2000**; *25*: 440–443.

21 Rogers DC, Fisher EM, Brown SD, Peters J, Hunter AJ, Martin JE. *Mamm Genome* **1997**; *8*: 711–713.

22 Johnstone JR, Alder VA, Johnstone BM, Robertson D, Yates GK. *J Acoust Soc Am* **1979**; *65*: 254–257.

23 Zheng QY, Johnson KR, Erway LC. *Hearing Res* **1999**; *130*: 94–107

24 Melcher JR, Guinan Jr. JJ, Knudson IM, Kiang NYS. *Hearing Res* **1996**; *93*: 28–51.

25 Steel KP, Bock GR. *Behav Neurosci* **1983**; *3*: 381–391.

26 Steel KP, Smith RJH. *Nature Genet* **1992**; *2*: 75–79.

27 Yoshida N, Hequembourg SJ, Atencio CA, Rosowski JJ, Liberman MC. *Hearing Res* **2000**; *141*: 97–106.

28 Patuzzi RB, Yates GK, Johnstone BM. *Hearing Res* **1989**; *39*: 177–188.

29 Dallos P, Cheatham MA. *J Acoust Soc Am* **1976**; *60*: 510–512.

30 Dallos P, Schoeny ZG, Cheatham MA. *Acta Otolaryng* **1972**; *302*(Suppl.): 1–46.

31 Dallos P. *Hearing Res* **1986**; *22*: 185–198.

32 Harvey D, Steel KP. *Hearing Res* **1992**; *61*: 137–146.

33 Steel KP, Harvey D. In *Development of Auditory and Vestibular Systems*, Romand R, (Ed.), Elsevier: Amsterdam, **1992**, 221–242.

34 Bock GR, Yates GK, Deol MS. *Neurosci Lett* **1982**; *34*: 19–25.

35 Horner KC, Lenoir M, Bock GR. *J Acoust Soc Am* **1985**; *78*: 1603–1611.

36 Lukashkin AN, Lukashkina VA, Legan PK, Richardson GP, Russell IJ. *J Neurophysiol* **2004**; *91*: 163–171.

37 Vazquez AE, Jimenez AM, Martin GK, Luebke AE, Lonsbury-Martin BL. *Hearing Res* **2004**; *194*: 87–96.

38 Holt JR, Gillespie SK, Provance DW, Shah K, Shokat KM, Corey DP, Mercer JA, Gillespie PG. *Cell* **2002**; *108*: 371–381.

39 Marcotti W, Johnson SL, Holley MC, Kros CJ. *J Physiol* **2003**; *548*: 383–400.

40 Geleoc GS, Holt JR. *Nature Neurosci* **2003**; *6*: 1019–1020.

41 Meyers JR, MacDonald RB, Duggan A, Lenzi D, Standaert DG, Corwin JT, Corey DP. *J Neurosci* **2003**; *23*: 4054–4065.

42 Taberner AM, Liberman MC. *J Neurophysiol* **2005**; *93*: 557–569.

43 Steel KP, Barkway C, Bock GR. *Hearing Res* **1987**; *27*: 11–26.

44 Steel KP, Bock GR. *Nature* **1980**; 288: 159–161.

45 Hardisty RE, Erven A, Logan K, Morse S, Guionaud S, Sancho-Oliver S, Hunter AJ, Brown SDM, Steel KP. *J Assoc Res Otolaryngol* **2003**; *4*: 130–138.

46 Steel KP, Moorjani P, Bock GR. *Hearing Res* **1987**; *28*: 227–236.

47 Rhodes CR, Parkinson N, Tsai H, Brooker D, Mansell S, Spurr N, Hunter AJ, Steel KP, Brown SDM. *J Neurocytol* **2003**; *32*: 1143–1154.

48 Martin P, Swanson GJ. *Dev Biol* **1993**; *159*: 549–558.

49 Sharpe J, Ahlgren U, Perry P, Hill B, Ross A, Hecksher-Sorensen J, Baldock R, Davidson D. *Science* **2002**; *296*: 541–545.

50 Avraham KB, Hasson T, Steel KP, Kingsley DM, Russell LB, Mooseker MS, Copeland NG, Jenkins NA. *Nature Genet* **1995**; *11*: 369–375.

51 Cable J, Huszar D, Jaenisch R, Steel KP. *Pigment Cell Res* **1994**; *7*: 7–17.

52 Colvin JS, Bohne BA, Harding GW, McEwen DG, Ornitz DM. *Nature Genet* **1996**; *12*: 390–397.

53 Steel KP, Barkway C. *Development* **1989**; *107*: 453–463.

54 Hunter-Duvar IM. *Acta Otolaryngol* **1978**; *351*(Suppl.): 3–23.

55 Pau H, Fuchs H, Hrabé de Angelis M, Steel KP. *Laryngoscope* **2005**; *115*: 116–124.

56 Kiernan AE, Ahituv N, Fuchs H, Balling R et al. *Proc Natl Acad Sci USA* **2001**; *98*: 3873–3878.

57 Kiernan AE, Erven A, Voegeling S, Peters J et al. *Mamm Genome* **2002**; *13*: 142–148,

2
Molecular Phenotyping: Gene Expression Profiling

Johannes Beckers

2.1
Why this Screen? Medical and Biological Relevance

In addition to identifying the function of every gene in the genome, the manner in which these gene functions are integrated into the complex network of molecular interactions in the cell will also need to be elucidated. Describing profiles of expressed genes in defined tissues and under distinct conditions is one important piece of this gigantic puzzle. Most biochemical processes within and between cells are put into effect by the interaction between proteins, or between proteins and their substrates. The proteome of a cell is the result of controlled biosynthesis, and hence largely but not exclusively, regulated by gene expression. In turn, gene expression can be regarded as a sensitive read-out of the biochemical state and the proteome of the cell. In this regard, the genome is the tool case of the organism. When and where these tools are applied during development, in homeostasis and disease is integral to gene function. The temporal and spatial logistics of gene expression is largely controlled by the regulation of transcription and translation. Feedback between the transcriptome and proteome is a highly complex procedure which is controlled by equally sophisticated mechanisms. The understanding of this functional regulation is still very incomplete and generally limited to isolated signaling or metabolic pathways. However, evidence is accumulating that such pathways are components of complex and interacting networks.

DNA-chips or DNA-microarrays provide the technology to monitor genome-wide gene expression at the transcription level. Gene-specific oligonucleotides or double-stranded cDNA fragments are immobilized in defined positions on a solid support and hybridized to complex mixtures of expressed nucleic acids. The current microarray spotters and spotting pins allow up to 40 000 probe-spots (also called features) to be fitted onto a standard chip of the size of a normal histological slide. An important advantage of using glass as a transparent, solid support is that it allows the simultaneous, competitive hybridization of test and reference samples labeled with distinct fluorescent dyes. Relative expression levels are analyzed directly by comparing each fluorescent signal of every feature. Production, hybridization, and scanning of such DNA-chips can be automated to a great extent thus allowing the use of high-throughput approaches.

Standards of Mouse Model Phenotyping. Edited by Martin Hrabé de Angelis, Pierre Chambon, and Steve Brown

ISBN: 3-527-31031-2

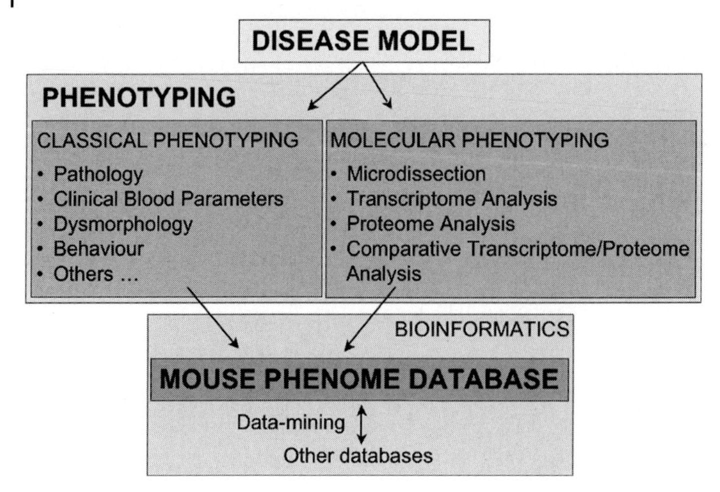

Fig. 2.1 Phenotyping is essential in the analysis of mouse mutants. In addition to classical phenotyping protocols, molecular phenotyping (transcriptome and proteome analyses) is an important tool in functional genomics and gene annotation. Transcriptional and post-transcriptional regulation can be distinguished by comparative analysis of the transcriptome and proteome. Sample homogeneity can be improved by the combination of innovative microdissection technologies with molecular phenotyping techniques, a procedure which poses an important challenge. The archiving of phenotypic data in mouse phenome databases serves as a virtual center. Phenome databases from different species are an important source for data-mining.

The recent technologies which enable genome-wide expression analyses to be carried out have been recognized as a complementary approach to the more classical phenotyping strategies. The characterization of changes in as many aspects as possible in an organism is a fundamental principal of a comprehensive phenotyping approach. The development of techniques to monitor gene expression at the genomic level makes phenotyping at a new molecular level possible [1, 2]. The feasibility of monitoring genome-wide gene expression makes it possible to achieve an unbiased assessment of changes in gene regulation that may be induced by a mutation, a drug or environmental changes. Expression profiling studies can detect underlying molecular phenotypes that may otherwise not be identified in standard phenotypic screens which focus only on external morphology, blood parameters, immunological alterations and so on. In particular, in combination with high-throughput mutagenesis methods, expression profiling improves the efficiency of the phenotype characterization of existing mutants and the isolation of new mutants (Fig. 2.1).

2.2 Examples: Diseases of Mouse and Man

Many of the early publications which described DNA-chip expression profiling, applied the procedure to the classification of tumors such as breast [3, 4] and prostate [5] cancers, leukemias, embryonal tumors [6], and inflammatory diseases such as rheumatoid arthritis [7]. The molecular phenotyping of tumor tissues has been used as an unbiased and systematic approach for tumor classification based on genome-wide expression data [8]. It enabled the tumors that are morphologically, histopathologically, and cytogenetically indistinguishable but which differ in their response to therapy, to be differentiated from one another [9]. Secondly, by defining the molecular characteristics it became potentially possible to specifically design drugs which can alter signaling activities to normal or closer to normal levels [10]. Several target-designed drugs, aimed at specific molecular pathways, are already on the commercial market. Two examples related to the classification of human leukemias which show how molecular data in the form of expression profiles can be used as a diagnostic tool and to select therapy, are given below.

Differentiation between acute leukemias derived from lymphoid precursors (acute lymphoblastic leukemia) and those derived from myeloid precursors (acute myeloid leukemia) is critical for successful chemotherapeutical treatment [11]. In the first example, using RNA samples from 38 acute leukemias and chips containing 6817 genes it was demonstrated that the expression levels of approximately 1100 of these genes were more likely to correlate with the class distinction between acute lymphoblastic and acute myeloid leukemia than to be random [12]. Independent new samples from heterogeneous sources of acute leukemias were assigned to one of the two classes based on the expression of 50 informative genes. Of 34 samples, strong predictions were made with 100% accuracy for 29. Interestingly, the arbitrarily chosen informative genes not only included markers of the hematopoietic lineage but also genes related to cancer pathogenesis, i. e. genes that code for proteins involved in S-phase cell cycle progression, chromatin remodeling, transcription, and cell adhesion as well as known oncogenes. Thus, expression profiling may also provide insights into cancer pathogenesis and pharmacology [12].

The second example provided evidence to show that the classification of cancers facilitates the design of therapeutic treatment for individual cases. It was shown that based on their RNA expression profiles, the morphologically indistinct group of diffuse large B-cell lymphomas (DLBCL) can be classified into at least two groups having similar molecular homogeneity. Based on the expression of B-cell marker genes, one group was related to germinal center (GC) B cells and the second subtype expressed genes that are indicative of *in vitro*-activated peripheral blood B cells. Interestingly, in this study GC B-like DLBCL patients had a 76% chance of survival for the 5 years following standard multi-agent chemotherapy, this was reduced to only 16% for patients with activated B-like DLBCL, whereas for the entire patient group the average chance of survival for 5 years post treatment was 52% [13]. This study demonstrated that applying more precise classifications of malignancies would lead to a more selective use of therapy [14, 15] on the one hand and on the other, that this technique could also be helpful in designing new drugs. By precisely identifying the therapeutically-induced molecular pathways in malignant cell lines which are activated in response to chemotherapy, it may become possible to design drugs that target only those pathways that are clinically beneficial. Such an approach could help circumvent the significant side-effects of chemotherapy.

Several publications have reported the use of DNA-microarray expression profiling for transcriptome analyses in mice. For example, the transcriptional response to aging in the mouse brain has significant similarities with that in human neurodegenerative disorders such as Alzheimer's disease [16, 17]. The differential gene expression in various regions of the brain and the response to seizure have also been analyzed and the results have provided evidence that particular differences in gene expression may account for distinct phenotypes in inbred mouse strains [18]. These reports and other work [19–21] have provided the proof-of-principle that despite the complexity of mammalian organs, expression profiling is a useful tool in identifying pathways associated with particular biological processes in the mouse model system.

2.3 Diagnostic Methods: History and State of the Art

Gene expression analyses have been performed on a gene-by-gene basis since the mid-1970 s typically using Northern blot analyses [22]. Since the development of an efficient method for the amplification of DNA fragments in the mid-1980 s reverse transcription PCR has become the most sensitive method for the detection of expressed genes [23]. There are currently two principle methods used for the simultaneous analysis of a large number of transcripts in a single experiment, i. e. serial analysis of gene expression (SAGE) and microarrays. SAGE is based on the finding that short sequence tags of 10 to 14 base-pairs are sufficient to identify the gene from which the cDNA tag has been isolated [24]. The frequency with which such SAGE tags are identified provides the expression level of the corresponding gene. To obtain this information SAGE tags are isolated from RNA pools, concatenated, cloned and finally sequenced. This technology has been successfully used for the evaluation of different transcript profiles. In particular, SAGE does not require any *a priori* knowledge about expressed sequences. It can detect novel genes in the transcriptome, provided that the SAGE tag is from a unique position within the expressed transcript. Microarray gene expression profiling, in contrast, is a hybridization-based technology. It is more commonly used than SAGE because the latter is technically rather challenging and requires a great deal of work. Microarrays for expression profiling employ a solid support to immobilize specific probes in defined positions. These microarrays are then hybridized to a complex mixture of expressed nucleic acids that are most commonly radioactively or fluorescently labeled. This is in marked contrast to Southern blot analyses (and also dot blot hybridization) where the complex mixture of expressed genes is immobilized on a membrane and hybridized to a specific and labeled probe. The first publication to report the application of the microarray concept to gene expression profiling, however, is rarely cited in reference to this technology [1]. In this seminal work a nylon membrane containing 4000 cloned cDNAs was used for the first time to profile relative gene expression in normal and neoplastic human large intestine biopsies and in colonic carcinoma cells differentiated *in vitro*. This work was of particular significance because the incorporation of such a large number of cloned cDNAs had not been attempted hitherto. Possibly because this publication focused on the biological questions at hand rather than on the application of a novel technique it is rarely recognized as the first application of microarrays. The most commonly used method for expression profiling microarrays, in particular for non-commercial arrays, employs glass (DNA-chips) as the solid support for probes and fluorescent dyes as the label for the target. A major advantage of this concept is the fact that it allows dual-color, competitive hybridizations on a single DNA-chip. This greatly facilitates direct comparison of expression levels in an experimental and reference sample, since perfect chip-to-chip reproducibility of spot sizes and morphology is not essential, and it is not necessary to calculate the absolute amount of probe contained in a spot. Custom-made DNA-chips are generally printed on slides of the size of a normal histological slide which, in contrast to membranes, allows for a high degree of automation in spotting, hybridization, washing, and scanning procedures. The proof-of-principle for this DNA-chip strategy was provided using poly-L-lysine-coated glass slides as support for 48 PCR-amplified cDNA probes [2]. This DNA chip was used to estimate the dynamic range of the detection method, and to confirm the regulation of specific genes in transgenic *Arabidopsis thaliana* and the regulation of genes between root and leaf tissue. The feasibility of extending the technology to genome-wide expression profiling had already been anticipated in this publication. Dual-color, competitive DNA-chip hybridizations are still most commonly used for microarray expression profiling with cDNA microarrays that are produced in academic laboratories. Commercial DNA-chips often use oligonucleotide microarrays. These may be manufactured by *in situ* synthesis on glass using a combination of photolithography and oligonucleotide synthetic chemistry. Such commercial arrays can be produced with a chip-to-chip reproducibility and density that is difficult to obtain using a spotting method. The reproducibility of these DNA-chips is generally regarded to be so good that using competitive, dual-color hybridizations is no longer advantageous. Instead, a single set of reference sample experiments may be sufficient to analyse expression profiles of more than one experimental sample, strongly reducing the number of reference RNAs needed in more complex experimental set-ups.

Since the mid-1990s DNA-chip expression profiling has become an indispensable tool in the analysis of molecular mechanisms underlying disease in humans and model organisms. Some early examples were given in the previous chapter. Today a plethora of papers has successfully described the use of expression profiling to classify tumor types and groups of patients at a molecular level, predict clinical outcome upon treatment, identify affected pathways and novel genes involved in biological processes and suggest targets for therapy. These findings from retrospective studies have generated great interest in the application of microarray-based expression profiling to human samples from real-time clinical studies. Despite these great expectations the application of DNA-chip expression profiling to clinical diagnosis is still elusive. However, a number of chip designs for expression profiling are in clinical trials [25]. In addition to its application in gene expression analysis, DNA-chip based technology has recently been approved for the first time for a clinic diagnostic purpose in the US and the European Union. This DNA-chip identifies allelic variations of two cytochrome P450 genes that determine the metabolic rates of several frequently prescribed drugs. This apparently first chip-based clinical diagnostic test aids physicians in selecting drugs and individualizing treatment doses according to the patient's genotype. Presumably, the arsenal of chip-based diagnostic products will grow in the coming years, as will the number of expression profiling applications.

2.4 Technical Requirements for Screening Protocols (Short): First and Second Line Approaches

Microarray expression profiling is a complex experimental procedure consisting of several consecutive steps: fluorescent or radioactive labeling of extracted RNA, hybridization, washing, scanning, image processing and data analysis. Each of these steps is a potential source of artefacts that may significantly influence the final result and the exact appraisal of differential gene expression. In parallel, the quality of expression profiling experiments also depends on the design and quality of the arrays themselves.

2.5 Logistics (Whom, When, How Many, Why)

2.5.1 Choice of Platform

The first consideration in microarray expression profiling is the choice between a commercial or a custom tailored, in-house platform. Commercial solutions are available which either cover the entire procedure from RNA extraction to expression data analysis or facilitate particular steps of the procedure: labeling of cDNA, the microarray itself, or data analysis. If the plan is to perform a rather limited number of experiments for a single analysis, it is most likely not advisable to establish a complete microarray facility. A comprehensive commercial solution is the most efficient method to use for a single expression profiling analysis. Microarrays for most model organisms are now available. These include whole genome arrays with up to 100 000 probes for almost every expressed sequence as well as thematically-focused arrays containing a few hundred probes. When planning to use expression profiling in systematic large-scale projects or as part of a centralized service unit, a microarray spotter and a suitable DNA-chip scanner are the minimum investments required. Commercial solutions for scanners generally include software for the identification of spots, background correction, flagging of irregular or missing spots, and normalization of array data. However, the identification of statistically significantly regulated genes from a set of microarray data will in most cases, require additional software. Some such statistical analysis tools are available in the public domain [26, 27], but are also offered on the free market. The latter may have the advantage of providing additional material for the interpretation of the statistical evaluation.

A major decision to be made when constructing a microarray is which sequences are to be arrayed. Sequences for probes can be identified from the public UniGene database for example [28]. This database automatically partitions GenBank sequences into a non-redundant set of gene-oriented clusters. Each UniGene cluster contains sequences that represent a unique gene, as well as related information such as the tissues in which expression of this gene has been detected in previous experiments and the chromosomal map location. For *Mus musculus* the UniGene database currently contains approximately 3.8 million sequences in clusters and close to 19 000 clusters that contain both mRNAs and expressed sequence tags (ESTs). In most cases UniGene clusters contain a number of clones that may be selected as probes for microarrays. However, selection of the right probes is an important and sometimes enigmatic task. Anyone with practical experience in hybridization methods such as Southern blot analysis or *in situ* hybridization methods may have found that the choice of the best working probe is to a certain extent a matter of trial and error. This is certainly also true for microarray hybridization experiments. Of course, the probe selection should include criteria such as uniqueness of the sequence, content of GC and AT bases, hypothetical melting temperature and so on. It is therefore a delicate task to select gene-specific probes, in particular when constructing a genome-comprehensive microarray. Since there is a decade of experience in DNA-chip expression profiling to consult, it is probably sensible to use microarray probe sets that have already been described. In some instances such probe sets have already been experimentally tested for their specific hybridization signals [29]. Microarray clone sets can be obtained from public resources or purchased either as bacterial clones or as oligonucleotides. It is highly recommended to choose a probe set that has been completely sequenced. The annotation of probes with gene names is often incorrect. Having the sequence of the probe gives the researcher the opportunity to test the probe sequence against the most recent up-date of the mouse genome sequence and to verify the uniqueness of the probe sequence. One advantage of oligonucleotide probe sets is that they can be spotted on arrays without prior amplification using PCR. In contrast, bacterial clone probe sets have the benefit of regeneration. Amplifying a genome-wide bacterial clone set is a major task requiring amplification by PCR from bacterial lysates and the purification and pooling of up to four repetitions to obtain the required concentration and quantity. Each PCR amplification should be tested on an agarose gel to validate the quality of the PCR product. Using liquid-handling robots the amplification of bacterial probe sets may be automated to a large extent.

PCR products or oligonucleotides are spotted by a robotic microarrayer onto glass slides (DNA-chips), or nylon or nitrocellulose membranes. The production of microarray membranes is technically less challenging. However, membranes may contain only up to a few hundred probes, and may also prohibit the use of competitive, dual-color hybridizations and are not well suited to the automation of hybridization and scanning processes. Both, chips and membranes can be purchased with surfaces that have been pre-treated to augment their surface charge, increase the adherence of the probe, and reduce background signals [30].

In addition to a microarray spotter, array scanner and an optional liquid-handling robot, a hybridization station may also be a useful investment. The latter will automate the parallel hybridization and washing and drying of several DNA-chips. This is not only a valuable addition to a DNA-chip facility if large-scale projects are being undertaken, it is also useful in largely increasing the chip-to-chip reproducibility. There are several hybridization stations for DNA-chips on the market. However, they vary greatly in their performance. For all major investments in establishing a microarray facility it is strongly recommended that they be tested in great detail in the researcher's own laboratory environment using their own microarray set-up.

2.5.2
Biological Samples

DNA-chip-based transcript profiling has been successfully used for the analysis of gene expression in mouse organs. For example, transcriptional responses to aging in the mouse brain [16, 17], differential gene expression in distinct brain regions and the response to seizure [18] and other analyses of mouse models [19–21, 31–33] have all demonstrated that despite the complexity of mammalian organs, expression profiling is a useful tool for identifying pathways associated with particular biological processes. In addition, systematic approaches have been initiated to analyse gene expression profiles in an extended collection of distinct mouse organs and to phenotype mutagenized mouse embryonic stem cells at the molecular level [34–36]. The feasibility of detecting transcriptionally affected organs among ENU-induced mouse mutant strains employing RNA expression profiling as a tool for molecular phenotyping has also been demonstrated [37].

A major advantage of the mouse model system, as compared to gene expression studies in human samples, is that in this system it is possible to analyze gene expression in homogenous genetic backgrounds. In addition, housing conditions can be standardized to a large extent. However, biological noise in gene expression levels will still be present when profiles are compared between organs of individual mice. Control of biological and experimental noise is therefore an important issue in the analysis of differential gene expression in complex tissues with heterogeneous cell types [29, 38, 39]. Natural variability in gene expression levels has been demonstrated for genes oscillating in a circadian rhythm, and for immune-modulated and stress-responsive genes [40, 41]. Thus, RNA expression profiling experiments require both biological and technical replicates to monitor noise in the data and to allow proper statistical analyses. In particular, biological replicates need to be analyzed in order to estimate the inter-individual variability in gene expression levels and to set up well designed experiments [38]. Several measures need be taken to reduce biological noise to a minimum, and parameters relating to age, gender (hormonal status), health and stress need to be considered. Controlling variables such as social status and food intake before necropsy will be beneficial with respect to biological variation in gene expression levels [37].

Nevertheless noisy genes do not, in general, compromise the identification of differentially expressed genes when replicate hybridizations are carried out with RNA samples from individual mice. Selecting genes from such data, based on their constant differential expression in all individuals strongly reduces the risk of identifying false positive genes. In this respect it is particularly important to select a pool of reference samples that does not introduce a bias into the expression profiling experiments. Controlling biological noise in reference or wild-type RNA samples may be achieved by performing hybridizations of individual wild-type samples against each other in addition to comparative mutant or experimental versus wild-type or reference hybridizations.

An approach suggested to increase throughput in DNA-chip expression profiling experiments is a combination of an increased number of experimental samples and sub-pooling strategies [42]. Reports describing the comparison of complete pools of differentially expressed genes with augmented numbers and individual samples [43] have shown that sub-pooling provides equivalent power, improved efficiency and cost-effectiveness [42, 44]. However, a direct side-by-side comparison of individual and pooled samples from the same RNA source suggests that, if biological variation in gene expression levels is an issue, pooling of samples is not an efficient method for the identification of differential expression. Using pooled samples a smoothing effect for genes with strong variability in expression levels between individuals was rarely observed [37]. Using only a single pooled sample from a limited number of RNA sources (as is occasionally required in biological experiments) true positives may not be distinguished from false positives.

2.6 Trouble Shooting

Various potential artefacts may be associated with microarray technology, which could lead to wrong conclusions if no independent verification is carried out to exclude them. Some attention was paid to this point in recent reviews [45], where the authors focused on the problem of reproducibility. A major pitfall of expression profiling is given by particular artefacts that can be highly reproducible and therefore may easily be interpreted as a true result. Here, we highlight some experimental aspects of this problem. We will not focus on bioinformatic and statistical aspects [46–48], or review experimental methods such as RNA extraction [49, 50] and the collection of biological samples. The standardization of these methods is a prerequisite for high quality expression profiling data, but satisfactory commercial solutions are available, for the isolation of RNA from essentially any source of biological material. For a concise guide to microarray experiments the reader is referred to published reviews and practical approaches [51].

Rather, we consider here the possible artefacts of microarray expression profiling experiments which come to light during the course of preparation and labeling of target RNA (the sample that will be hybridized to the immobilized probe), during hybridization, and artefacts that may come into play during post-hybridization image processing and those that may result from the production of microarrays.

The issues discussed below mostly concern users of custom-made cDNA microarrays. For the major commercial DNA array platforms the reader is referred to internet-based discussion groups and trouble-shooting guides published on the web.

2.6.1
Preparation of Hybridization Target

To be used in microarray expression profiling experiments, samples of mRNA need to be detected after hybridization. The two most popular approaches exploit the detection of fluorescent or radioactive labels. We will not discuss here other approaches such as surface plasmon resonance [52, 53] due to their current lack of practical use in expression profiling experiments. However, surface plasmon resonance enables monitoring of hybridization kinetics [54], which could help to estimate cross-hybridization (see below). Therefore, this method may have potential for future improvement of microarray technology.

Both fluorescent and radioactive approaches require the target to be labeled prior to hybridization. The fluorescent approach enables the simultaneous use of two or more labeled samples within the same microarray (Fig. 2.2). This method may, for example, be used to include an internal control in the expression profiling experiments, e. g. a wild-type target or a general RNA standard, labeled with one fluorescent dye versus a mutant sample labeled with another dye. Such a comparison of an experimental sample with an internal control eliminates some artefacts associated with array reproducibility, such as size and shape of the spots and the amount of probe transferred. Another advantage of the fluorescent approach is the improved spatial resolution of fluorescent scanners as compared to phosphorimagers by more than two orders of magnitude. This permits the use of microarrays with feature sizes of 20 μm and containing up to hundreds of thousands of elements [55].

Artefacts in radioactive approaches may arise from the interference of spots with strong hybridization signals in adjacent features when they are more densely spotted. However, some publications have argued that this poses no serious problem [56]. A major advantage of radioactive approaches is their higher sensitivity, i. e. the ability to detect transcripts of low copy number. Therefore, efforts are being made to increase the sensitivity of fluorescent approaches.

One way of improving the sensitivity of fluorescent approaches is by increasing the number of fluorescent groups per target molecule. Instead of labeling the 5' or 3' ends of the targets, fluorescently labeled dCTP (or other dNTPs) can be incorporated by reverse transcription of mRNA pools [57]. This procedure is usually referred as direct labeling (Fig. 2.3). However, reverse transcriptase processes unlabeled and labeled dNTPs with variable sequence-dependent efficiency. This efficiency is also different for alternative fluorescent labels, such as Cy3 or Cy5. This is known to result in a dye-dependent bias of reverse transcription yield and labeling efficiency [45]. To avoid this problem aminoallyl-labeled dNTPs are often used for reverse transcription. Fluorescent dyes are then coupled to the incorporated amino groups in a second labeling step. Since the same aminoallyl dNTPs are used for both fluorescent dyes, a dye-specific bias is avoided during reverse transcription. A recent protocol for this indirect labeling method is available from The Institute of

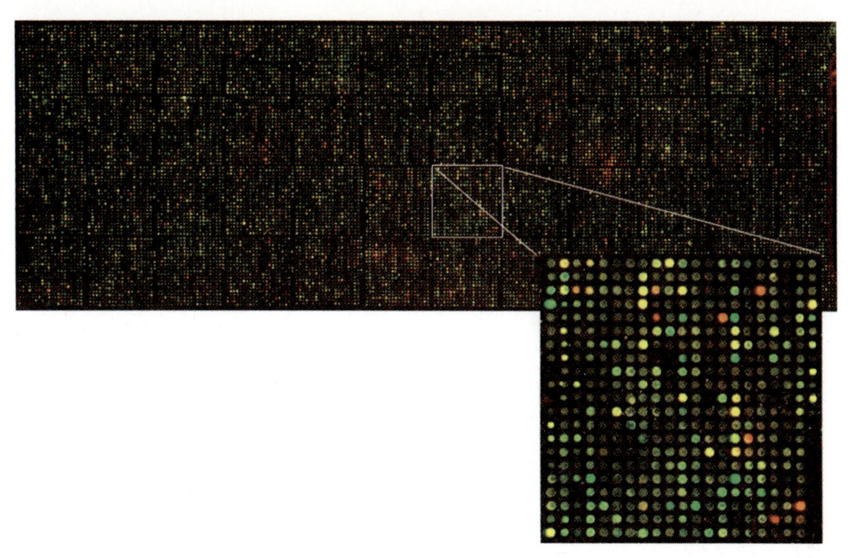

Fig. 2.2 Scanned image of a glass DNA-chip after dual color, competitive hybridization. Red staining indicates overexpression, and green, underexpression. The DNA-chip was printed on a slide of the size of a normal histological slide and contains approximately 21,000 PCR amplified probes.

Genomic Researches (http://atarrays.tigr.org/). Indirect labeling was shown to produce less dye-specific bias and a better correlation of Cy3 versus Cy5 data as compared to direct labeling procedures [58].

The sensitivity of fluorescent approaches can be further increased by more than 10-fold by the incorporation of fluorescent dendrimers to target cDNAs [59]. Commercially available dendrimers are three-dimensional DNA structures with several hundred fluorescent labels per molecule. In expression profiling experiments labeled dendrimers are designed such that they have free arms that are complementary to primer sequences that are incorporated into cDNAs derived from the target mRNAs during reverse transcription. Experimental and reference cDNA pools are produced with various primers complementary to dendrimers with different fluorescent labels. Both dendrimers are then added to the hybridization mixture and hybridized to the 5' ends of the cDNAs. As with the use of aminoallyl dNTPs, this method also excludes the potential bias of direct incorporation of fluorescent labels into the target cDNA during reverse transcription. The dendrimer method was evaluated in comparison to direct and indirect labeling protocols in many aspects and was found to be comparable and sometimes even superior [58, 60]. However, one essential limitation was discovered: there is a weak response to changes in concentrations of cDNA pools labeled with one particular dye [60]. This response was especially weak for highly expressed genes. For example, a four-fold dilution of Cy-5 labelled target cDNAs resulted in less than a two-fold change in the signal ratio. A possible explanation for the lack of a quantitative response may

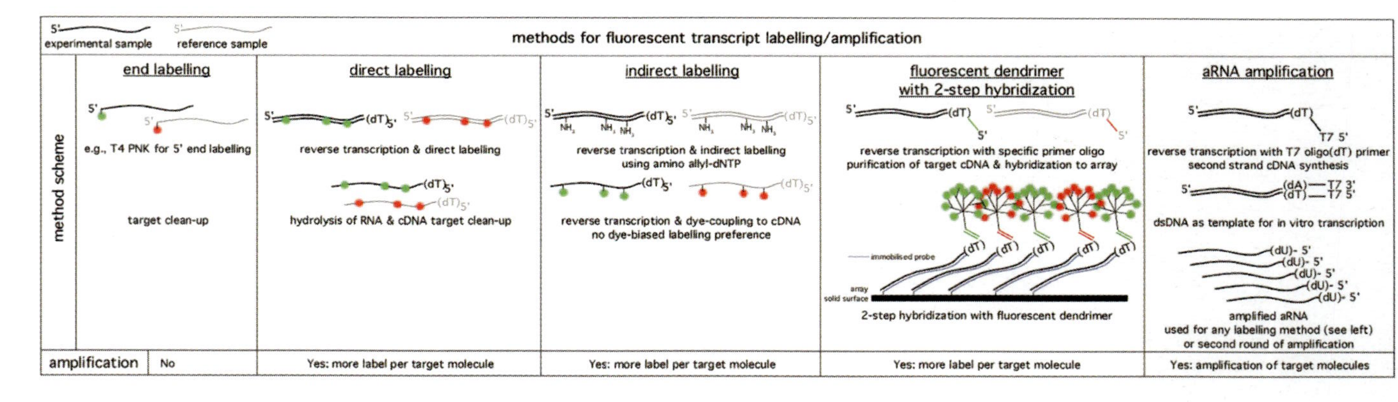

Fig. 2.3 Schematic illustration of the most commonly applied methods for labeling and amplification of target RNAs.

be found in the limitations of the kinetics with which dendrimers hybridize to the specific primer sequences of the target cDNAs.

Several applications of gene expression profiling require the use of limited amounts of biological material. This is the case for example, when biopsies of human patients or specific embryonic tissues of animal models are being analyzed. Whereas the methods described above are generally applied to amounts of total RNA in the range of 1 to 50 μg, microdissected tissues may yield total RNA in the range of a few ng to 1 μg. Such minute amounts of biological material require a reproducible and efficient way of amplification of the starting material. At least two methods are currently used: one method is based on the use of T7-promoter activity to amplify ds-cDNA, the other approach uses RT-PCR. The first technique has been suggested to result in a linear amplification in the range of two to three orders of magnitude. Such linear amplification methods are also used in some of the major commercial DNA-chip arrays [61]. It was argued that amplification introduces virtually no bias in expression profiling. Subsequently contradictory results were published on the reliability of this method. Some research groups [62] found that this bias may be considerable and suggested specific modifications of this technology to minimize artefacts resulting from amplification. In contrast, it was argued by others that transcript representation is "faithfully preserved" in both linear and exponential RT-PCR amplification [63]. It was also found that linear amplification using oligo-dT primers yielded a better specificity than direct reverse transcription labeling with Cy3-dUTP or linear amplification using random primers and mixtures of random and oligo-dT primers [64]. These contradictory results suggest that the methods for amplification-labeling in expression profiling experiments are not yet robust and not easily reproducible in different laboratories. The establishment of such techniques, therefore, requires a very careful examination of standards and controls.

Since none of the target preparation methods described above guarantee artefact-free results, (e. g. exclusion of gene-dependent dye-specific bias) it is common practice to include so-called color flip (or dye swap) hybridizations in the experimental design. This means that dual-color competitive hybridizations are repeated such that the fluorescent labels are swapped between reference and experimental target in independent hybridization experiments. Only genes with dye-independent response in color flip experiments should be considered as candidates for differential gene expression.

2.6.2 Critical Issues of Chip Hybridization

Microarray expression profiling relies on a fundamental property of nucleic acids: their ability to form duplexes (to hybridize) with complementary nucleic acids. However, they can also hybridize with only partially complementary nucleic acids, thus forming mismatched duplexes that may be rather stable [65, 66]. This phenomenon is usually referred to as non-specific hybridization or cross-hybridization and is one of the major concerns in microarray experiments.

One method of controlling this source of artefacts is the careful design of probes using bioinformatics tools [67]. Theoretical considerations such as avoiding repetitive sequences and conserved functional domains of paralogous genes have been suggested as criteria for the selection of specific probes. The applicability of such strategies depends on the completeness of sequence information. However, experimental approaches to the reliable verification of probe specificities may be required. This may be achieved by including at least some *a priori* non-specific probes as negative controls in the array (e. g. plant probes on an array of mouse probes). This strategy allows the assessment of the overall stringency of the hybridization rather than the specificity of individual probes. In the case of significant hybridization with such *a priori* non-specific probes, the hybridization stringency should be increased. The latter depends on the particular hybridization platform, i. e. the type of support (solid surface or gel matrix), the type of probe immobilization (covalent or non-covalent) and the type of immobilized probes (oligonucleotides with equalized melting temperatures or cDNAs from commercially or academically available libraries).

Most of the current microarray experiments are based on the immobilization of probes on a solid surface. In this approach the surface is modified with chemical groups (such as aldehyde, epoxy or amino groups) that either covalently or non-covalently bind the (chemically modified) probes. Alternatively, using gel immobilization such chemical groups are positioned within the three-dimensional structure of the gel, which in turn is covalently attached to the solid surface. Such approaches were introduced by several research groups [68, 69] and later became commercially available. It was argued [70] that gels offer a three-dimensional environment for immobilized probes and hybridizing targets. Hybridization in gel is similar to hybridization of nucleic acids in solution, which has been well studied over recent decades. In addition, thermodynamic parameters were measured *in situ* for oligonucleotides immobilized in polyacrylamide gels which enables an estimation of the hybridization thermodynamics and efficiency [71]. The well-understood physics of hybridization in solution (and gel matrices) provides a better theoretical basis for controlling the parameters required for specific hybridization procedures.

Depending on the chemical group used to bind nucleic acid probes, immobilization may be covalent or non-covalent. One of the most popular immobilization techniques relies on non-covalent binding of non-modified nucleic acids on a positively-charged glass surface [51]. The positive charge is typically provided by amino groups (poly-L-lysine, gamma-amiminopropylsilane, etc.). This method is simple, inexpensive and results in relatively strong hybridization signals. This is probably due to the compensation of the negative charges of the probe by the positive charges of the substrate and the elimination of electrostatic repulsion.

Importantly, there is very little experimental data concerning the true physical mechanisms of probe immobilization and duplex formation on a solid surface. On the one hand, oligonucleotides immobilized on a glass surface can discriminate single base-pair mismatch hybridizations [72]. At the same time it was demonstrated [73] that such probes do not form ordinary double helical structures with hybridized targets. It was calculated that the duplex between the target and the immobilized probe is approximately 1.5 times longer than the B-form of duplex DNA and that the identified duplex structure cannot be attributed to any of the known structures of dsDNA. Therefore, the question arises: what other non-canonical structures could be formed with the immobilized probe? And in particular whether such duplex structures may in some way favor non-specific hybridizations? Indeed, it was found [74] that hybridized targets can form triplex structures with arrayed probes on positively-charged surfaces, similar to triplex structures in homologous recombination events facilitated by RecA protein. Interestingly, the authors point out that the binding of RecA protein has been shown to elongate the duplex DNA by a factor of ~ 1.5 [75]. The duplex elongation on a positively-charged surface was confirmed independently (see above) and it is still unclear what other consequence of such abnormal behavior will be discovered. A fluorescent signal in the center of probe spots, which cannot be attributed to any kind of hybridization, was also reported to be seen on such slides [76].

A common application of covalent probe immobilization uses amino-modified probes (cDNAs or oligonucleotides) and aldehyde-modified glass surfaces [57]. In some commercial applications amino-modified probes are covalently immobilized on gel-coated slides. In this case the sensitivity and specificity of 50-mer oligonucleotides were assessed [77]. It was found that cross-hybridization may be considerable if the hybridizing target has more that 75 % overall homology with the 50-mer probe or if it contains stretches of contiguous complementary sequences more than 15 base-pairs long. The specificity of immobilized probes has also been evaluated based on hybridization kinetics from 4 to 48 h of hybridization [78]. It was found that a non-specific target hybridizes faster to arrays of *in situ*-synthesized 60-mers than specific target molecules. This implies that longer hybridizations result not only in stronger but also in more specific signals. Recently, probe specificity was also assessed by means of post-hybridization fractionation curves for cDNA probes covalently immobilized on a glass surface [29]. Fractionation curves (hybridization signal intensity versus the formamide concentration in the washing solution) were generally found to have a stepped profile with the position of the step corresponding to the stability of the duplex. For non-specific hybridization the step was observed at lower washing stringency than that for specific hybridizations. In contrast to the incorporation of *a priori* non-specific probes, fractionation curves and the kinetics method described above are experimental tools used to assess the specificity of each individual probe of a microarray. The iterative use of these methods provides an essential tool for establishing experimentally-verified sets of probes with tested specificity.

Historically, the first microarrays for expression profiling were arrays of cDNAs [1, 2]. It was believed that using such long cDNA probes as compared to short probes such as oligonucleotides, a higher specificity could be achieved. The competitive approach applied by some of the major commercial DNA-chip platforms exploits light-directed highly parallel *in situ* synthesis of oligonucleotide probes [79]. Taking advantage of the high parallelism, this approach allocates up to 40 probes for each gene. Half of them are designed to perfectly match the target sequence and the other half includes mismatch hybridizations. The hybridization signal for a particular gene is calculated based on intensity differences between perfect match and corresponding mismatch hybridizations. This method should improve the reliability of expression profiling data but suffers from some difficulties when signals from mismatch hybridizations are stronger than those from perfect match hybridizations. This theoretically well-controlled method is complicated in practice by protocols involving several enzymatic reactions followed by the fragmentation of cRNA, which is necessary to achieve strong hybridization signals with short probes immobilized on a solid surface. This can in turn results in some bias regarding linear target amplification as discussed in the previous chapter.

It is also worth mentioning that dual-color competitive hybridization underestimates the actual ratio of target concentrations. This effect can easily be measured by increasing the concentration of one cDNA pool in competitive dual-color hybridizations [64, 80]. This phenomenon is unlikely to be attributed to cross-hybridization since the cross-hybridizing target concentration will increase to the same extent as the specific target. Recently, the degree of underestimation was assessed by comparison with RT-PCR and an empirical correction was suggested [81].

2.6.3
Image Processing and Array Design

To retrieve useful and reliable information from microarray experiments it is essential to correctly process the scanned images of hybridized arrays. Spots of hybridized target must be identified against the background of noise and the fluorescent signal must be properly quantified. This task is further complicated by irregularities in size, shape, and position of spots. Microarray image processing software tools can generally solve these problems. One example of this type of software is the GenePix software supplied with Axon microarray scanners, which was also used in many publications reviewed in this chapter [29, 63, 76, 80–82]. Its success may in part be due to the fact that it is designed for a reliable microarray scanner, but may also be due to a convenient interface and reliable algorithms. It is essential in microarray expression profiling experiments that not every spot should be quantified and used in further statistical analysis. For example, low intensity spots cannot provide accurate values of intensity ratios in dual-color hybridizations. Such low intensity spots should rather be marked or flagged as unreliable and should be excluded from further consideration of differential gene expression. High intensity and small size spots should also be flagged as unreliable since they are unlikely to result from hybridization signals, rather they are most likely caused by noise or dust particles. Another widely used image processing software, ImaGene, has additional features enabling more flexible flagging, and has also been used in some of the work cited in this chapter [83, 84]. More advanced algorithms for spot locating and flagging have been proposed [83] based on computation of quality indices accounting for signal to noise ratios, spot size, local background variability, extremely high local background and spot saturation. Another simple and powerful method of improving the quality of microarray expression data (also used by GenePix) is to take into account the correlation of mean and median intensities of spots [82]. This allows the effective elimination of spots with extremely irregular shape and spots contaminated with, e. g. dust particles.

When high quality microarray images and data can be achieved, consistent results are obtained with the different image processing software tools with or without the quality scoring considered above [83]. In practice, problems appear when the quality of arrays and/or images is not optimal. Therefore, it is not surprising that more reliable results are obtained with repetitive data for each gene [85]. In the case of multiple spotting of particular probes on a single array, the coefficient of variation can be used as penalty score for this particular probe or gene, and it was found that the quality of the data on the expression level of each gene is inversely related to its coefficient of variation [84]. By measuring the coefficient of variation it was demonstrated that the absence of Denhardt's solution in hybridization buffer results in a lower quality of data. Denhardt's solution is recommended by many hybridization protocols (e. g. by the TIGR's protocol mentioned above) and is used to prevent non-specific sorption of fluorescently-labeled target on the surface of slides, especially if these are polylysin coated. It was also demonstrated that the log-transformed variance of intensity ratios depends linearly on the quality score [83], i. e. the spread of replicate data is generally due to poor quality of spots and can be controlled by image processing software.

In essence, these observations suggest that data interpretation is reliable if expression data is of high quality. Therefore, the obvious complementary approach to the use of sophisticated tools in post-hybridization image analysis is to improve the quality of spots. Indeed some improvement in spotting procedures [82] has led some authors to claim that they "find it unnecessary to generate duplicate spots on the same slide to ensure the reproducibility of the data". In this context it should be mentioned that a simple and useful method to improve spot morphologies is the addition of betain to the spotting buffer [86]. Betain is strongly hygroscopic, so that solutions containing betain do not dry out at ambient humidity. This prevents spotted droplets from drying and helps to eliminate irregularities in spot morphology due to the formation of salt crystals.

It is evident from the considerations above that image processing and meticulous spot quality control are crucial to the reliability of microarray data. However, relatively little attention is being paid to this aspect by bioinformaticians compared to the issues of subsequent data analyses such as fluorescent channel normalization, detection of differentially expressed genes, clustering, etc. At the same time several statistical procedures *de facto* struggle with the consequences of image-processing artefacts. Unification of bioinformatics research on image processing, data normalization and further statistical analysis could, in our opinion, lead to more advanced procedures, relying on the assignment of qualitative reliability of the data for particular spots and genes rather than excluding "unreliable" data.

2.7 Short-term Outlook

Without doubt gene expression profiling using microarray techniques has become an efficient and versatile tool for the analysis of gene expression on the genomic scale. Many publications have proven the usefulness of the technology although it sometimes suffers from artefacts that may be highly reproducible. The respective experimental platforms and technical details used for expression profiling experiments are still a matter of choice. One reason for this diversity of methods in each step of DNA-chip experiments appears to be that some of the fundamental knowledge about the physics of nucleic acid hybridization on solid surfaces is not yet available. We did not attempt to recommend any particular platform but endeavoured to provide guidelines to facilitate the choice from the available methods and to discuss potential future directions for the solution of related problems.

New solutions will probably come from the emerging applications of nanobiotechnology [87]. One such nano-device with great potential for future applications in gene expression studies should be mentioned: quantum dots (QDs), colloidal semiconductor nanocrystals with a CdSe core and ZnS shell, have exciting physical properties that may eventually make them the method of choice over the fluorophore Cyanine tags which are most frequently used in expression profiling experiments today. The absorption spectra of QDs are very broad and their emission is confined to a narrow, symmetrical band that is characteristic of the nano-particle size. They are highly luminescent and stable against photobleaching. DNA-hybridization studies have proven their applicability to the detection of nucleic acids [88]. QDs may be fabricated for any emission spectrum and excited at a single wavelength. Furthermore, defined quantities and ratios of different color QDs have been packed into microbeads and subsequently linked to distinct nucleic acid probes. The relative ratios of colors label the identity of the probe. Such microbeads with three different color QDs have been detected at the single bead resolution in DNA hybridization studies [88]. Coding nucleic acid probes with QD-loaded microbeads or with fluorescent microbarcodes has evoked the vision of flow systems with tag-by-tag recognition instead of monitoring hybridization to a probe in a fixed position on a microarray [89, 90].

New technologies are being developed or improved to monitor proteomes at similar throughput rates. At least two prerequisites make this effort a major challenge: first, the proteome is at least one or two orders of magnitude more complex than the transcriptome. Second, proteins are biochemically highly diverse and cannot be amplified as easily as nucleic acids. Despite these difficulties research is being undertaken to improve the sensitivity and reproducibility of 2D-gel electrophoresis and to automate the subsequent identification of proteins. Pre-fractionation of protein samples and the availability of comprehensive sets of protein-specific antibodies and antibodies against characteristic functional domains of protein families will be an integral part of proteome analysis. When it becomes possible to integrate transcriptomics and proteomics with techniques such as laser microdissection to isolate homogenous populations of cells from complex tissues, then for the first time, it will be feasible to comprehensively analyze gene expression in the context of the molecular network of the cell. Such a holistic approach of molecular analysis would have important synergistic effects on the analysis of regulatory interdependencies that determine the molecular phenotype of the cell; it would also allow the distinction to be made between transcriptional and post-transcriptional regulation in a comprehensive approach.

References

1 L. H. Augenlicht et al. **1987**, *Cancer Res.* 47, 6017.
2 M. Schena, D. Shalon, R. W. Davis, P. O. Brown **1995**, *Science* 270, 467.
3 M. Nacht et al. **1999**, *Cancer Res.* 59, 5464.
4 L. J. van't Veer et al. **2002**, *Nature* 415, 530.
5 J. Elek, K. H. Park, R. Narayanan **2000**, *In Vivo* 14, 173.
6 S. L. Pomeroy et al. **2002**, *Nature* 415, 436.
7 R. A. Heller et al. **1997**, *Proc. Natl Acad. Sci. USA* 94, 2150.
8 R. Wooster **2000**, *Trends Genet.* 16, 327.
9 M. Bittner et al. 2000, *Nature* 406, 536.
10 C. Debouck, P. N. Goodfellow **1999**, *Nat. Genet.* 21, 48.
11 J. F. Bishop **1999**, *Med. J. Aust.* 170, 39.
12 T. R. Golub et al. **1999**, *Science* 286, 531.
13 A. A. Alizadeh et al. **2000**, *Nature* 403, 503.
14 D. Pinkel **2000**, *Nat. Genet.* 24, 208.
15 D. M. Roden, A. L. George **2002**, *Nat. Rev.* 1, 37.
16 C. K. Lee, R. Weindruch, T. A. Prolla **2000**, *Nat. Genet.* 25, 294.
17 C. K. Lee, R. G. Klopp, R. Weindruch, T. A. Prolla **1999**, *Science* 285, 1390.
18 R. Sandberg et al. **2000**, *Proc. Natl Acad. Sci. USA* 97, 11038.
19 J. D. Porter et al. **2001**, *Proc. Natl Acad. Sci. USA* 98, 12062.
20 W. G. Campbell et al. **2001**, *Am. J. Physiol. Cell Physiol.* 280, C763.
21 F. J. Livesey, T. Furukawa, M. A. Steffen, G. M. Church, C. L. Cepko **2000**, *Curr. Biol.* 10, 301.
22 E. M. Southern **1975**, *J. Mol. Biol.* 98, 503.
23 R. K. Saiki et al. **1985**, *Science* 230, 1350.
24 V. E. Velculescu, L. Zhang, B. Vogelstein, K. W. Kinzler **1995**, *Science* 270, 484.
25 M. E. Burczynski et al. **2005**, *Curr. Mol. Med.* 5, 83.
26 B. M. Bolstad, R. A. Irizarry, M. Astrand, T. P. Speed **2003**, *Bioinformatics* 19, 185.
27 C. Li, W. H. Wong **2001**, *Proc. Natl Acad. Sci. USA* 98, 31.
28 D. L. Wheeler et al. **2003**, *Nucleic Acids Res.* 31, 28.
29 A. L. Drobyshev et al. **2003**, *Nucleic Acids Res.* 31, E1.
30 A. Schulze, J. Downward **2001**, *Nat. Cell Biol.* 3, E190.
31 T. S. Tanaka et al. **2000**, *Proc. Natl Acad. Sci. USA* 97, 9127.
32 A. R. Espanhol et al. **2003**, *Mol. Cell Biochem.* 252, 223.
33 H. L. Keen et al. **2004**, *Physiol. Genomics* 18, 33.
34 D. J. Symula, Y. Zhu, J. C. Schimenti, E. M. Rubin **2004**, *Mamm. Genome* 15, 1.
35 H. Bono et al. **2003**, Genome Res. 13, 1318.
36 E. Matsuda et al. **2004**, *Proc. Natl Acad. Sci. USA* 101, 4170.
37 M. Seltmann et al. **2005**, *Mamm. Genome* 16, 1.
38 G. A. Churchill **2002**, *Nat. Genet.* 32(Suppl.), 490.
39 A. L. Drobyshev, M. Hrabe de Angelis, J. Beckers **2003**, *Curr. Genomics* 4, 615.
40 C. C. Pritchard, L. Hsu, J. Delrow, P. S. Nelson **2001**, *Proc. Natl Acad. Sci. USA* 98, 13266.
41 K. Oishi et al. **2003**, *J. Biol. Chem.* 278, 41519.
42 X. Peng et al. **2003**, *BMC Bioinformatics* 4, 26.
43 D. Agrawal et al. **2002**, *J. Natl Cancer Inst.* 94, 513.
44 C. M. Kendziorski, Y. Zhang, H. Lan, A. D. Attie **2003**, *Biostatistics* 4, 465.
45 X. Li, W. Gu, S. Mohan, D. J. Baylink **2002**, *Microcirculation* 9, 13.
46 M. Vingron, J. Hoheisel **1999**, *J. Mol. Med.* 77, 3.
47 J. Quackenbush **2001**, *Nat. Rev. Genet.* 2, 418.
48 C. Sabatti, S. L. Karsten, D. H. Geschwind **2002**, *Math. Biosci.* 176, 17.
49 D. S. Millican, I. M. Bird **1998**, *Methods Mol. Biol.* 105, 315.
50 A. E. Krafft, B. W. Duncan, K. E. Bijwaard, J. K. Taubenberger, J. H. Lichy **1997**, *Mol. Diagn.* 2, 217.
51 P. Hegde et al. **2000**, *Biotechniques* 29, 548.
52 S. L. Beaucage **2001**, *Curr. Med. Chem.* 8, 1213.
53 A. P. Turner **2000**, *Science* 290, 1315.
54 A. W. Peterson, R. J. Heaton, R. M. Georgiadis **2001**, *Nucleic Acids Res.* 29, 5163.

55 R. J. Lipshutz, S. P. Fodor, T. R. Gingeras, D. J. Lockhart **1999**, *Nat. Genet.* 21, 20.
56 A. W. Machl, C. Schaab, I. Ivanov **2002**, *Nucleic Acids Res.* 30, e127.
57 M. Schena et al. **1996**, *Proc. Natl Acad. Sci. USA* 93, 10614.
58 J. Yu et al. **2002**, *Mol. Vis.* 8, 130.
59 R. L. Stears, R. C. Getts, S. R. Gullans **2000**, *Physiol. Genomics* 3, 93.
60 E. Manduchi et al. **2002**, *Physiol. Genomics* 10, 169.
61 E. Wang, L. D. Miller, G. A. Ohnmacht, E. T. Liu, F. M. Marincola **2000**, *Nat. Biotechnol.* 18, 457.
62 L. R. Baugh, A. A. Hill, E. L. Brown, C. P. Hunter **2001**, *Nucleic Acids Res.* 29, E29.
63 N. N. Iscove et al. **2002**, *Nat. Biotechnol.* 20, 940.
64 A. Relogio, C. Schwager, A. Richter, W. Ansorge, J. Valcarcel **2002**, *Nucleic Acids Res.* 30, e51.
65 Y. Li, G. Zon, W. D. Wilson **1991**, *Biochemistry* 30, 7566.
66 Y. Li, G. Zon, W. D. Wilson **1991**, *Proc. Natl Acad. Sci. USA* 88, 26.
67 A. Krause, S. A. Haas, E. Coward, M. Vingron **2002**, *Nucleic Acids Res.* 30, 299.
68 G. Yershov et al. **1996**, *Proc. Natl Acad. Sci. USA* 93, 4913.
69 N. P. Gerry et al. **1999**, *J. Mol. Biol.* 292, 251.
70 N. E. Broude, K. Woodward, R. Cavallo, C. R. Cantor, D. Englert **2001**, *Nucleic Acids Res.* 29, E92.
71 A. V. Fotin, A. L. Drobyshev, D. Y. Proudnikov, A. N. Perov, A. D. Mirzabekov **1998**, *Nucleic Acids Res.* 26, 1515.
72 Y. Belosludtsev et al. **2001**, *Anal. Biochem.* 292, 250.
73 S. V. Lemeshko, T. Powdrill, Y. Y. Belosludtsev, M. Hogan **2001**, *Nucleic Acids Res.* 29, 3051.
74 S. J. Shi, A. Scheffer, E. Bjeldanes, M. A. Reynolds, L. J. Arnold **2001**, *Nucleic Acids Res.* 29, 4251.
75 T. Nishinaka, A. Shinohara, Y. Ito, S. Yokoyama, T. Shibata **1998**, *Proc. Natl Acad. Sci. USA* 95, 11071.
76 M. J. Martinez et al. **2003**, *Nucleic Acids Res.* 31, e18.
77 M. D. Kane et al. **2000**, *Nucleic Acids Res.* 28, 4552.
78 H. Dai, M. Meyer, S. Stepaniants, M. Ziman, R. Stoughton **2002**, *Nucleic Acids Res.* 30, e86.
79 D. J. Lockhart et al. **1996**, *Nat. Biotechnol.* 14, 1675.
80 H. Yue et al. **2001**, *Nucleic Acids Res.* 29, E41.
81 T. Yuen, E. Wurmbach, R. L. Pfeffer, B. J. Ebersole, S. C. Sealfon **2002**, *Nucleic Acids Res.* 30, e48.
82 P. H. Tran et al. **2002**, *Nucleic Acids Res.* 30, e54.
83 X. Wang, S. Ghosh, S. W. Guo **2001**, *Nucleic Acids Res.* 29, E75.
84 G. C. Tseng, M. K. Oh, L. Rohlin, J. C. Liao, W. H. Wong **2001**, *Nucleic Acids Res.* 29, 2549.
85 M. L. Lee, F. C. Kuo, G. A. Whitmore, J. Sklar **2000**, *Proc. Natl Acad. Sci. USA* 97, 9834.
86 F. Diehl, S. Grahlmann, M. Beier, J. D. Hoheisel **2001**, *Nucleic Acids Res.* 29, E38.
87 P. Fortina, L. J. Kricka, S. Surrey, P. Grodzinski **2005**, *Trends Biotechnol.* 23, 168.
88 M. Han, X. Gao, J. Z. Su, S. Nie **2001**, *Nat. Biotechnol.* 19, 631.
89 X. Gao, S. Nie **2005**, *Methods Mol. Biol.* 303, 61.
90 M. J. Dejneka et al. **2003**, *Proc. Natl Acad. Sci. USA* 100, 389.

3
Screening for Bone and Cartilage Phenotypes in Mice

Helmut Fuchs, Thomas Lisse, Koichiro Abe, and Martin Hrabé de Angelis

3.1
Introduction

The Egyptian pharaoh Tutankhamen (pharaoh between 1334 and 1325 BC) died in his late teens and remained at rest in Egypt's Valley of the Kings for over 3300 years. In his tomb his child was found, and X-ray analysis confirmed scoliosis in the fetus (Harrison 1979; OMIM 181800 Scoliosis, Idiopathic; OMIM 148900 Klippel–Feil syndrome). In another example, six native Americans who lived between 3000 and 5000 years ago in North America exhibited erosive polyarthritis characteristic of rheumatoid arthritis (Rothschild et al. 1988). And interestingly, these might be the most ancient examples of skeletal diseases in humans. At present in Europe, nearly one-quarter of adults suffer from musculoskeletal pain at any one time in their lives (Woolf et al. 2004). It is apparent that skeletal diseases have unwaveringly threatened the health-related quality of life among humans, but despite tremendous efforts in bone research and development to date, most molecular mechanisms involving skeletal diseases remain unclear. There are many reasons why the mouse is perhaps the best tool with which to learn, understand and expand the repertoire of human skeletal diseases. In this chapter, we will focus on the methods used to identify skeletal phenotypes in genetically-altered mice.

3.1.1
The Skeleton

The skeleton accounts for about 15% of the total body weight, and thus it is the third largest organ system in the human body. The skeleton consists of bone and cartilage, whereby adult humans have 206 individual bones. Biochemically, two-thirds of bone is inorganic substances, and one-third is organic material. The inorganic component, calcium hydroxyapatite ($Ca_{10}(PO_4)_6(OH)_2$), is the mineral that gives bone combined strength and hardness. The osseous tissue can be structured into periosteum, the outer substancia corticalis (compacta), and the inner substancia spongiosa. The bones are interconnected by joints, which allow a wide range of movements. Most cartilage is located at the circumference of bones facing the joint.

Standards of Mouse Model Phenotyping. Edited by Martin Hrabé de Angelis, Pierre Chambon, and Steve Brown

ISBN: 3-527-31031-2

This articular cartilage forms the smooth gliding surface of joints such as the knee and hip, which permits free locomotion.

The functions of the skeleton are to provide a scaffold for the body since it has to carry most of the body weight, and to give the body stability, which is needed as a basis for mobility. Another important role of the skeleton is facilitation of locomotion together with muscle. Furthermore, the skeleton serves as a store for body minerals and provides mechanical protection for inner organs. All in all, the skeleton has to be hard and stable as well as flexible.

3.1.2 Skeletal Development in the Embryo

The skeleton is distributed in many locations in the whole human body. Progenitor cells of the skeleton are derived from different origins during embryogenesis: axial skeleton from somites, skull from neural crest cells, and limb skeleton from the lateral mesoderm. The unique properties of the skeletal developmental process control the shape of bones by local signaling in the embryo. Despite different origins, the differentiation processes of the skeleton have similarities. For example, the mesenchymal cells of different origins properly migrate to locations in the embryo, and then aggregate to form prechondral condensations. The cells in the condensations proliferate and differentiate into chondrocytes. During these processes, joint formation also proceeds at the boundary between two condensations or within a single condensation. The prospective joint regions become more dense and flattened, thereby ensuring that future cell death within these regions will create the joint cavities. In skeletal development, this phase is generally categorized as "patterning", since the cartilage templates prefigure the future bone locations. And this patterning process tightly interacts with the developmental signaling cascade such as the morphogen gradient. The classical grafting experiments in the chicken embryo predicted the existence of activity in the specific region of the limb bud. For instance, if the posterior distal margin of the limb bud is grafted onto the anterior distal margin of another embryo, a duplicated structure in the left–right axis of the limb will form. This region is called the zone of polarizing activity (ZPA). Skeletal "patterning" defects in human genetic diseases such as polydactylism and syndactylism often exhibit similar phenotypes to the chick grafting experiments, suggesting disturbed developmental signals might cause similar phenotypes in human. (For an overview of the genetic control of skeletal development, and bone and joint formation, see Karsenty (2001) and Kingsley (2001) respectively; for detailed insights into the early stages of limb patterning and chondrogenesis, see Pizette and Niswander (2001), Mariani and Martin (2003)).

3.1.3 Growth and Maintenance of Bone and Cartilage

After prepatterning of the avascular cartilage template, the chondrocytes of the centers of condensation stop proliferating and become hypertrophic. The hypertrophic chondrocytes direct mineralization of the extracellular matrix (ECM), and

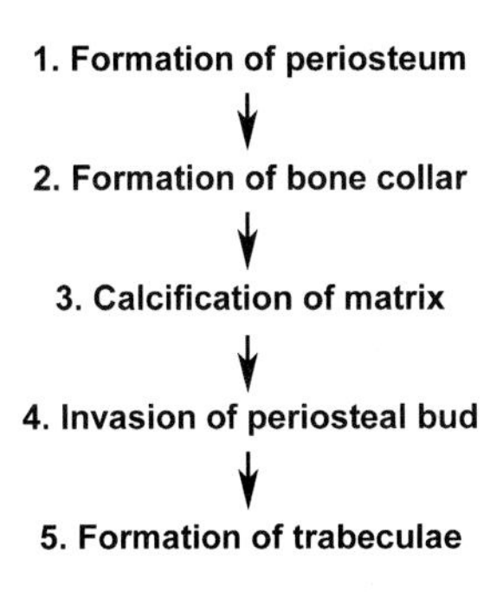

Fig. 3.1 The periosteum harbors a layer of undifferentiated cells which later become osteoblasts (step 1). Osteoblasts secrete osteoid i. e. a collagen I protein matrix, against the shaft of the cartilage model, thus becoming the support for the new bone (step 2). Chondrocytes in the primary center of ossification begin to grow (hypertrophy). Collagen secretion ceases, and other proteoglycans and hypertrophic chondrocytes begin secreting alkaline phosphatase essential for mineral deposition. Nutrients can no longer diffuse if the matrix becomes calcified and the chondrocytes subsequently die, thereby creating cavities within the bone (step 3). A periosteal bud consisting of blood vessels, lymph vessels and nerves, invades the resultant cavities (step 4). The vascularization carries hemopoietic cells, osteoblasts and osteoclasts to the cavity, whereby the hemopoietic cells later form the bone marrow. Osteoblasts manage the calcified matrix as a scaffold, and begin to secrete osteoid, thus forming the bone trabecula (step 5). Osteoclasts, on the other hand, break down cancellous bone to form the medullary (bone marrow) cavity.

vascular invasion then occurs only in the ECM surrounding the hypertrophic chondrocytes due to the loss of vascularization inhibitors present only in proliferating chondrocytes. By blood circulation, hematopoietic cell-derived chondroclasts migrate and start to digest cartilaginous ECM to make the bone marrow cavity. The hypertrophic chondrocytes also direct perichondrial cells to become osteoblasts, and undergo apoptotic cell death. The perichondrium is a layer of dense connective tissue which surrounds the cartilage. It consists of two separate layers that include the outer fibrous layer and the inner chondrogenic layer. Once vascularized, the perichondrium becomes the periosteum. The periosteum contains a layer of undifferentiated cells which later become osteoblasts. The osteoblasts invade the cartilage and position themselves to construct true mineralized ECM forming a bone collar. This site of vascular invasion and osteoblast gathering is called the primary spongiosa. The balance between chondroclasts and osteoblasts, destruction and construction of the ECM, changes the cartilage template to mineralized bone matrix. This process is called endochondral ossification (Fig. 3.1). At the ends of bones, chondrocytes continue to proliferate forming the second ossification center and lengthen the bone collar. The region between the primary spongiosa and the secondary ossification center is called the growth plate. After colonization of hematopoietic stem cells in the bone marrow, osteoclasts are derived from hematopoietic stem cells and dominate chondroclasts. This phase including endochondral ossification can be categorized as “growth and mineralization”.

Finally, a process of growth and remodeling, the balance between osteoblastic synthesis and osteoclastic resorption, results in proper skeletal function in adults. In humans, bone formation until puberty is controlled by genetic factors, the supply of calcium and vitamin D and by the amount of physical activity. After puberty the control and maintenance is taken over by hormones and nutritional balance. At the age of about 35 to 40 years, the formation of the skeleton is complete, and a peak bone mass is reached. From this point and further, 0.5 to 1.5% of the bone mass is resorbed per year (see Morriss-Kay et al. 2001, Kronenberg 2003) for general information regarding the genetic control of the proliferation–differentiation balance during bone formation; Ornitz (2001) regarding the regulation of chondrocyte growth and differentiation, and Suda et al. (2001) for detailed information about the molecular basis of osteoclast differentiation and activation).

3.1.4
Diseases Involving Cartilage and Bone

From the complex process of skeletal development featured above, disorders of the skeleton can be manifold. They can be divided into four classes which include: (1) developmental patterning defects (e.g. polydactylism and syndactylism), (2) metabolic and growth defects (e.g. osteogenesis imperfecta, osteomalacia), (3) modeling and remodeling defects (e.g. osteoporosis, osteopetrosis), and (4) aging and immune system defects (e.g. arthritis, sponsirosis, ruptured disks; see Fig. 3.2). The etiology and pathologic basis of skeletal diseases are diverse in nature. Table 3.1 lists functionally subcategorized modes of genetic defects of cartilage and bone, whereby the primary genetic cause of many examples within the above disease categories have already been identified in both human and mouse (see Superti-Furga et al. (2002); Shum et al. (2003); Zelzer and Olsen (2003) for a more complete and detailed listing of known skeletal genes).

Developmental patterning defects such as polydactyly, syndactyly or brachydactyly have already been alluded to in the previous section. In human, defects of the growth and mineralization phase cause growth retardation and/or malformation of bones such as bending or fractures. Rickets and osteomalacia are defects in ECM mineralization occurring during childhood and adult age, respectively. These diseases are most often related to a lack of vitamin D or its metabolism. Osteogenesis imperfecta (OI) is caused by abnormal collagen organization and exhibits bone abnormalities similar to those seen in rickets and osteomalacia. Achondroplasia is caused by a defect in the growth plate, a major cause of dwarfism with bone growth anomalies.

The balance between osteoclastic destruction and osteoblastic construction is called bone remodeling. Bone mass increases rapidly during childhood, which suggests osteoblast activity dominates over osteoclasts. It reaches a peak in the third decade and is termed the peak bone mass or peak bone density. Bone mass is stable for the next 10 to 20 years, and then becomes unstable, presumably due to osteoclast dominance over osteoblast activity. The phase after the peak bone mass is categorized as "remodeling". Normal bone mineral density (BMD) values are con-

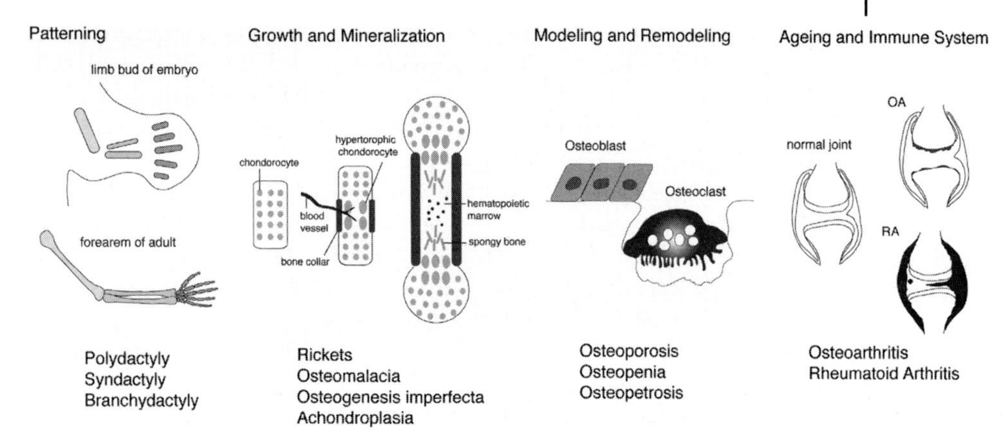

Fig. 3.2 Skeletal development and related human diseases.

Table 3.1 Modes of genetic defects of cartilage and bone in human and/or mouse.

Functional category	Example(s)
ECM and other related proteins	perlecan, COMP, lysyl hydroxylase, MMPs, collagens
Metabolic pathway	DTDST, cathepsin-K, TNSALP, TCIRGI, CIC-7, PEX7
Growth factors	BMP-4,5, GH, PTHrP, GDF-5, CSF-1, OPGL
Signal transduction cascade	FGFR-1,2,3,23, c-Src, CASR, VDr, DLL3, IHH, SOST
TF and nuclear proteins	cbfa-1, msx-1,2, hoxa-2, hoxd-13, c-fos, SOX9, Gli3, TWIST
Oncogenes and tumor suppressor genes	EXT1,2, SH3BP2
DNA and RNA processing/ metabolism	ADA

ECM, extracellular matrix; MMPS, matrix metalloproteinases; DTDST, carrier family 26 sulfate transporter, member 2; BMP, bone morphogenic protein; GH, growth hormone; PTHrP, parathyroid hormone related protein; GDF, growth differentiation factor; CSF-1, colony stimulating factor 1; OPGL, tumor necrosis factor ligand superfamily, member 11; FGF, fibroblast growth factor; hox, homeobox gene; msx, homeobox, msh-like; cbfa, core-binding factor; c-Src, Rous sarcoma oncogene; COMP, cartilage oligomeric matrix protein; TNSALP, tissue non-specific alkaline phosphatase; TCIRGI, osteoblast proton pump unit; CIC-7, chloride channel 7; PEX7, peroxisomal receptor/importer; CASR, calcium sensor/receptor; VDr, 1,25-α-dihydroxy-vitamin D3 receptor; DLL, delta-like; IHH, indian hedgehog; SOST, sclerostin; Gli3, GLI-Kruppel family member 3; TWIST, helix–loop–helix transcription factor; EXT, exostosin; SH3BP2, c-Abl-binding protein; ADA, adenosine deaminase; TF, transcription factor.

sidered to be all values within one standard deviation (SD) of the mean of a young adult person. If the value decreases by more than 1 SD, but not by more than 2.5 SD, the diagnosis is oesteopenia, and values decreasing by more than 2.5 SD are indicative of osteoporosis (WHO Technical Report Series 843, 1994). Osteoporosis is a common multi-factorial disorder of reduced bone mass manifesting clinically as fragility fractures. Low BMD itself, however, does not cause pain or deformity. When the bone mass is already severely reduced, fractures frequently occur in the vertebrae, hip and forearms notably. In a healthy person, bone turnover is in equilibrium. Osteoporosis patients exhibit either reduced bone formation or increased bone resorption. As a consequence, the skeleton is less stable, and cannot provide as much as support as it does in healthy people, which results in an increased risk for fractures.

In old age, maintenance of the skeleton is important for health-related quality of life. However, 50% of people over 65 years of age suffer from movement disability due to arthritis. Arthritis is an inflammatory condition that causes pain, swelling, and stiffness in the joints. The chronic nature of arthritis results in long-term physical disability. Osteoarthritis and rheumatoid arthritis are the most frequent joint diseases in the world. Osteoarthritis is a degenerative disease of the articular cartilage during old age. Patients exhibit degeneration of the synovium and cartilage of affected joints, mostly the hip and knee. The affected area causes pain upon weight-bearing since the cartilage of the joints no longer protects the bone surface (Fig. 3.2). Rheumatoid arthritis is a symmetric inflammatory arthritis that targets multiple joints of hands, feet, wrists, elbows, ankles, shoulders, and knees. It is characterized pathologically by inflammatory infiltration of the synovium and the formation of rheumatoid pannus, which is a progressive overgrowth of the synovium with the capacity to erode the articular surface (Fig. 3.2). The chronic nature of this disease results in cartilage destruction, bone erosion, ultimate joint deformity, and a loss of joint function. These features are not related to bone remodeling, and can thus be categorized as occurring and involving the "aging and immune systems" respectively.

Many of the diseases share symptoms of pain and/or inflammation, and can involve limitation of motion, disability and even death. Many skeletal diseases are late-onset diseases and cause severe long-term complications. As the population ages, the economic toll of medical treatment is predicted to increase dramatically in the future (Linden 2003). The Bone and Joint Decade (www.bonejointdecade.org) was launched by the World Health Organization (WHO) in Geneva, Switzerland to reduce such increasing social and financial costs to society by raising awareness and advancing the understanding of musculoskeletal diseases. The European Bone and Joint Health Strategies Project defines osteoarthritis, rheumatoid arthritis, osteoporosis and back pain as major disease targets (Bone & Joint Decade). Detailed information about clinical disorders of bone resorption can be found in Russell et al. (2001), and the report by Mundlos (2001) provides an overview of models and mechanisms of defects in human skeletogenesis.

3.1.5
The Mouse as a Model for Skeletal Diseases

In genetics, most scientists favor sophisticated genetic systems of tiny organisms such as the fruit fly or worm since they reproduce rapidly with large numbers of offspring. However, one obvious difficulty in studying skeletal diseases in *C. elegans* or Drosophila is that the conservation of regulatory pathways between humans cannot be relied upon, as the skeleton was acquired late during evolution. For instance, insects obtained an exoskeletal system, whereby they harness hard outer chitin without bones. It is therefore necessary to identify disease models using vertebrate organisms such as the mouse. The mouse is well suited to genetic analysis because of its relatively short generation time and its ability to produce large litters in the laboratory when compared to other mammalian model organisms. A variety of inbred mouse strains, of which every locus within the genome is homozygous, have been used in genetic studies of bone density. For example, Beamer et al. (1999) used both C57BL/6J and CAST/EiJ mice strains as low and high bone density strains respectively, to search for quantitative trait loci (QTL) and identified four candidates.

In the mouse, forward and reverse genetics approaches (i. e. phenotype to gene and gene to phenotype respectively) are readily available (Jackson and Abbott 2000). In the forward genetics approach, more than 120 spontaneous mutant lines with skeletal defects have been identified and maintained in the mouse genetics community (Lyon et al. 1996). By the positional cloning strategy, some of the mutations have been identified and the contribution to human skeletal diseases confirmed (Chalhoub et al. 2003; Ho et al. 2000; Malkin et al. 2005). Further efforts to increase the number of models for human diseases such as large scale mouse mutagenesis projects have been launched to obtain and share mutant resources worldwide (Hrabé de Angelis et al. 2000, Nolan et al. 2000). In the Munich screen, we have isolated skeletal mutants as mouse models for osteogenesis imperfecta, inflammatory arthritis, and low bone mass density (unpublished data, manuscripts in preparation). In the reverse genetics approach, the development of embryonic manipulation technology has enabled the production of gain of function (transgenic) and loss of function (knockout) alleles of any gene in the genome, whereby abnormalities in patterning, bone remodeling and joints in transgenic and knockout mice have been reported (Peacock et al. 2002; Superti-Furga et al. 2002).

In addition to advantages of these genetic approaches, the genome sequence of the mouse is available on various public databases. Recent developments in bioinformatics technology of the genome sequence reduce the routine and laborious experimental work carried out by biologists. For instance, in forward genetics and QTL analysis, the increase in polymorphic markers such as microsatellites and single nucleotide polymorphisms (SNPs) between inbred strains facilitates genetic mapping, and from sequence information of critical intervals from mapping data one can quickly "pick up" candidate genes for mutation analysis. In the reverse genetics approach, information regarding exon–intron structure and restriction enzyme sites from the genome sequence can be used in the construction of gene targeting vectors. Thus, the combination of genomics and genetics in the mouse enables direct and fast access to molecular mechanisms of complex biological systems. In addition, mouse phenome databases (MPD aretha.jax.org/pub-cgi/phenome/mpdcgi and PhenomicDB www.phenomicdb.de) provide a comprehensive search tool for phenotype evaluation and phenotyping centers provide a wealth of biological information to the international scientific forum pertaining to mouse lines of special interest (Gailus-Durner et al. 2005). In this context, detailed phenotyping technology will become more important in the future for the analysis of mutant mice produced either by the forward or reverse genetics approach.

3.2
Screening Protocols

3.2.1
Morphological Analysis

Morphological analysis of the mouse is the simplest method of obtaining first-line information about skeletal malformations. For most protocols no special equipment is needed. For example, kinky and curly tail mutants can easily be detected at an early stage by visual inspection alone. There have been many protocols developed by different groups to screen for morphological abnormalities in mice. In many of them, skeletal parameters are combined with other parameters, for example for behavior or neurology. In most cases this is because the group was interested not only in bones, but also in other areas of research. But even if it is skeletal parameters which are of primary interest, the study of additional parameters relating to behavior is very important for the interpretation of the results, as this approach may help to detect subtle abnormalities. For example, abnormal gait in a mouse may be indicative of a limb abnormality. In another case, we were able to detect the beginnings of knee joint hardening as a result of the abnormal struggling behavior and positioning of the legs, when the mouse was suspended by its tail.

These types of test have the advantage of being non-invasive and do not require the use of X-rays or other harmful chemical (e. g. narcosis) or physical treatments. By careful observation and tactile investigation of the animals even hidden phenotypes may be uncovered. The disadvantages of this test are that certain phenotypes which are only detectable by X-ray analysis for example, would certainly be missed, and in many cases an abnormality may be suspected but the unequivocal determination of the phenotype is not possible using this type of test. Thus, these techniques are primarily useful for application in ENU-screens, where the priority is to detect inherited phenotypes using a non-invasive method.

There are many protocols available, such as the so-called SHIRPA (Irwin 1968; Rogers et al. 1997) protocol or a limb-specific protocol. We will present here a modified protocol according to Fuchs et al. (2000). We recommend that this protocol is used on mice which are 8 weeks of age or older. In younger mice, certain subtle phenotypes will remain hidden.

As soon as the first phenotypes are identified, storage of the findings in databases is useful. We recommend the use of the terms found in Mammalian Phenotype Ontology (http://www.informatics.jax.org/searches/MPform.shtml) for describing the phenotypes. This will facilitate the identification of mutants with similar phenotypes at a later stage and will also facilitate comparisons between different laboratories, uploading data, and searching mutant resource databases.

3.2.1.1 **Protocol**

- Take a cage of animals out of the rack and put it on the laboratory desk
- Allow the mice to settle and open the lid of the cage
- Observe the mice in the cage without disturbing them
- Record any type of abnormality

Comments: by observing the mice in the cage, small or large mice can easily be identified. Mice with skeletal problems might move in an unusual manner (e. g. slower movements or abnormal gait). Diseased mice might sit offsite from their cage mates or may be reluctant to come into contact with their cage mates.

- Catch one mouse and put it on your hand
- Check the (ear-) labeling of the mouse and then observe the way the mouse is sitting on the hand (positioning, vertebral column, head shape, pelvis elevation, tail movement)

Comments: healthy mice sit on the hand, and then start to explore the environment. If the hand is not in a horizontal position, the mouse will try to achieve an upright position. Trembling and a fuzzy coat may indicate that the mouse is suffering from any number of diseases which may result from bone abnormalities. Any malformations of the skull should be identified; sometimes microphtalmia is associated with hydrocephalus. Abnormal movements of the tail can give information about malformations of the tail vertebrae. If vertebrae of the tail are malformed, other vertebrae might also be malformed.

- Suspend the mouse by its tail and note the positioning of hind and forelimbs as well as the degree of struggling

Comments: Suspending the mouse by the tail will show whether the limb joints are functioning normally. This is also a good occasion to see whether the limb bones are of normal length. A healthy mouse will position its hind legs in a V-shape and will stretch the front limbs towards the ground. After a few seconds, it will start to struggle and try to climb back onto the observer's hands.

- Check the tail for vertebral abnormalities by moving along the tail with two fingers

Comments: By moving the fingers along the tail, any vertebral malformation can be easily detected.

- Holding the animal by its neck, inspect the mouse from the ventral side
  - Skull
  - Basket
  - Limbs
  - Shape and number of digits on hind and front limbs

Comments: In most cases of abnormality the hind limbs are more severely affected than the front limbs.

- Force the mouse to move its limbs to test for correct reactions
- Open the mouth of the mouse and check the mandibles and teeth

Comments: Tooth malformation (e. g. long teeth) may indicate skull abnormalities.

- Suspend the mouse by the tail and drop it into a new cage.

Comments: Abnormal landing of the mouse may indicate limb abnormalities.

- Continue in the same way with the next mouse in this cage

With regard to these observations, the genetic background of the investigated animals should be taken into consideration.

3.2.2 X-Ray Analysis

3.2.2.1 General

From the outside the skeleton is an invisible tissue, but ever since Whilhelm Röntgen's discovery of X-rays, this has become a powerful technique for the visualization of skeletal diseases. The X-ray, having a wave length in the range of 10 nm to 100 pm, penetrates soft tissues, but is strongly attenuated by the bony counterparts. Radiography is a common technique used to produce photographs by exposing a photographic film or other image receptor to X-rays. X-ray analysis is a reasonable technique with which to obtain information about developmental disorders and to a minor extent metabolic bone disorders. Within a few minutes the whole mouse skeleton can be visualized and a complete appraisal of the appearance and presence of all bones and joints can be made.

Conventional X-ray applications common within the clinical setting can diagnose and show a myriad of conditions such as congenital bone diseases (e. g. scoliosis, mucopolysaccharidoses, OI, diabetes, fibrodysplasia ossificans progressive, arachnodactyly), joint dislocations, infections of the bone (e. g. osteomyelitis and cellulites), degenerative conditions (e. g. damage due to prosthesis or osteoarthritis) and metabolic bone diseases (e. g. Paget's disease, Rickets, hyperparathyroidism, acromegaly). For example in progressive fibrodysplasia ossificans (FOP; de la Pena et al. 2005; Kaplan and Smith 1997), a rare heritable disorder of connective tissue attributed to altered BMP-4 signaling, congenital malformations of the big toes and progressive heterotopic endochondral osteogenesis (i. e. bony formations) in anatomic and temporal patterns are easily discernable using conventional X-ray imaging. In the appendicular skeleton, both digital and plain film radiography can readily localize fractures, fusions and provide a qualitative assessment of BMD. In line with the latter, semi-quantitative indices have been successfully developed to assess BMD (e. g. by combined cortical thicknesses; Grampp et al. 1993). However, the degree of bone loss needs to be significant before osteopenia is visible by radiography. Incorporation of X-ray analysis into the screen for mouse skeletal phenotypes aims to identify similar types of the aforementioned genres of disorders by way of characterizing the shape, lengths and integrity of the skeletal system.

3.2.2.2 Imaging

X-rays can be used to darken X-ray films, or can be applied in digital systems such as X-ray scanners or sensitive cameras. The use of X-ray films is the traditional method of performing X-ray analysis, yielding high quality resolution. The disadvantage of films is the need for a dark room, developing machine to process the films, and the high costs of consumables.

Digital imaging of X-rays overcomes the limitations of X-ray films. The X-ray images are directly transferred to a computer for digital archiving. This saves space, and also time when endeavoring to locate the desired images. The disadvantages of digital systems are the large financial investment for the equipment, and in terms of the scanner, the lower resolution when compared to X-ray films. As X-ray scanners are still quite slow, the exposure time to radiation is around double compared to X-ray films.

Because of the limitations of scanners, recent developments have focused on the replacement of scanners with X-ray sensitive cameras. These are much faster than scanners, and thus the time needed to take the image and the exposure time of animals is dramatically reduced. Also, X-ray cameras overcome the limitations in resolution associated with scanners. Unfortunately, the investment costs for the camera are still high; an additional limitation of most cameras is that they are not able to perform whole body scans of mice.

3.2.2.3 Analysis of Images

X-ray analysis is a powerful tool for the investigation of various scientific questions, and can be applied to the mice at any age. The analysis of X-ray images should be accomplished in both a qualitative and quantitative manner. First, the researcher should carry out a systematic visual check of the images for the presence and shapes of all bones and joints. In addition, bone density can be crudely evaluated using clear images that indicate anomalies such as bone transparencies or image scatter. In a second step, quantitative analysis can be carried out using a ruler or ruler tools supplied by the software which accompanies the digital imaging machine. One of the disadvantages is the anisotropic nature of the X-ray scans. Thus, the measurements should either be standardized or analyzed repeatedly. It is very important to always place the mouse in the same position (e. g. by the use of a mouse bed), and to clearly define the start and end points of the measurements.

X-ray analysis should be performed using an X-ray cabinet so as to reduce the radiation exposure of the experimenter. When using digital analysis systems, the researcher should be aware that the mouse bed might be on a different level to the scanner or camera. This will result in a projection of the mouse, and a correction factor for the metric data will need to be determined. Care should be taken when using X-ray analysis for time-series experiments since repeated exposure to radiation will lead to radiation-induced anomalies. An alternative to X-ray analysis should be considered for male mice scheduled for any subsequent breeding since the radiation might induce chromosomal aberrations and deletions in the sperm (Russell et al. 1998). It is recommended that mice should not be mated until at least 8 weeks after the X-ray analysis.

3.2.2.4 Protocol

X-Ray Film

- Prepare the X-ray cassette with the X-ray film
- Anesthetize the mouse (e. g. intraperitoneal injection of Ketamin/Xylacin)
- Test to see whether the mouse is asleep with a toe pinch
- Put the mouse on the cassette
- Fix the digits of the mouse with tape (make sure the digits do not overlap)
- Set the radiation level of the X-ray cabinet to 25 kV
- Put the mouse in the X-ray cabinet
- Take precautions regarding the X-ray cabinet and surrounding working areas and conditions
- Turn on the X-ray source and wait for 25 s
- Switch off the radiation source
- Take out the cassette and put the mouse back into its cage
- Put the cage onto a heating plate as long as the mouse is asleep to prevent loss of body heat
- Control the mouse until it has re-awakened
- Take the cassette to the dark room and develop the X-ray film
- Analyze the film

Comments and hints: the settings may need to be adjusted depending on the type of X-ray cabinet and X-ray film used (i. e. the settings quoted above are for Faxitron).

Digital Device

- Start the imaging software of the scanner or camera
- Anesthetize the mouse (e. g. intraperitoneal injection of Ketamin/Xylacin)
- Test to see whether the mouse is asleep with a toe pinch
- Put the mouse on the mouse bed and fix it according to the recommendations of the supplier
- Set the radiation level of the X-ray cabinet to 25 kV
- Take precautions regarding the safety of the X-ray cabinet and surrounding working conditions
- Turn on the X-ray source and wait until the software has completed the imaging process
- Switch off the radiation source
- Put the mouse back into its cage
- Put the cage on a heating plate as long as the mouse is asleep to prevent loss of body heat
- Control the mouse until it has re-awakened
- Analyze the image

Comments and hints: the settings may need to be adjusted depending on the type of X-ray cabinet and the software scanner/camera system used (i. e. the settings are for Faxitron).

3.2.3 DXA-Analysis

3.2.3.1 General

DXA (dual energy X-ray absorption) analysis is one of the most commonly used bone density modalities in human diagnostics for the detection of osteoporosis and defects of bone mineralization. The object is exposed to two distinct X-ray energy levels, whereby the regions of interest are then systematically scanned. DXA takes advantage of the variable absorption of X-rays by different organ systems in order to discriminate between the skeleton and lean and fatty tissues (Sorenson et al. 1989).

It is well-accepted that bone mass, density, size and shape are all governed by a complex interplay of genetic and environmental factors (Dequeker et al. 1987; Lund et al. 1997; Moller et al. 1978; Slemenda et al. 1992). There are a number of traditional or state-of-the-art techniques that approximate or assess apparent (powder) bone content or degree of mineralization in small animals (Tab. 3.2). Each of these methods has advantages and drawbacks. DXA is one of the most widely used of the current bone densitometry modalities in the biological and clinical sciences of

Table 3.2 Methods for the assessment bone density, content, and integrity.

Method	Reference(s)
Plain film radiography	NA
Ash fraction	Hernandez et al. 2001; Meunier and Boivin 1997; Tablante et al. 2003
Archimedes' principle	Arjmandi et al. 1996
DXA	Akhter et al. 2000; Beamer et al. 1996, 1999, 2002; Bohlooly-Y et al. 2004; Hamrick 2003; Kasukawa et al. 2003; Phillips et al. 2000; Nagy and Clair 1999
pQCT	Bohlooly-Y et al. 2004; Butz et al. 1994; Gu et al. 2002; Hamrick 2003; Kasukawa et al. 2003
MRI	Kroger et al. 1995; Taicher et al. 2003
µCT	Boone et al. 2004; Chung et al. 2004; Guldberg et al. 2004; Jiang et al. 2000; Li et al. 2005; Ritman 2004; Recinos et al. 2004; Wachsmuth and Engelke 2004
Ultrasound	Pressel et al. 2005

pQCT, peripheral quantitative computed tomography; MRI, magnetic resonance imagery; µCT, micro computed tomography; NA, not applicable.

small animals. The principles of bone densitometry using DXA have been reviewed by Blake and Fogelman (1996); and DXA utilization in inbred strains of mice for bone research has been reviewed by Beamer et al. (2002). This section will focus on the steps, benefits and limitations of DXA in animal models as a tool for skeletal phenotyping.

3.2.3.2 **Advantages**

The main advantage of DXA is its proven and reproducible ability to measure bone and body composition parameters rapidly, precisely and accurately (Baroncelli et al. 1998). Bone mineral density (BMD; g/cm^2) is a major DXA parameter that provides a number of opportunities for retrospective, cross-sectional and longitudinal observations between age-matched controls in the assessment of a mutation, transgene and/or feasible bone therapeutic intervention (see Huang et al. 2005; Phillips et al. 2000). In DXA, mineral radiation is propagated to determine the bone density of the whole body or regions of interest (ROI) with a radiation intensity of $< 1/20$ of that of a chest X-ray. Furthermore, this lower X-ray dosage and stable calibration translates into a cost-effective option for radiological testing. DXA analysis is more sensitive (i. e. having a more improved spatial resolution) than ordinary radiographs, more accurate than radiograms, and can diagnose bone loss at earlier stages, for example in osteoporosis. Importantly, DXA requires shorter whole body scanning times in mouse. In addition, DXA can accurately quantify the effects of overlying soft tissues in determining lean and fat mass content, two key metabolic values. These advantages make DXA a technology that is germane to mouse applications such as large-scale mutation screens. Lastly, screening with DXA may unravel or reveal other aspects of bone mass maintenance that may include new determinants in the maintenance of blood calcium levels, hematopoiesis, development of the growth plate, strength, weight and aging (Harada and Rodan 2003).

3.2.3.3 **Disadvantages**

One of the major disadvantages of DXA analysis of bone metabolism is that it only reports the status of mineralization as a consequence of a therapy or mutation for example, but not of any specific cellular turnover mechanism in bone. For example, investigators monitoring perturbed conditions accounting for reduced BMD, as assessed by DXA, are ignorant of the fact of whether reduced osteoid and/or deficient organic matrix mineral calcification (collagen : mineral ratio) in a particular disease model is part of the effect or not. DXA alone cannot identify whether the quantity of bone is deficient while the quality is normal, or vice versa as exemplified by animal models of osteogenesis imperfecta (OI). As a result, it is recommended that biochemical bone markers are taken into account in interpreting DXA results (see Section 3.2.4). Likewise, in a situation where the normal equilibrium between bone formation and resorption is lost with resorption predominating, DXA would not be able to elucidate whether there is resultant loss of secondary trabeculae, although the primary trabeculae may be reinforced due to the increased stress they are subjected to. This is due to technical limitations, whereby on the one hand DXA provides an integrated summation of cortical and cancelleous bone measurements but on the other is incapable of segregating more metabolically active bone from the less active cortical region in areas such as the vertebral body and proximal femur. Furthermore, the intricacies of the synovial region hinder the attainment of reliable DXA results even at high resolutions. Of course, the major disadvantage of DXA is the lack of true volumetric BMD expressed as grams per cubic unit (centimeters), thereby excluding the third-dimension thickness component from the data. Since DXA values are not true volumetric calculations, the affected body thickness (i. e. the effect of the test condition on weight) must be subtracted in order to achieve strict bone comparisons. To compensate for this effect, Srivastava et al. (2003) applied an estimation regression model using baseline weight data as a normalization step during their skeletal phenotype screen (for principles on BMD normalization, see Nielsen (2000), Nielsen et al. (1993, 1998, 1999) and Prentice et al. (1994)).

3.2.3.4 Small Animal Applications

The relevance to humans of BMD data obtained by DXA was in predicting the future risk of fracture since it was an objective and direct measure of bone mass (Black et al. 1992; Cummings et al. 1993; Glüer et al. 1993; Hui et al. 1989; Peacock et al. 2002). Since then, the application of DXA to small animal has become more widespread in an attempt to investigate the early response, maintenance and outcomes of therapeutic drugs, growth factors, and genetic manipulation, to name only a few, on skeletal regulation (Tab. 3.3). To date, DXA analysis is being continuously applied in both human and animal osteoporosis research in a similar, yet more progressive manner than the classical approaches (Bohlooly-Y et al. 2004; Bonnick 2000; Han et al. 2005; Melton et al. 2005; Morrison et al. 1994; Roy et al. 2005; Vis et al. 2005).

For example, DXA analysis has been extensively applied in the monitoring of the effects of therapeutic intervention. In an *in vivo* model of osteoporosis, ovariectomized (OVX) mice showed bone and body composition resilience when put on a diet of soy isoflavone, which is one of the phytoestrogens (Wu et al. 2004). In a separate study Sun et al. (2003) looked at OVX mice to assess the effects of dietary fish oil (FO) on osteoporosis, where FO decreased bone loss as measured by DXA due to the inhibition of osteoclastogenesis which was measured independently in culture. Using DXA Samuels et al. (2001) came to the conclusion that parathyroid hormone (PTH) and 17 beta-estradiol (E2), when administered in combination, have additive effects on the mass of the long bone of female mice, possibly reflecting cross-talk between the two agents during bone formation.

3.2.3.5 Precision and Accuracy

The precision and accuracy of DXA is commensurate with its widespread application in all areas of clinical and biological sciences as documented in a number of correlation studies validating the technology. For instance, Nagy and Wharton (1999) and Nagy et al. (1999) performed correlation experiments in order to

Table 3.3 Recent applications of DXA to small animals.

Feature	Reference
Longitudinal study of oim/oim OI mouse model	Philips et al. 2000
Longitudinal study of GH-deficient lit/lit mice	Kasukawa et al. 2003
Musculoskeletal effects in GDF8 knockout mice	Hamrick 2003
Fracture healing	Li et al. 2005
Hormone signaling	Bohlooly-Y et al. 2004
BMP regulation	Rittenberg et al. 2005
Monitoring rat type 2 diabetes	Hayes et al. 2000
Depicting mouse strain-specific bone parameters	Akhter et al. 2000 Beamer et al. 1996,
Mouse-ENU skeletal phenotyping	Srivastava et al. 2003
Identification of peak bone density QTLs and elucidating bone turnover	Beamer et al. 1999; Gu et al. 2002; Klein et al. 1998; Klein et al. 2001; Shimizu et al. 1999
Therapy	Samuels et al. 2001; Sun et al. 2003; Wu et al. 2004

oim, osteogenesis imperfecta murine; GH, growth hormone; GDF, growth differentiation factor; BMP, bone morphogen; ENU, ethylnitroso-urea; QTL, quantitative trait loci.

successfully relate DXA-derived BMC values to ash weight. Complementary to this work was the study by Nagy and Clair (1999) who applied gravimetric and chemical extraction techniques to positively correlate DXA body composition measurements with the biochemical standards. The prediction of lean mass was the focus of the report by Nunez et al. (2000), who validated the fat-free mass index obtained by DXA for *in vivo* body composition measurements in mice. With regard to small animals in particular, there are different systems on the market, but for the most part, correlation of the results is very high. Coleman et al. (1999) showed that both the PIXI and traditional DXA technology were able to predict *in vivo* bone mass precisely, and correlated well during BMD and BMC comparisons at multiple sites in the mouse skeleton.

3.2.3.6 Considerations

Most available DXA systems do not have a breach system to prevent the operator from crossing the X-ray beam during a scan, or radiation escaping from the analysis platform. Thus special operating instructions must be adhered to (e. g. clearance of space, removal of jewelry) even though the dose of DXA radiation from the system is relatively low; in DXA the X-ray dose for the mouse is about 100 to 1000 times lower than that used in X-ray analysis. In mice many factors such as age, influence the quality of the DXA data obtained. Most systems only work reliably when the mice have reached a minimum weight. The critical weight for a whole body scan in most systems is between 15 and 25 g. Within the first 100 days, there is a strong, nearly linear increase in the bone tissue per day. Therefore individual variations in growth phases will have strong gradual influences on the values obtained and as the mice age the individual variations in BMC and BMD will increase. Thus it is critical to estimate the age at which the population variance is minimized during the linear growth phase in order to collect bone parameters such as BMD.

As already mentioned, mouse BMC and BMD levels depend on age and weight, but also on genetic background and gender. Thus, it is important to analyze male and female data separately. In addition, the data should also be analyzed in a standardized format for age and weight. As already discussed, we suggest that animals between the ages of 100 and 150 days should be selected for comparison of age/sex-matched animals. The collected data of these animals can then be analyzed for the raw BMC and BMD values as well as for weight-corrected data. The simplest method for the standardization of the data is to divide BMC and BMD data by the weight of the animal (Carter et al. 1992; Compston 1995).

3.2.3.7 **Protocol**

- Calibrate the system according to the manufacturer's instructions
- Anesthetize the mouse (e. g. intraperitoneal injection of Ketamin/Xylacin)
- Test to see whether the mouse is asleep with a toe pinch
- Directly place and position the mouse on the analysis platform
- If necessary fix tail and limbs with tape
- Take precautions to avoid contamination by radiation
- Start a scout scan to test whether the ROI is covered by the scan
- Start the measure scan
- When the scan is finished and the radiation source turned off, put the mouse back in its cage
- Put the cage on a heating plate as long as the mouse is asleep to prevent a fall in its body temperature
- Control the mouse until it has re-awakened

3.2.4 Biochemical Bone Markers

3.2.4.1 **Clinical Utility of Biochemical Markers of Bone Turnover in Small Animals**

Significant advances in the understanding of bone and mineral metabolism and the pathophysiology of associated pathological disorders have dramatically increased the demand for routine clinical measurements of bone and mineral analytes. The purpose of using bone markers within the skeletal screen is to identify phenodeviants of systemic metabolic bone diseases as early as possible and with the use of non-invasive techniques. Markers of bone metabolism have an important and potential role in the detection of phenodeviants at risk for dysregulated bone turnover (Allen 2003; Watts 1999). Metabolically active and dynamic bone tissue undergoes constant remodeling or "turnover" which is characterized by two processes, bone resorption and formation, and mediated by two types of bone cells, osteoclasts and osteoblasts respectively. Activated osteoclasts degrade "old" bone during bone resorption, while osteoblasts fill the resorption cavities with "new" bone during bone formation within bone remodeling units. In general, the resorption phase occurs more rapidly than the bone formation phase during which the cavity is refilled. In a healthy adult mouse, bone resorption and formation are tightly coupled within a bone remodeling unit signifying the remodeling process.

Bone mass depends on two factors: (1) the coupling of bone resorption and formation in a remodeling unit, and (2) the number of remodeling units activated in a given area of bone. Bone loss results from a persistent imbalance or uncoupling of the resorption and formation phases in a given bone remodeling unit, coupled with an increased activation velocity of these remodeling units. This consequential loss of bone can be monitored statically by way of DXA over an extended period of time, or dynamically by using biochemical markers. Biochemical markers alone are not advocated as a means of diagnosing bone loss. Biochemical markers of bone turnover present a means with which to monitor the short-term, initial compliance to a mutation or a diseased state in an objective manner. Nevertheless, bone turnover measurements can be helpful in identifying phenodeviants with significantly increased early-onset or small-change bone turnover, particularly bone resorption, even in the face of a normal or nominal change in BMD. In retrospect, since bone mass accounts for approximately 80 % of the variance in bone strength, low BMD is an important risk factor for osteoporosis and fracture (Boskey et al. 1999). Therefore, BMD assessments should be used in conjunction with biochemical markers of bone turnover for a full and more effective analysis of disease.

A mutation or anomaly of a strict bone component, e. g. type I collagen, could also exacerbate and complicate the consequences of bone loss attributed to secondary increased bone turnover. Certain diseases can be grouped together and display characteristic marker profiles via mechanisms that increase/decrease bone turnover and hormonal regulation associated with a particular pathology (Tab. 3.4). Table 3.4 shows the relevant parameters which should be measured in first-line screening for various skeletal catabolic/anabolic phenotypes. For the most part, the listed disease states are complex disorders with an hereditary basis influenced by many conditions. Based on this table, changes in bone metabolism may be detected using only the less expensive markers such as tAP and urinary hydroxyproline. Serum tAP and urine calcium excretion, although less specific, may give useful measures of bone turnover and the consequences of a disorder. Causes of bone loss include endocrine disorders, fracture healing, certain medications, lifestyle behaviors and skeletal complications of malignancies (e. g. previous fractures, myeloma). It must also be stressed that the biochemical profile featured in Tab. 3.4 is generalized. For example, depending on the nature of the mutation, genetic background, sex, age, and environmental factors, the results obtained using biochemical markers may be significantly different. Nevertheless, the performance of currently available markers in identifying Paget's disease, postmenopausal osteoporosis, hyperthyroidism, primary hyperparathyroidism, hypoparathyroidism, osteomalacia and rickets, acromegaly and Cushing's syndrome, osteogenesis imperfecta, arthritic disease, chronic renal failure, and cancer have all been extensively reviewed. The results of OI blood chemistry in general are highly variable. Seeliger et al. (2003) noted no differences in the clinical chemistry of serum including alkaline phosphatase activity, in a type I OI model in dog. Furthermore, Shapiro et al. (1996) provided a synopsis of various cases of OI with varying clinical chemistry outcomes. As an example relevant to mice, serum markers of bone metabolism (PTH, TRAP, and OC) have been successfully used in beta-arrestin2(–/–) mice suggesting the arrestin-mediated regulation of the differential effects of PTH on cancellous and cortical bone (Bouxsein et al. 2005).

3.2.4.2 **Mouse Markers of Bone Turnover/Metabolism and Hormonal Regulation**

The biochemical consequences of osteoclasts and/or osteoblasts during bone remodeling can be assessed by measuring their enzymatic activity or by measuring the levels of bone matrix metabolites released into circulation. Table 3.5 features a list of bone marker assays and indicates those bone markers which give positive cross-reactivity with mouse and other companion animals. During bone resorption, breakdown products such as deoxypyridinoline (Dpd) and the N-terminal telopeptide (NTX) from type I collagen digested by osteoclasts, are released into the bloodstream and excreted in a non-metabolized form in urine. Tartrate-resistant acid phosphatase (TRAP) is another common biochemical enzymatic marker for resorption whereby isoforms occur in osteoclasts, platelets, and erythrocytes. Bone formation markers are released into the circulation during resorption as well during the synthesis of new bone protein matrix and include bone-specific alkaline phosphatase (BSAP), type I procollagen C-terminal propeptide (PICP) and osteocalcin (OC). Measurements of all these products are indicative of the rate of bone turnover. Tests for several hormonal markers of bone metabolism are currently available for use in the research and clinical settings. The endocrine effect in metabolic bone conditions can either be primary or secondary. PTH and FGF-23 levels can be reasonably monitored in serum as an indicator of the degree and direction of calcium maintenance in complex disease conditions such as osteomalacia or primary hyperparathyroidism. The new line in ELISA kits makes it feasible to undertake these assays on a routine basis. Bone formation and hormonal markers are usually measured in serum, whereas markers of bone resorption are typically measured in urine. However, tests for measuring bone resorption markers in serum are also readily available. The urinary markers of bone resorption are derived from type I collagen. Although pyridinoline (Pyr) is found in type I collagen, it is also found in type II collagen of cartilage. Therefore, it is recommended that the helical peptide assay is used for specific detection of breakdown products of type I collagen. Table 3.5 lists each marker, its purpose, and its species specificity. The reviews by Allen (2003) and Watts (1999) provide descriptions of all other bone marker assays.

Table 3.4 Biochemical marker profile of human bone diseases.

Diagnosis	Differential Diagnosis	Serum											Urine Excretion			
		Regulation		Formation			Resorption			Hormonal						
		Ca^{2+}	PO_4^{3-}	(BS)AP	OC	PICP	SP	ICTP	DH	PTH	CT	GH	Ca^{2+}	PO_4^{3-}	HYPRO	cAMP
Primary OP[1]	M. Cushing[1], malabsorption	■	■	↑■	↑/↓	■	↑■	↑[1]	■	■	■	□	↑■	↓■	↑■[1]	□
ROD (-malacia)	Secondary HPT[2]	↓■	↑	↑	↑	↑■	↑	↑	↓	↑	□	□	↓	↓	↑■	□
Osteomalacia[3]	ROD, Blout's disease[4] (burnt-out rickets)	↓■/ ■	↓■/ ↓M ■F4	↑/ ↑M ■F4	↑/↓	↑	↑■	↑	↓	↑	□	□	↑■	↓	↑■	↑
Primay HPT	Secondary and tertiary HPT[5]	↑	↓■	↑	↑	↑	↑	↑	↑	↑	↑■	□	↑■	■	↑	↑
M. Paget[6]	Neoplasia arthritis	↑■	■	↑	↓/□[6]	↑	↑■	↑	□	□	□	□	■	■	↑	□
Acromegaly[7]	Diabetes mellitus, arthritis	↑■	↑	↑	↑	?	■	?	□	□	□	↑	↓	■	↑	□
OI[8]	Marfan syndrome	↑■	■	↑	↑	↓	↑/↓[7]	↑/↓[7]	□	?	■	?	?	?	↑	□[8]
Osteoclastic neoplasia	Primary tumor, primary HPT, M. Paget	↑	↑■	↑■	↑	■	↑	↑	□	■	□	□	↑	↑■	↑	■
Osteoblastic neoplasia	Secondary HPT, malabsorption	↓	↓■	↑	↑	↑	↑	↑	□	□	■	□	↓■	↓■	↑	□

The practical significance of each parameter should be interpreted with caution. Conditional differences in age, genetic background, sex, diet (calcium intake), diurnal and seasonal rhythms, physical activity (stress), light exposure, and menstrual cycle, to name but a few, can all influence bone turnover and the results of bone marker tests. The profiles shown in this table were adapted from Bilezikian et al. (1996), and Seibel et al. (1999), and papers cited therein.

↑ ↑ ↑, level of difference; ■, reference level; □, not applicable; ?, unknown to the authors; ↑ / ↓, variable; OP, osteoporosis; ROD, renal osteodystrophy; HPT, hyperparathyroidism; OI, osteogenesis imperfecta; (BS)AP, bone specific alkaline phosphatase; SP, total acid phosphatase; DH, 1,25$(OH)_2D_3$/calcitriol vitamin D hormone; PTH, parathyroid hormone; CT, calcitonin; OC, osteocalcin; GH, growth hormone; HYPRO, hydroxyproline; cAMP, cyclic adenine monophosphate; PICP, type I procollagen C-terminal propeptide; ICTP, cross-linked telopeptide of type I collagen; M, male; F, female; RIA, radioimmunoassay; IRMA, immunoradiometric assay; tAP, total AP; TRAP, tartrate-resistant acid phosphatase.

1) Primary osteoporosis is categorized as postmenopausal or senile; M. Cushing's syndrome is hyperadrenocorticism; ICTP and HYDRO increase only in high bone-turnover diseases.
2) Secondary kidney clinical chemistry method suggested.
3) Adult rickets due to malabsorption, vitamin D deficiency with secondary hyperparathyroidism; direction of effect may depend on the compensatory parathyroid activity.
4) Congenital tibia vara characterized by bowing or torsion of the tibia, and breaks in the medial tibia metaphysis as seen on a plain radiograph (Giwa et al. 2004).
5) Secondary calcium suppression method suggested.
6) Resultant enlarged and deformed bones, etiology unknown; there is no known explanation for the discrepancy between increased tAP and decreased osteocalcin levels.
7) Hyperpituitarism; the most common etiology is a benign hGH (human growth hormone) -producing pituitary adenoma; secondary clinical GH method suggested.
8) The clinical chemistry profile for OI can vary dramatically depending on the severity (mutation) of the phenotype (Whyte 1999).

Table 3.5 Serum-/urine-based clinical assays of bone metabolism in mouse.

	Assay	Purpose	Company	Detection method	Sample	Cross reactivity	Cost (sample/price)
Formation	CPII	Monitor the synthesis of cartilage components (type II collagen)	IBEX Technologies Inc.	ELISA	Serum	Mouse, rat, human, dog, rabbit, guinea pig, horse, cow	96/690 €
	Osteocalcin, mouse	Bone formation marker	Immutopics Inc.	IRMA	Serum	Mouse	100/480 €
	VEGF-A, mouse	Angiogenesis factor	Reliatech Inc.	ELISA	Serum	Mouse	96/450 €
Mineral metabolism/hormonal	FGF-23 Intact mouse	Monitor hypophosphatemic; indication of calcium metabolism	Immutopics Inc.	ELISA	Serum or plasma	Mouse	96/690 €
	Osteopontin, mouse	Monitor sepsis and metastatic cancers; calcium deposition and bone metabolism	Osteomedical Group Inc.	ELISA	Serum	Mouse	96/690 €
	Calcitonin, rat	Monitor the regulation of calcium and phosphate levels	Immutopics Inc.	IRMA	Serum	Mouse, rat	100/590 €
	PTH Bioactive, rat	Maintenance of ionized calcium in blood	Immutopics Inc.	ELISA	Serum or plasma	Mouse, rat	96/620 €
	PTH Intact, mouse	Maintenance of ionized calcium in blood	Immutopics Inc.	ELISA	Serum or plasma	Mouse, rat	96/620 €
	PTH Intact, rat	Maintenance of ionized calcium in blood	Immutopics Inc.	IRMA	Serum or plasma	Mouse, rat, human, cow, pig, rhesus monkey	96/620 €
Resorption/collagen degradation	Osteoprotegerin, mouse	Monitoring of resorption (osteoclastogenesis)	Osteomedical Group Inc.	ELISA	Serum	Mouse	96/690 €

Free s-RankL	Differentiation marker of osteoclasts	Osteomedical Group Inc.	ELISA	Serum	Mouse, rat	96/690 €
Mouse TRAP	Bone resorption	Osteomedical Group Inc.	ELISA	Serum	Mouse	96/545 €
CTX Rat (CrossLaps)	Monitor bone resorption by the degradation of C-terminal telopeptides of type I collagen	Nordic Bio-science Inc.	ELISA	Serum or urine	Mouse, rat	96/770–685 €
Pyrilinks D (Dpd)	Monitor unmetabolized deoxy-pyridinoline crosslink (Dpd) of bone type I collagen, a specific marker of bone resorption	Quidel Inc.	ELISA	Urine	Mouse, rat, human, squirrel, dog, rabbit, guinea pig, horse cow, pig, rhesus monkey, sheep	96/580 €
Total Dpd crosslink	Measure total Dpd	Quidel Inc.	ELISA	Serum or urine	Mouse, rat, human, dog, guinea pig, horse, cow, rhesus monkey	96/350 €
Helical peptide	Monitor type I collagen helical peptide 1(I) 620–633, a marker for collagen degradation	Quidel Inc.	ELISA	Urine	Mouse, rat, human, rabbit, guinea pig, cat, dog, pig, goat, sheep, cow, horse, Rhesus monkey, baboon, chimpanzee	96/690
Pyrilinks (Pyd+Dpd)	Measure of total pyridinium crosslink as an indicator of type I bone collagen resorption	Quidel Inc.	ELISA	Urine	Mouse, rat, human, guinea pig, rabbit, pig, dog, sheep, horse, Rhesus monkey	96/580
Pyridinoline (Pyd)	Indicator of type I collagen resorption	Quidel Inc.	ELISA	Serum	Mouse, rat, human, baboon, cat, cow, dog, guinea pig, horse, Rhesus monkey	96/690 €

3.2.4.3 Variability/Sensitivity/Sample Choice

Baseline and screening results using biochemical markers should be interpreted with caution. Standardization of each bone marker is obligatory before use in animal screening. Inherent variability is associated with each bone marker and false conclusions may be drawn from a single value assessed twice or more if the variability is too high. Two types of variability are associated with bone marker measurements: analytical and biological. Analytical variability is related to the performance characteristics of a given assay, and is a function of the analyte itself combined with the design and quality of the assay methodology. The analytical performance (inter- and intra-assay variability) should be established before the assay is carried out. Biological variability includes variability resulting from seasonal and environmental differences, for a single animal over a single day (diurnal), from day to day, and from animal to animal within a given population. Inter- and intra-breed variability in skeletal growth and remodeling is unavoidable, but strict constraints on sample collection procedures (i. e. collection time, temperature, storage) should reduce the amount of noise on marker measurements when compared to age/sex-matched controls. Efforts should be made to establish and minimize variability by collecting age/strain/sex-specific baseline data over the course of a day, between days, and preferably between seasons and animal housing facilities in a large cohort in a repeated-measures design if possible. Subsequently, taking at least two sample measurements at the determined time-point should reasonably allow for animal to animal differences within a given population. Longitudinal studies utilizing bone markers would be informative and physiologically relevant, although extremely expensive in a high volume screen for phenodeviants. In terms of which samples to collect, there is generally less biological variability associated with serum samples, although urine collection is easier and less invasive for the animals. In deciding which type of sample to use, convenience, expense, and assay availability have all to be considered.

When used optimally mouse assays exhibit adequate sensitivity for the identification of phenodeviants of systemic metabolic bone diseases. The sensitivity depends on the bone metabolite or molecule to be detected, metabolic condition and on the detection method (e. g. ELISA or RIA). For example, more benign skeletal conditions such as osteoarthritis may be characterized by undetectable traces of circulating markers, thereby making a particular bone assay unsuitable. The detection spectrum and details provided by the manufacturer should be consulted before adhering to any particular assay.

3.2.4.4 Which Markers Should be Used During the Screen?

The serum and urinary assays described above are relatively simple and cost-effective methods only when compared to high definition µCT imaging and histomorphometrical analysis carried out on a large scale. Due to the current and substantially high prices associated with many of these bone-specific kits, their use is impractical in primary screening for metabolic/anabolic mutants for example, and thus only recommended for secondary screening and beyond. One exception is the total Dpd kit, whose cost is half that of most other urine assays, but its specificity is much lower. Nevertheless, the clinical chemical profiles of known skeletal disorders can be used to assist and potentially identify affected mice using more cost-effective strategies. Several trends are obvious and prevalent (e. g. increases in AP and collagen metabolite excretion, increases/decreases in Ca^{2+} and PO_4^{3-} levels), thus warranting obligatory inclusion in the primary screen. In addition, electrolytes and creatine levels are monitored to assess the animal's body fluid balance and kidney function respectively. tAP is a relatively inexpensive non-specific bone marker for bone turnover and has the potential to detect alterations in either osteoblast differentiation or proliferation, making it a practical tool for high volume screening. Serum osteocalcin is a sensitive marker of osteoblastic activity since osteocalcin is a non-collagen protein found only in bone and dentin. Therefore in mouse, this assay could be substituted for BSAP analysis as a bone-specific formation marker in the secondary screen, bearing in mind the lack of any commercial mouse BSAP ELISA kit known to the authors to date. In retrospect, a primary screen which entails several methods spanning each individual remodeling status would be ideal, but impractical, expensive and unrealistic. Instead, a primary screen monitoring the levels of calcium, phosphate, tAP, and a type I collagen excretion metabolite is recommended, whereby the events of bone formation, regulation and urine resorption are systematically and comparatively evaluated. A follow-up secondary screen reconfirming the status of bone formation, and an evaluation of the status of serum resorption and the hormonal status are further recommended (see Fig. 3.5).

3.2.5
Advanced Small Animal Imaging Techniques

There are many biomedical imaging techniques which give accurate results in the assessment of body composition parameters, and cortical and trabecular bone integrity in arbitrary orientations to varying degrees. These methods, for the most part, are highly automated, objective, and non-user-specific thus providing an unbiased comparison between the control and altered states. In addition, they are non-destructive, thereby allowing further biomechanical, radiographic and/or chemical testing on the same sample. On the other hand, these techniques are time-consuming, and are thus reserved for either the secondary or tertiary skeletal phenotyping screens. Below we provide a brief overview and highlight several common techniques, their advantages and disadvantages, and current applications using small animals, in particular the mouse.

3.2.5.1 pQCT

Background, Advantages and Disadvantages Peripheral quantitative computed tomography (pQCT) is a method of assessing bone parameters using multiple cross-sectional X-rays to reconstruct a volumetric model of bone density distribution. The most common pQCT equipment on the market is manufactured by Stratec Medizintechnik GmbH, Germany, and certain models have been adapted specifically for smaller rodents. pQCT is a widespread, non-invasive technique used in small animals to provide separate estimates of trabecular and cortical BMD and BMC. Axial and appendicular skeletal development, degree of bone loss or gain, and response to therapy can all be assessed with high accuracy using this method (Gasser 1995). pQCT can compute bone density in terms of a true physical measurement (g/cm^3). pQCT can also be used determine the axial and polar moment of inertia, and a strength strain index, which are estimates of mechanical strength and are dependent on both bone geometry and density. In the cross-sectional analysis of bone area pQCT can also distinguish between periosteal modeling and cortical bone. One disadvantage of the pQCT set-up is its inability to monitor mouse lumbar vertebral bodies *in vivo*, therefore *ex vivo* measurements are used to reduce associative errors. In other words, for *in vivo* monitoring of bone density, mass and architecture, pQCT analysis is restricted to locations on the appendicular skeleton and tail vertebra. In addition, the largest source of error in pQCT systems is the fat within bone marrow; this can be overcome by performing scans at two different X-ray energy levels to improve the accuracy. pQCT is also limited by the degree of in-depth trabecular bone analysis that can be achieved. Other technical issues which need to be taken into account when using pQCT equipment are the slice thickness and spatial resolution that can be obtained to reflect the smaller bone size of mice.

Technical and Practical Considerations When comparing mutants or possible therapeutic effects, as examples, precise and standardized acquisition regions and distances should be determined. The appropriate anesthesia (isofluran or narcosis) should be applied in one dose, and special holders for stable limb positioning during the scan should be used. Before applying pQCT, proper quality assurance steps (i. e. calibration) should be taken. The procedures for setting up the scanner are beyond the scope of this chapter. Threshold-based contour separation procedures are standard for the detection of outer bone for example, in small rodents. Moreover, adjustment of the threshold value will depend on the age of the mice. Likewise, the data analysis procedures are too elaborate to describe in detail, but it should be borne in mind that the absolute gain or loss in BMC can be calculated and attributed to either an endosteal or bone size effect by cross-comparing other parameters, for example. Further guidance on the interpretation and normalization of results can be found in the report by Helfrich and Ralston (2003) and in the manufacturers' guides. For tissue comparison, the proximal tibia metaphysis has been the region of choice in small animals due to the presence of cancellous bone and its metabolic status and reaction to diseased or therapeutic states. On the other hand, the distal femur metaphysis is hardly ever used for analysis because its cross-section is too small and differentiating cancellous from cortical bone in this region is virtually impossible in mice. If other appendicular locations are to be considered, then the degree of regional bone loss that can be detected with pQCT must be estimated. For this problem and more elaborate trabecular measurements, high definition µCT or µMRI analysis is recommended.

Mouse Applications Recent examples of the use of pQCT in mice can be found in the work of Bohlooly-Y et al. (2004) and Keller and Kneissel (2005) who quantified BMD in a MCHR1-defiicient osteoporotic mouse line and showed inhibition of

SOST transcription in bone in PTH-treated mice respectively. Srivastava et al. (2003) applied pQCT in a secondary mouse ENU screen as a follow-up or validation step to confirm phenodeviants from the primary DXA screen. The use of pQCT is recommended for the secondary skeletal phenotyping screen mainly on the grounds of cost and radiation burden as compared to DXA.

3.2.5.2 μCT

Background It was in the early 1970s when computed tomography (CT), also known as computed axial tomography (CAT), was introduced by Hounsfield and Cormack (Di Chiro and Brooks 1979) who shared the Nobel Prize for Medicine in 1979, as a novel biomedical imaging technique. Since then, CT has advanced both human and animal biomedical imaging. CT is a tomographic imaging method whereby digital processing is used to create a 3-D reconstruction of internal organs, arteries and bone from a large series of 2-D X-ray data taken around a single axis of rotation. The 3-D extension is accomplished via several mathematical methods such as Radon transform and algebraic reconstruction techniques (ART), whereby the computational costs and additional post-filtering techniques affect which methods to employ initially. Here we discuss the measurable parameters, advantages and disadvantages, and the current μCT applications in small animals, while the actual μCT system and detailed data analysis (e. g. microfocus X-ray tubes, stage, detectors, software, etc.) will not be discussed here.

Advantages and Parameters Currently, μCT is the method of choice for bone microanalysis based on absorption of high energy X-rays, as it is the most accurate method of assessing volumetric trabecular bone mass and density in mice. One advantage of high definition μCT analysis is that it is highly correlated with the "gold standard" histomorphometrical method of evaluating the degree of mineralization in bone (Muller et al. 1996), and it arose and was adapted from industrial applications for biomedical uses such as bone research. At present, this micro-technique is more commonly applied *in vitro* due to high radiation levels and acquisition times, although newer machines and software offer the possibility of *in vivo* examination. Advances in μCT (nanoCT) using high intensity and tight collimation synchrotron radiation can achieve spatial resolutions of 1–2 μm (i. e. at the cellular level), thus providing another dimension and the capacity to assess resumption cavities in animal models featuring high bone turnover rates (Jiang et al., 2000).

μCT can accurately determine a plethora of true bone parameters in a model-independent manner. The bone morphometric parameters of μCT are: TV (VOI, volume of interest; mm^3), BV (object volume; mm^3), BV/TV (relative volume; %), BS (object surface; mm^2), BS/BV (specific surface; mm^{-1}), Tb.Th (trabecular thickness; mm), Tb.N (trabecular number; mm^{-1}), Tb.Sp (trabecular space; mm), Tb.Pf (trabecular pattern factor; mm^{-1}), SMI (structure model index), DA (degree of anisotropy), E.Conn (Euler connectivity), and E.Conn.D (Euler connectivity density; mm^{-3}). For a brief clarification: the BV fraction is the volume of bone divide by the total volume of the bone sample; the Tb.N is the average number of trabeculae that

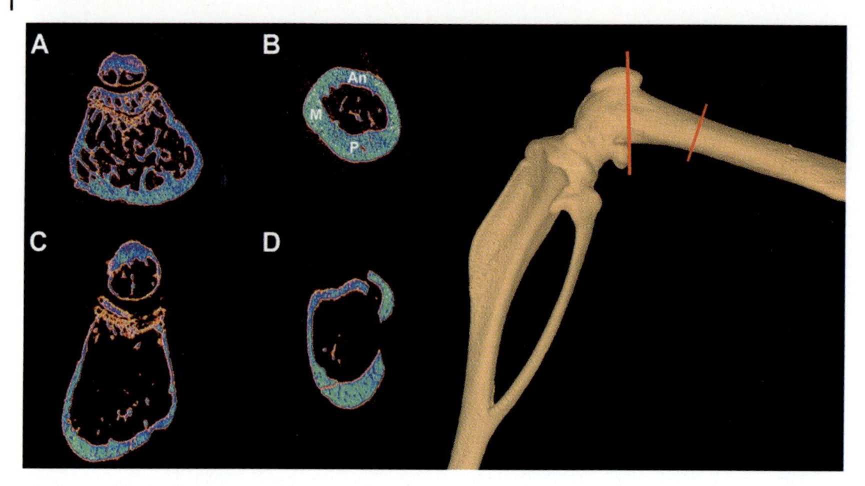

Fig. 3.3 High resolution μCT of mouse femora (6.3 μm pixels). Cross-sectional analysis of the distal femur metaphysis (A) and shaft (B) of wild-type C3HeB/FeJ 16-week-old mice. (C and D) comparative analysis of the corresponding regions in an OI mouse model on the same genetic background. In panel A, 27 trabecular objects were selected yielding an average object size of 0.097 mm occupying a relative area of 17.8 %, as opposed to panel C where four objects were selected yielding an average object size of 0.058 mm covering a relative area of 0.9 %. Panel B and D compare the cortical thicknesses yielding anterior values of 0.28 and 0.18 mm, posterior values of 0.45 and 0.37 mm, and medial values of 0.19 and 0.09 mm between wild-type and OI samples respectively. An, anterior; P, posterior; M, medial.

occur per millimeter in a line drawn through the bone sample in a particular direction; the Tb.Th is the average thickness of the trabecula in a particular direction; the Tb.Sp is the average space between two trabeculae in a certain direction; and connectivity is a measurement of the number of equivalent voxels next to each other. Some of these parameters have relevance in μMRI analysis, although standardization is still in development (see later). A general overview of the volumetric data processing method is given in the Appendix.

Disadvantages The disadvantages of using μCT analysis are the high radiation load and long acquisition times. Also, the fluid and live tissue status (e. g. bone marrow and synovial fluid) cannot be assessed using this method. In addition, the high costs of the hardware and software license limit easy access and usage of dedicated μCT research workstations.

Mouse Applications μCT has been applied extensively in mouse for the study of osteoporosis and in phenotypes of transgenic and/or gene knockout mice models of human skeletal disease (Fig. 3.3). Kawano et al. (2003) used μCT analysis for the quantification of mineralization in androgen knockout mice, showing a clear reduction in trabecular volume. In the skeletal phenotyping of the EP4 receptor, one of the subtypes of the prostaglandin E(2) (PGE(2)) receptor, μCT imagery revealed osteopenia, thinner cortices, reduced trabeculae, and a deteriorated trabecular network in KO mice (Li et al. 2005). Jiang et al. (2000) used µCT to evaluate ovariectomy- (OVX) induced osteopenia in rats and its treatment with agents such as estrogen and sodium fluoride. In another animal study using OVX rats treated with alfacacidol, a prodrug of D-hormone (calcitriol; Dambacher et al. 2004; Ito et al. 2003), Tb.Sp was decreased, and BV/TV, Tb.Th, and Tb.N were all increased in a concentration-dependent manner. µCT analysis in small animals will continue to be a vital part of research into bone metabolism and anabolism.

3.2.5.3 µMRI

Background Magnetic resonance imaging (MRI), also called magnetic resonance tomography (MRT), uses non-hazardous radio waves and a strong magnetic field rather than X-rays to provide remarkably clear and high resolution pictures of internal organs and tissues such as porous trabecular bone and articular cartilage to demonstrate pathological or other physiological alterations of living tissues. MRI was developed from the principles of nuclear magnetic resonance. The word *nuclear* is almost universally omitted in order to avoid the negative connotations associated with radiation exposure. Other MRI-related methodologies such as functional MRI, magnetic resonance spectroscopy, magnetic resonance angiography, and diffusion MRI will not be discussed here. The basic principles of MRI and imaging of MRI datasets are discussed in the Appendix.

Advantages The main advantage of MRI compared to other commonly used imaging methods is that it is non-ionizing, and has no reported damaging side-effects. Recently, quantitative µMRI techniques have been compared and validated, although with limitations, between DXA, pQCT, µCT and histomorphometrical methods using animal (Ichikawa et al. 2004; Jiang et al. 2000; Taicher et al. 2003; Weber et al. 2005) and human (Link et al. 1998; Majumdar et al. 1998; McWalter et al. 2005) bone samples for relevant body composition parameters, and bone measurements such as BMD, BMC and trabecular characterization, thereby revealing another modality in bone research. µCT can neither be used for analysis at the molecular level, nor for the measurement of muscle or blood fat. On the other hand depending on the system, quantitative MRI (qMRI) can be used rapidly and without anesthesia in mice to measure fat (and temperature), lean mass and free amount of fluids irrespective of whole body weight. The key benefit of the MRI system in body composition measurements is the precision and accuracy it offers when compared to DXA, where in some comparisons the precision was found to be about 25 times better. This reduces the need to administer chemical compounds (i. e. less toxicity) and produces more detailed measurements of body composition in mice.

The major reports comparing (µ)CT to (µ)MRI for bone parameters suggest very few differences, thus implying that (µ)MRI can be feasibly used in the same way that (µ)CT is currently being applied in bone research. However in one human comparison (Link et al. 1998), texture measurements using MRI combined with BMD significantly increased the correlation with CT using a multivariate-regression model, indicating that MRI techniques provide additional information regarding bone strength and quality. The authors also suggest that the CT images represent projected patterns rather than a true image of trabecular structure as obtained using MRI owing to the superior spatial resolution of the latter. However, the authors were not using a dedicated research CT system for the comparisons that yielded a spatial resolution of only 400 µm. In addition, the authors did not discuss methods of choosing the threshold in order to reliably compare the scanned data to the real data obtained from a second form of measurement. The simplest method would be to use a generally accepted measurement for the BV parameter for example, and then choose a threshold value for both MRI and CT that would result in the same BV value, the spatial resolutions can then be compared.

qMRI has been compared with DXA in the measurement of bone parameters. Taicher et al. (2003) compared the two methods using bovine tibia samples and showed a high correlation for BMD and BMC calculations; the precise method for determining the BMD and BMC values using µMRI is described in their report (Taicher et al. 2003). The assessment of trabecular bone structure and discrimination between individual trabeculae with quantitative high-field µMRI provides information pertaining not only to BMD and BMC, but also to the trabecular microarchitecture, thus adding a further dimension to the evaluation of bone strength. Bone tissue is a solid material that has a low mobile proton density, therefore direct MRI measurements of bone tissue is relatively unsuccessful. However, trabecular bone can be analyzed indirectly by observing the surrounding bone marrow instead. It is hypothesized that near the boundaries of two physical phases of different magnetic susceptibility, the magnetic susceptibility of bone is smaller than that of the bone marrow. Thus, the lack of homogeneity in the spaces of the bone marrow can be seen in the static magnetic field, this results in a measurable alteration of the marrow relaxation times (T2) and properties. In addition, alteration of the strength of the MRI field (i. e. by using a high-efficiency coil) may improve the acquisition of high-resolution images of bone.

Disadvantages A major disadvantage of quantitative µMRI in bone research is that the research and development phase is still in its infancy. For example, the spatial resolution of µCT is much better understood than that of the current µMRI range. In other words, the higher spatial resolution of µCT gives better contrast due to the sharper definitions of the boundaries between bone and marrow. This problem could be solved by overcoming several of the technical limitations of µMRI, such as requirements for higher signal-to-noise ratio, stronger and more linear magnetic field gradients and more homogenous and stable static magnetic fields. Other physical limitations of µMRI include the width of the spectral signal line and the chemical shift measurements. Another disadvantage is the long acquisition times of µMRI in bone analysis, which makes multiple scanning impractical. µMRI analysis in mice cannot be used to estimate certain histomorphometric parameters since there is no proof or clear distinction between calcified and non-calcified trabeculae using this method. Also, if bone marrow is to be analyzed, great care must be taken to prevent dehydration of the marrow during *in vitro* evaluation to avoid border artifacts. The current status will provide the impetus for future progress and improvements in MRI fidelity toward standardizing the entire procedure for application to bone.

Small Animal Applications To date, MRI has widespread applications in the routine assessment of soft tissues in mice and humans. Bone analysis using μMRI is quite a new technique for both humans and small animals, and the results should be interpreted cautiously. There is potential for *in vivo* μMRI analysis of synovial joint/fluid in arthritic or inflammation-induced mouse models. Jiang et al. (2000) used μMRI in OVX rats to show changes in trabeculae structure which were not detected by DXA. They also showed decreased cartilage thickness, subchondral osteosclerosis and osteophytes in a rat model of induced osteoarthritis; these changes were only partially detected using radiographs. Gardner et al. (2001) used μMRI to monitor morphological changes in mouse trabecular bone induced by a microgravity environment. Ichikawa et al. (2004) reported the use of a 9.4-T μMRI system in the assessment of early bone development in mice. Although the method was useful in assessing cartilage maturation, the resolution was greatly inferior to that achieved by histological procedures. However, the advantage of using this μMRI system to evaluate bone development is that each tissue such as water (which is lost during histological preparations), collagens, bone marrow, and fat has a characteristic proton density and status. The conclusion therefore, was that μMRI analysis in embryonic bone development could be adjunctive, although impractical on its own, which would negate its advantage of being non-invasive. In the context of functional μMRI, Mayer-Kuckuk et al. (2005) monitored labeled bone precursor cells in the bone marrow cavity of mouse femora in order to evaluate bone marrow cell types during bone maintenance and regeneration. Interestingly, there is potential for streamlining μMRI by using a custom-built seven-bed mouse holder for rapid 3-D acquisition (Dazai et al. 2004), unfortunately however, this system has yet to be applied to bone research.

3.2.6 Whole-mount Skeletal Preparations

Skeletal preparation is a simple histological technique whereby alizarin red and alcian blue stain bone and cartilage with red and blue respectively. Alizarin red sulfonate is a bidentate chelating agent for calcium ions when used at the appropriate pH. In the mouse embryos, chondrification starts around 12.5 dpc, and ossification begins about 14.5 dpc (Patton and Kaufman 1995). This two-color staining enables the observation of the endochondral ossification process in order to assess any cartilage or bone malformations. The results include 3D information about the skeleton and samples can be stored for long periods in glycerol. However, since animals must be sacrificed for this analysis, further phenotype analysis is not possible. In addition, to obtain good staining results, each step takes from a few days to over 1 week.

Reagents

- A-R solution: 0.1% Alizarin red in 96% ethanol
- A-B solution: 0.3% Alcian blue in 70% ethanol
- For the working stain: mix 50 ml each of the A-R solution, A-B solution and acetic acid plus 850 ml of 70% ethanol

Protocol

- Kill adult mice with ether or CO_2 (do not damage the skeleton) and remove as much of the skin, muscle, and internal organs as possible
- Fix in 96 % ethanol for a minimum of 1 week
- Fix in acetone for a minimum of 1 week
- Stain in the working stain for a minimum of 1 week with continual agitation
- Destain in 20 % glycerol, 1 % potassium hydrate at 37 °C for 1–2 days
- Continue destaining at room temperature for 5–6 days
- Clear in 50 % and 70 % glycerol each for a minimum of 1 week
- Keep in glycerol and observe the skeleton by microscopy

3.2.7 Histomorphometry

Background The dynamics of bone osteogenesis is routinely analyzed by histomorphological–histometrical and immunohistochemical techniques to assess the differentiation status of bone deposition and growth. Histomorphometric examination provides information about localized bone turnover and remodeling, and structure at the cellular level which cannot be obtained by other investigative means such as high definition μCT, biochemical markers, or even paraffin sections when mineralization and antigenicity have been lost. Although there are correlations between μCT and histomorphometric evaluation (Won et al. 2003) for certain parameters such as bone volume, the cellular profile obtained by histomorphometric analysis is the defining “gold standard”.

Advantages, Parameters, Applications The major advantage of histomorphometry is the plethora of dynamic parameters of bone development and maintenance that can be analyzed. Keep in mind that conventional sections are viewed in two dimensions in which 3D parameters can be extrapolated using stereological formulas, although prone to errors due to anisotropy. The indices in Tab. 3.6 are either primary or derived in accessing bone remodeling events. It should be borne in mind that the primary measurements of perimeter, area, and number are normalized by the amount of tissue present, therefore they can only be compared when a clearly demarcated area or perimeter within a section is analyzed. For histomorphometrical purposes, cancellous bone should be viewed in the corticoendosteal and mid-cancellous regions, and the selection of these regions should be consistent by using proportional or fixed distances from the outer periosteum. A similar approach in consistency should be applied to cortical bone regions. Factors that can affect the interpretation of histomorphometric data include sample preparation/staining procedure, observer bias, and criteria standardization for histological features. Some of these factors can be eliminated using special software (see later). Histomorphometry can be used in the elucidation of the pathogenic mechanics of OI bone fragility, for example, in terms of the effects of thinner collagen fibrils on mineral crystal size and concentration, and whether the mineral is located outside the collagen fibrils.

Table 3.6 Primary and derived histomorphometric indices.

Entity	Measured Parameters
Cortical	B.Ar/T.Ar, Ct.Ar, Ct.Th, O.Ar/T.Ar or O.Ar/B.Ar, O/B.Ar, O.Wi, Ob.Pm/B.Pm
Spongy tissue	Tb.Th, Tb.N, Tb.Sp
Mineralization/ dynamic metabolism	M.Pm/B.PM, MAR, E.De, E.Ar, E.Pm/B.Pm, Oc.Pm/B.Pm, E.Le, N.Cv/B.Pm or /T.Ar, Oc/T.Ar, Mlt, Omt, BFR/B.Pm or /B.Ar, Rm.P, FP, FP(a+), QP, RP, Rv.P, Tt.P
Multiple structures	TV or T.Ar, BV or B.Ar, BS or B.Pm

Some of the listed parameters are individual 3D referents (e. g. TV). Some cortical parameters can be applied to the trabelcular field (e. g. Ob.Pm/B.Pm).
B.Ar/T.Ar, bone area (%); Ct.Ar, cortical area (μm^2); Ct.Th, cortical thickness (μm); O.Ar/T.Ar or O.Ar/B.Ar, osteoid area (%); O/B.Ar, osteoid number (cells/mm^2); O.Wi, osteoid width (μm); Ob.Om/B.Pm, osteoblast perimeter (%); Tb.Th, trabecular thickness (μm); Tb.N, trabecular number (/mm); Tb.Sp, trabecular separation (μm or mm); M.Pm/B.PM, mineralization perimeter (%); MAR, mineral apposition rate (μm/day); E.De, erode depth (μm); E.Ar, erode cavity area (μm^2); E.Pm/B.Pm, erode perimeter (%); Oc.Pm/B.Pm, osteoclast perimeter (%); E.Le, erosion length (μm); N.Cv/B.Pm or /T.Ar, cavity number (No./mm or /mm^2); Oc/T.Ar, osteoclast number (cells/mm^2); Mlt, mineralization lag time (days); Omt, osteoid maturation period (days); BFR/B.Pm or /B.Ar, bone formation rate (μm^2/μm/day or %/year); Rm.P, remodeling period (days); FP, formation period (days); FP(a+), active formation period (days); QP, quiescent period (days); RP, resorption period (days); Rv.P, reversal period (days); Tt.P, total period (days); TV or T.Ar, tissue volume/area (μm^3 or μm^2); BV or B.Ar, bone volume/area (μm^3 or μm^2); BS or B.Pm, bone surface/perimeter (μm^2 or μm).

Disadvantages The major disadvantages of histomorphometric analysis are the amount of time needed to prepare the sample, and the variable topology and structural characteristics of each sample analyzed since histomorphomtetry is a 2-D analysis of anisotropic features (i. e. certain micro-architectural deviations such as the SMI or DA cannot be calculated or correlated). A major problem of mouse histomorphometry is that the small size of the bone samples makes adequate trimming difficult, thus leading to "pockets" within the sections which cannot be completely infiltrated by the plastic resin.

Below is a generalized MMA (4 °C) protocol with an approximate processing time of 6 days.

Protocol

- Fix specimen accordingly; for example 70 % ethanol for 1 h at a low temperature (LT)
- Transfer into 95 % ethanol for 1 h at LT

- Transfer into three changes each of 100% ethanol for 1 to 6 h
- Immerse in methacrylic acid methyl ester infiltrating solution (see notes) and leave overnight with slow rotation
- Transfer into two changes of methacrylic acid methyl ester infiltrating solution for 24 and then 72 h each with slow rotation
- Surround the embedding molds in crushed ice and fill the molds with embedding solution to the outer rim (see notes). Make sure that the specimen is orientated in the direction that the block is to be sectioned. Place a label on the block giving relevant information. Molten wax can be used to seal the rim if necessary
- Apply low pressure at the correct temperature using an exikator if further infiltration is necessary
- Complete polymerization will take approximately 3 h

Technical Notes

- The infiltrating solution is an 80 : 20 solution of methacrylic acid methyl ester to dibutyl phthalate (plasticizer)
- The embedding solution contains 5 g benzyl peroxide (catalyst dried at 37 °C for 1 h) dissolved in 100 ml of infiltrating solution using a slow rotor. Subsequently, 15 µl of N,N-dimethylaniline (accelerator) is adder per ml of embedding solution and mixed by gentle inversion and then placed into appropriate molds
- Alternative polymerization methods (e. g. UV)
- Complete polymerization may take several days during which time shrinkage may occur, and polymerization above 37 °C can lead to the formation of air bubbles. Re-embedding samples is possible, but requires time and the manipulations are difficult.

Technical Considerations: Sectioning, Staining, Decalcification, Resins, Fluorochromes, Equipment The cost of adapting a current microtome/histological system to facilitate the plastic embedding of samples can be considerable. Such costs include the amount of plastic used, chromatography column (i. e. for destabilization), blade re-sharpening, and accessory costs. The universal motorized microtome (Leica RM2165) has yielded consistent and excellent quality bone sections for histomorphological analysis. A special blade/knife support should be used with the tungsten-carbide hard metal disposable blades or reusable knives. When using a rotary microtome, it is recommended to use a D profile 16-cm tungsten carbide cutting edge upon resin sectioning. This set-up will ensure 3–5-µm thick sections prepared at room temperature. It is important to keep the knife and block moist using a 30% ethanol solution during sectioning. Sections are then transferred to slides and carefully positioned using a 96% ethanol solution. The resin sections should be collected using adhesive polylysine-coated slides, covered with polyethylene foils (Heraeus Kulzer), and then compressed overnight or longer at 42 °C using a slide presser from Leica. Sections labeled with fluorochromes should not be deplasticized, left unstained, or mounted using any commercial fluoromount. Mayer's hematoxylin should be used for counterstaining for 30 s to 1 min. The same non-decalcified sections can be used for histochemical and immunohistochemical purposes.

The most common special stains used on non-decalcified bone sections are the Villanueva–Goldner stain for mineralization (Kawano et al. 2003; Villanueva and Mehr 1977) and the Masson trichrome stain (Sigma–Aldrich) for collagen differentiation (Phillips et al. 2000). The Villanueva osteochrome bone stain can be purchased from Polysciences, Inc. (Warrington, PA, USA; www.polysciences.com). Other stains used routinely on non-decalcified bone include Von Kossa/Van Gieson and toluidine blue. The advantage of the former is in conjunction with an osteoclast stain, resorption of the black mineralized bone is clearly depicted. Toluidine blue can be used to identify mineralized bone, front, and reverse line simultaneously. Certain considerations must be taken into account during tissue preparation such as the type of fixative (e. g. Bouin for mineralization; sucrose-buffered formalin for collagen; 1.4 % paraformaldehyde, 0.02 M phosphate buffered saline, 5 % sucrose solution for alkaline phosphatase; 70 % ethanol for fluorochromes) and method of plastic polymerization (e. g. specific light used for polymerization can bleach certain fluorochromes). All fixation steps should be performed via perfusion using a peristaltic pump with a heparin flush. As a general rule, it is advisable to restrict blocks to a thickness of 3 mm and no thicker than 5 mm for routine diagnostic work.

tAP activity and localization within bone sections was adequately demonstrated using the AP leukocyte kit by Sigma–Aldrich together with the manufacturer's protocol. Likewise, the cytochemical acid phosphatase leukocyte kit by Sigma–Aldrich revealed the presence of strong tartrate-resistant acid phosphatase (TRAP) activity in femur–tibia sections of a murine model of OI (Fig. 3.4). Finally, the alkaline phosphatase–anti alkaline phosphatase (APAAP) method can be used for immunohistochemical detection of antibody using non-decalcified sections. The difficulty in distinguishing bone when using decalcified samples must be taken into account, otherwise any analysis would be deemed semi-automated using a color threshold technique.

Decalcified samples can reveal the major features of histological organization of bone (e. g. lamellae, chondrocytes, osteoblasts, osteocytes, osteoclasts, trabeculae, and matrix) and/or cellular events by way of enzymatic or antibody stains following paraffin section microtomy where possible. Ethylenediaminetetraacetic acid (EDTA; 0.5 M), commercial Decal® (Decal Chemical Corporation USA) solution or 8 % formic acid can all be used for decalcification. Samples should then be examined using either radiological or chemical methods to determine the extent of decalcification. Following decalcification, all structural components of bone tissue, regardless of their prior degree of mineralization, appear indistinguishable in sections stained with hematoxylin and eosin (H&E).

It is often necessary and crucial to quantify the amount of mineralized bone in a sample, in particular in metabolic bone diseases. Currently, there are few available embedding resins which are suitable for both morphological and immunohistochemical analyses of mineralized tissue. Methyl methacrylate (MMA; Fluka), the resin of choice for non-decalcified bone histology, can be used in bone immunohistochemistry if the usual, highly exothermic polymerization procedure is avoided as this destroys tissue antigenicity. A report by Erben (1997) provides detailed protocols outlining the dehydration and MMA processing schemes using rat samples,

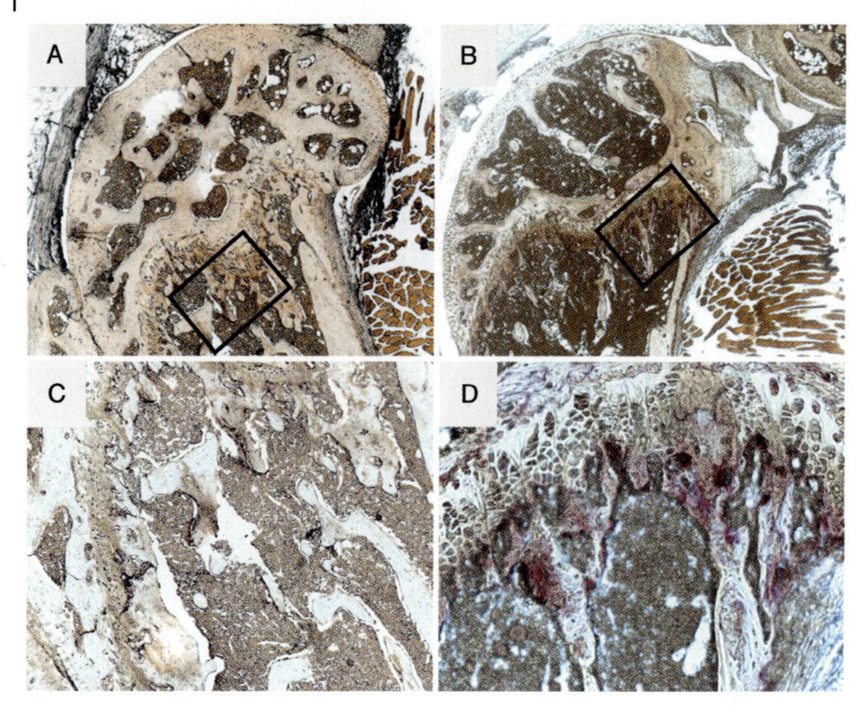

Fig. 3.4 TRAP enzymatic staining. The TRAP enzymatic assay carried out on femur sections of control (panels A and C) and OI (panels B and D) mice. Panels C and D are higher magnification images of regions within the overviews. Increased areas of bone resorption were revealed by way of increased TRAP activity in the OI line. The thicknesses of trabeculae and cortical bone, and the fullness of the secondary ossification center can be seen to be much greater in the control animal.

while Helfrich and Ralston (2003) describe applications in humans and rodents. In addition, Helfrich and Ralston (2003) also give a detailed description of the calculation procedure for all the primary and derived histomorphometric indices (Tab. 3.6).

MMA offers a variety of advantages such as better penetration compared to water-soluble methacrylates such as glycolmethacrylate (GMA) and complete removal of the resin from tissue sections resulting in acceptable staining and morphological details. One disadvantage of conventional MMA is the complete loss of enzyme activity and protein antigenicity in the tissue due to the exothermic radical chain reaction and the increase in temperature during polymerization. Alternatively, polymerization can be initiated using blue or UV light photons as this does not destroy the enzyme activity or antigenic determinants. In addition, modifications of conventional MMA techniques such as lowering the temperature threshold for polymerization, have improved the quality of the staining procedure (Erben 1997; see step six in protocol section of 3.2.5.5). Consequently, most current practices involve cutting samples in half and processing each half in separate resins when more than one type of analysis is required. In addition this “trimming” step is obligatory for complete and satisfactory infiltration when utilizing bone samples from mice or any small animal. As an alternative to plastic resin embedding for the generation of bone sections, samples can be deep frozen using cellulose–hexan (Kawamoto 2003; see Section 3.2.5.6). This method bypasses the tedious preparative steps of resin embedding, although it has the major drawback of requiring an automatic cryostat for the collection of bone sections on an adhesive film. Provided that an efficient fixation procedure is employed, there is excellent preservation of the morphology and antigenicity of frozen sections prepared by the Kawamoto method as compared to resin or paraffin embedding techniques.

Rates of bone mineralization apposition and formation can be measured at the cellular and tissue levels by quantitative histomorphometry after staggered intravital fluorochrome double labeling (e. g. tetracycline hydrochloride and/or calcein, Sigma–Aldrich; 30 µg or 15 µg/g body weight respectively), and at the organ level by ^{47}Ca-kinetics employing different multi-compartment models or the continuous expanding calcium pool model. Fluorochromes attach to the calcium molecules that bind to the mineralization fronts in bone formation sites, and the dynamic range of the growth profile must be arbitrarily determined before the collection of tissues. In non-decalcified sections, fluorochromes can be visualized by their characteristic fluorescence at 365 nm UV or by blue light excitation. A reduced mineralization rate may be due to reduced bone formation (e. g. due to hypothyroidism or glucocorticoid treatment) or compromised mineralization of the produced osteoid (e. g. during osteomalacia).

Small hacksaws, bone trimmers, and a micro-saw are all used during the trimming step prior to infiltration. Especially difficult regions to infiltrate with resin include the encased knee joint, and trimming before resin processing is obligatory in this case. Any standard pressure system such as the exikator (< 0.2 bar) or regulated Univapor 150 (Germany) can be used to improve infiltration of the resin at low temperatures.

Software Histomorphometry is greatly enhanced and optimized by using state-of-the-art analytical software for static and polarized indices, and labeled (epifluorescence) surfaces. Such programs include OsteoLab (Biocom, France; www.biocom.fr) and BIOQUANT OSTEO (BIOQUANT Image Analysis Corporation, USA; www.bioquant.com). These systems are specially designed to assist biologists to explore and analyze bone slides following the standards and guidelines of bone histomorphometry (Parfitt et al. 1987). Some of the structural (e. g. cortical, spongy, and skeletal), static (e. g. osteiod, resorption, and osteoclasts), and dynamic (e. g. epifluorescence) factors that can be measured in bone sections using such software packages are listed in Tab. 3.6. It is advisable to evaluate a demo-version of any software prior to the purchase of software licenses. Bone lamellation patterns in variable (severity) OI cases can be easily analyzed under polarized light and quantified using such program packages (Rauch and Glorieux 2004) as can trabelcular struts, number of nodes, and bone strength and integrity.

3.2.8 Miscellaneous

A number of specialized and highly relevant techniques are available for the analysis of bone disorders. These methods are recommended for either the secondary or tertiary levels of the bone phenotyping screen (Fig. 3.5), and are only briefly described here. Detailed protocols and steps in data analysis can be obtained from the references cited.

IS-PCR, IS-RT-PCR, Microarray Unlike conventional RT-PCR, differential display PCR, or microchip array applications, *in situ* PCR (IS-PCR) and *in situ* reverse transcription PCR (IS-RT-PCR) allow the cellular localization of genes to be identified directly on the bone section. IS-PCR is used mainly to monitor either viral DNA or single copy genes within cells (Komminoth et al. 1992), while IS-RT-PCR is commonly used to monitor the expression of genes in entire bone sections (Mee et al. 1996). Detailed *in situ* protocols using bone samples can be found in Helfrich and Ralston (2003; chapter 15). Raouf and Seth (2002) used high-density mouse microarray gene chips to study the expression profile of over 8700 genes during *in vitro* osteoblast maturation (i. e. from proliferation to mineralization stages). The authors' intent was to elucidate the molecular phenotype of osteoblast differentiation in order to identify novel differentially expressed genes within signaling pathways (e. g. the c-Jun N-terminal kinase (JNK) mitogen-activated protein (MAP) kinase pathway). Culture applications have always been difficult and the effects of media conditions (e. g. serum, vitamins) on the expression profile of genes affected by serum-response elements or stress factors must be taken into account. The most common source of osteoblasts for molecular phenotyping experiments are derived from mouse calvaria. Array applications using intact bone are extremely difficult because of the intrinsically low RNA content of bone. The conflicting effects of hematopoietic progenitors (i. e. stroma cells) in the bone marrow can drastically overshadow any observable direct signal from bone. Flushing out the marrow from the bone can still leave residues which will give substantial noise. The key step for successful microarray analysis is the quality of the starting RNA material and the aforementioned *in situ* techniques provide methods for adequate RNA extraction.

Apoptosis Detecting apoptosis within bone cells is an exciting area of bone research that requires specialized techniques (Croce et al. 2004; Roux et al. 2005; Weinstein 2001). The loss of cells by apoptosis is considered to be a normal step in the life cycle of bone cells. Apoptosis is assumed to regulate the number of osteoblasts at the bone surface, manage endochondral bone formation by chondrocytes, and remove osteoclasts at the end of bone resorption. In addition, glucocorticoids are known to induce apoptosis of bone cells in osteoporosis, leading to the development of specialized techniques to study apoptosis (Eberhardt et al. 2001). Unfortunately the study of cellular events in mineralized bone is complicated by the difficulty of obtaining thin sections, and by the contrasting material properties of calcium-phosphate and non-mineralized components such as bone marrow in the same tissue. Some of the techniques used to study apoptosis in bone include: morphological analysis using toluidine blue staining, viability using the LDH enzyme assay, nick translation to view cleaved DNA, and the detection of conventional apoptotic markers such as effector caspases and phosphatidylserine.

Frozen Sectioning The Kuwamoto (see Section 3.2.5.5) and polyvinyl alcohol (PVA; see Helfrich and Ralston 2003, chapter 17) methods for embedding frozen sections are alternative techniques to either wax or plastic resin embedding for immunohistochemical analysis when using bone samples. Results using the Kuwamoto method on non-decalcified small rodent samples have been published, but in contrast, the PVA method has been used only on neonatal ribs of postmortem human infants and it is unknown whether the method can be extended to small animals. The advantage of frozen sections is that their preparation does not include harsh fixation steps which may prevent access of antibodies to antigen. The disadvantage is that an expensive slow drive, high-torque motor cryostat with an automatic speed control and section capturing techniques (e. g. adhesive films, vacuum) is needed to obtain good-quality thin sections. As usual, choices of antibodies and control antibodies, secondary antibodies and chromogen or flurogen all need to be optimized on a case-by-case basis in order to view the cells.

Calvarial Analysis *In vivo* calvarial analysis and preparations have been described in rodents (Boyce et al. 1989; van't Hof et al. 2000), and are useful in understanding local bone remodeling events and osteoclast formation. This application can be used in the analysis of the effects of disruptive hormonal pathways and/or drugs on bone metabolism *in vivo*. The important steps in this procedure are tissue processing (i. e. resin embedding), TRAP/von Kossa/light green and Goldner's trichrome staining, and the analysis of the results.

LSCM, TEM, SEM Laser scanning confocal microscopy (LSCM) has been used for screening large numbers of primary osteoclasts, for example, to assess cell adhesion and resorption sites, and to track exogenous proteins (e. g. antibodies) during bone resorption without recourse to tissue sectioning (Nesbitt et al. 2000). Transmission electron microscopy (TEM) has been applied to bone research quite extensively and has focused on cell-to-cell/matrix interactions and sample composition. Specifically, TEM has been used to show cellular vacuolation and differences in collagen striation (i. e. an indication of cross-linking defects) in a case of OI in dog (Seeliger et al. 2003) and Ehlers–Danlos Syndrome (Giunta et al. 2002) respectively. Special attention should be paid to all perfusion/immersion fixation, post-fixation, trimming/sectioning, and staining steps. Scanning electron microscopy (SEM) is used to obtain structural, analytical and associative information from bone samples. By using a magnification range of 10- to 10,000-fold, anorganic bone overviews detailing functional and structural aspects can be established. Proper bone preparation is the key to securing reasonable images from which all organic material has been removed. Eberhardt et al. (2001) used SEM to show resorptive pits in glucocorticoid-treated samples of subarticular bone. Croce et al. (2004) used a multitude of methods (MTT assay, confocal microscopy, SEM,

TEM) to understand the dynamics of cell adhesion and proliferation of fibroblasts in collagen matrices. Specifically, SEM was used to define and characterize how fibroblasts aligned on the collagen matrix in relation to one another, and how they produce cytoplasmic projections that contact or adhere to the superficial collagen fibers *in vitro*.

Mechanical Bending The mechanical strength of bone is governed by the relative amount and characteristics of the mineral (e. g. mass, density) and the integrity of the organic matrix (type I collagen), as well as the spatial arrangement of bone at the microscopic and macroscopic levels (i. e. micro-architecture and anatomy respectively). The mechanistic connection between type I collagen and bone strength has been demonstrated using OI murine models (Boskey et al. 1999) and other clinical cases affecting type I collagen processing as a result of mutations in the COLIA1 and COLIA2 genes. The site of fracture in a bone is determined by the distribution of applied forces. Forces can be applied either directly or indirectly to the bone. The great majority of fractures are caused by indirect forces applied at some distance from the actual fracture site. There are almost always simultaneous combinations of forces at work in order to create a fracture. Tension, rotation, bending, or compressive forces (weight-bearing or axial loading) all play a part in where the fracture will occur, what its shape will be, and the extent of damage. Initially, when the femur is subjected to stress, for example during the three-point bend test (Instron Inc., USA), the bone resists the applied force by bending or stretching and enters a zone of elastic deformation. The femur then stretches to its limit or yield point. Beyond the yield point, any additional stress applied is no longer proportional to the stress the femur can withstand. At this point the femur is nearing its breaking point or zone of plastic deformation. Failure (fracture) occurs when the bone can no longer withstand the applied force, and the three-point bend test measures these parameters. The three-point bend test is commonly used to evaluate cortical bone and can be combined modally with other geometric and material density analyses to provide a global picture. Compression tests on cancellous bone are uncommon due to its small size, anisotropic nature and variable geometry/structure. In addition, the results are affected by inconsistencies in loading orientation, storage conditions, temperature, and frictional differences between bone samples and the apparatus. In a report by Misof et al. (2005), treatment of oim/oim mice with alendronate, a bisphosphonate, did not result in any improvement in cortical width and strength, or any other mechanical property as monitored by histomorphometry or the femoral three-point bend test respectively.

3.3
Order of the Tests

First and foremost before instigating a skeletal phenotype screen, the results expected from the primary and subsequent screens must be clearly distinguished. Furthermore, a relationship between the results obtained from each tier must be rationalized before proceeding to the next level. The aim of the primary tests is to

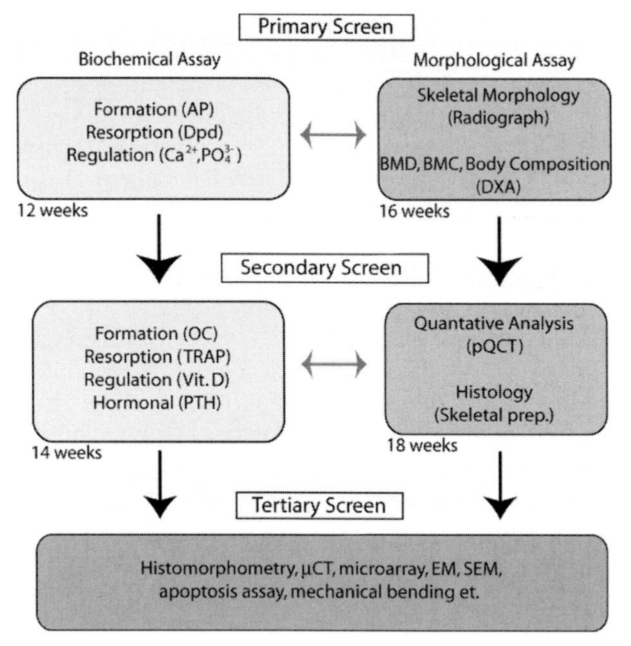

Fig. 3.5 Proposed skeletal screening strategy. In the primary screen, if changes in serum AP or Dpd levels are observed, confirm bone- and/or osteoclast-specificity respectively, with the secondary screen. If any changes in the biochemical regulatory parameters are noticed, confirm secondary hormonal effects and calcium maintenance anomalies using PTH and vitamin D markers respectively. If the bone turnover status is still not clear, broaden the repertoire of bone markers further or perform histomorphometrical analysis using special stains. If changes in BMD are noticed in the primary screen, confirm vBMD with the secondary screen using pQCT. If dysmorphological features are shown by radiography, proceed to skeletal preparation depending on the type of screening protocol. If vBMD is confirmed, proceed to high definition μCT analysis for further analysis of cancellous bone. Special interests include for example, measurements of collagen-to-mineral ratio and growth/mineralization rates that require histomorphometry in the tertiary screen.

obtain first-line information about the potential of a mouse as a model organism for human skeletal diseases. Thus, these tests should be non-invasive, and quick and easy to perform covering the key parameters pertaining to skeletal metabolism. This will enable the researcher to analyze many potential candidate parameters and the same animals in multiple tests, thus maximizing the amount of information gained from a limited number of animals. The second-line experiments are more focused on special phenotypes ascribed to human diseases and in most cases, these tests are more time consuming, require more expensive equipment and/or consumables, and the animals may have to be sacrificed.

The proposed screening tier in the isolation of bone phenodeviants is shown in Fig. 3.5. We recommend the following order of tests to ensure cost effectiveness, the logical scope of skeletal parameters and no interference from one test to another (e. g. due to narcosis, bleeding). First, an anatomical observation for visible, binary anomalies of the mice at the age of 8 to 10 weeks should be carried out. At the age of 10 to 14 weeks, blood samples should be taken via a retro-orbital sinus puncture and the blood-based parameters determined. Urine should also be collected, and both urine and plasma can be aliquoted and refrigerated for 1 h or stored at –70 °C for future use. Blood and urine are taken at these ages before the mice are subjected to radiographic tests to avoid the influences of narcosis. Blood and urine collection times must be standardized to avoid circadian variations based of baseline experiments for each biochemical marker. For confirmation of values differing from the expected values, a second blood sample should be taken 2 weeks later and the spurious parameter(s) determined again. Second-line blood-based parameters can also be measured using stored samples. At the age of 15 to 20 weeks, the animals should be examined by X-ray, and thereafter while they are still narcotized, the animals should be subjected to DXA analysis. Comparison of the results will lead to further steps. Subsequently, both pQCT and μCT analyses will be necessary for bone characterization on a volumetric and dynamic level. When animals are readily available (i. e. whether the animal is derived from the ENU vs. phenotyping screens), skeleton preparation will reveal information about cartilage and bone formation. Histomorphometric analysis is recommended when the effects on cellular dynamics are being investigated. Other special interests will ensue depending on the nature of the phenotype on a case-by-case basis. Archiving of samples (e. g. sperm or tissues) should be organized in a logical manner.

3.4 Conclusion

Disorders of the skeleton are manifold, and therefore require a phenotype screen capable of tackling a diverse array of bone anomalies. Here we have described and refined an operative and feasibly cost-effective musculoskeletal phenotype screen in the hope of discovering novel genes and the nature of their function during skeletal regulation and further characterizing established animal models of human skeletal diseases. The screen incorporates diverse methodologies in capturing skeletal phenotypes related to growth, mineral density, bone size/mass, bone turnover/metabolism, and dysmorphology, not to mention all the auxiliary data obtained from the clinical chemical and radiological analyses. Although the proposed screen is diverse in nature, it will not capture all skeletal phenotypes due to assay sensitivity barriers, genetic backgrounds (i. e. hypomorphy), subtleness of phenotypes, and the systemic and dynamic representations of bone markers and radiological analyses respectively.

Osteoporosis is listed by WHO as one of the 10 most crucial diseases, thus giving impetus to the development and refinement of strategies for finding suitable candidates for an animal model of osteoporosis. Osteoporosis is most common among Asian and European populations, while the African population is virtually unaffected by the disease. Osteoporosis patients exhibit anomalies in bone turnover, leading to bone fractures which limit mobility and quality of life. Currently there are no reliable therapies that either halt or prevent osteoporosis. The identified candidates would provide a model base for testing new and more potent bisphosphonates, and antisense gene or cell therapies. The techniques used in the diagnosis of human osteoporosis form part of our proposed mouse screen. Understanding and pursuing novel genetic factors which may regulate bone metabolism in humans is one of the aims of studying this complex skeletal disease.

Most of the high resolution imaging techniques have only been available for a few years, and recent developments have made it possible to use them in mouse genetics. Progress and refinement in technological research and development will facilitate the identification of bone phenodeviants. Higher precision, speed and sensitivity will assist already established methodologies in identifying outliers (e. g. X-ray cameras, DXA, etc.), while new techniques will be able to identify novel pathways/candidates (e. g. bone marker antibodies, *in vivo* labeling techniques, PET etc.). Recent improvements in *in vivo* imaging technologies at the cellular level are remarkable. Temporal and spatial organization, development and interaction of certain molecules within bone-related cells will lead to a better understanding of dynamic processes in homeostasis and disease. In addition, increasing the number of parameters that are screened in order to "pick-up" more candidates is plausible, although not cost effective.

The two most frequent applications of comprehensive mouse screening protocols are large scale projects such as ENU–mutagenesis screens and phenotyping centers (e. g. German Mouse Clinic www.mouseclinic.de, Institute Clinic Souris www-mci.u-strasbg.fr, Toronto Center for Phenogenomics www.phenogenomics.ca/index.html). The main aim of an ENU project is the establishment of new mutant lines with a phenotype of interest among a large population of "normal" animals. Phenotyping centers are more comprehensive, targeted and open to mutants of different origin. In contrast to ENU projects, the goal is not to establish new mutant lines with a particular phenotype, but to assign new phenotypes to already-established mouse mutant lines, thereby generating an integral picture of the phenotype.

The application of comprehensive and standardized phenotyping methods will lead to a more systemic understanding of mouse models. Large data sets of every mouse model will be available and accessible. This is a new standard of phenotyping and will be essential in system biology approaches to building molecular models of diseases.

The next step in the area of advanced phenotyping of mouse models will be to adjust certain environmental conditions to alter the phenotypic outcome. The establishment of standardized phenotyping approaches for mice under defined challenge conditions is probably one of the most important prerequisites for the identification of novel genes or gene functions. Changes in e. g. nutrition, activity/exercise, air composition, social environment and exposure to pathogens will reveal unexpected reactions in mutant lines which will be valuable tools in the field of bone and cartilage research.

We thus anticipate that the proposed phenotype screen will contribute to the international scientific community engaged in bone research, a body of information which will expand the growing list of skeletal-related genes and functions and ultimately lead to a better quality of life for many patients.

List of Abbreviations

AP	Alkaline phosphatase
BMC	Bone mineral content
BMD	Bone mineral density
DXA	Dual energy X-ray absorption
ECM	Extracellular matrix
ELISA	Enzyme linked immunosorbent assay
ENU	Ethyl-N-nitrosourea
μCT	Micro computed tomography
MMA	Methyl methacrylate
MRI	Magnetic resonance imaging
OC	Osteocalcin
OI	Osteogenesis imperfecta
Oim	Osteogenesis imperfecta murine
OMIM	Online Mendelian Inheritance in Man
OVX	Ovariectomy
QTL	Quantitative trait loci
PCR	Polymerase chain reaction
PQCT	Peripheral quantitative computed tomography
PTH	Parathyroid hormone
ROI	Region of interest
TAP	Total alkaline phosphatase
TRAP	Tartrate-resistant acid phosphatase
VOI	Volume of interest
WHO	World Health Organization

References

Akhter MP, Iwaniec UT, Covey MA, Cullen DM, Kimmel DB, Recker RR **2000**, Genetic variations in bone density, histomorphometry, and strength in mice, *Calcif. Tissue Int.* 67: 337–344.

Allen MJ **2003**, Biochemical markers of bone metabolism in animals: uses and limitations, *Vet. Clin. Path.* 32(3): 101–113.

Arjmandi BH, Alekel L, Hollis BW, Amin D, Stacewicz-Sapuntzakis M, Guo P, Kukreja SC **1996**, Dietary soybean protein prevents bone loss in an ovariectomized rat model of osteoporosis, *J. Nutr.* 126: 161–167.

Baroncelli GI, Bertelloni S, Ceccarelli C, Saggese G **1998**, Measurement of volumetric bone mineral density accurately determines degree of lumbar undermineralization in children with growth hormone deficiency, *J. Clin. Endocrinol. Metab.* 83(9): 3150–3154.

Beamer WG, Donahue LR, Rosen CJ, Baylink DJ **1996**, Genetic variability in adult bone density among inbred strains of mice, *J. Bone Miner. Res.* 18: 397–403.

Beamer WG, Shultz KL, Churchill GA, Frankel WN, Baylink DJ, Rosen CJ, Donahue LR **1999**, Quantitative trait loci for bone density in C57BL/6J and CAST/EiJ inbred mice. *Mamm. Genome* 10: 1043–1049.

Beamer WG, Donahue LR, Rosen CJ **2002**, Genetics and bone using the mouse to understand man, *J. Musculoskelet. Neuronal Interact.* 2(3):.225–231.

Bilezikian JP, Raisz LG, Rodan GA. **1996**, *Principles of Bone Biology*, Academic Press Inc.

Black DM, Cummings SR, Genant HK, Nevitt MC, Palermo L, Browner W **1992**, Axial and appendicular bone density predict fractures in older women, *J. Bone Miner. Res.* 7: 633–638.

Blake GM and Fogelman I **1996**, Principles of bone densitometry. In *Principles of Bone Biology*, Academic Press Inc., 1313–1332.

Bohlooly-Y M, Mahlapuu M, Andersen H, Astrand A, Hjorth S, Svensson L, Tornell J, Snaith MR, Morgan DG, Ohlsson C **2004**, Osteoporosis in MCHR1-deficient mice, *Biochem. Biophys. Res. Commun.* 318: 964–969.

The Bone & Joint Decade, European action towards better musculoskeletal health, ISBN 91–975284–0-4.

Bonnick SL **2000**, Monitoring osteoporosis therapy with bone densitometry: a vital tool or regression toward mediocrity? *J. Clin. Endocrinol. Metab.* 85(10): 3493–3495.

Boone JM, Velazquez O, Cherry SR **2004**, Small-animal X-ray dose from micro-CT, *Mol. Imaging.* 3(3): 149–158.

Boskey AL, Wright TM, Blank RD **1999**, Collagen and bone strength, *J. Bone Miner. Res.* 14(3): 330.

Bouxsein ML, Pierroz DD, Glatt V, Goddard DS, Cavat F, Rizzoli R, Ferrari SL **2005**, Beta-Arrestin2 regulates the differential response of cortical and trabecular bone to intermittent PTH in female mice, *J. Bone Miner. Res.* 20(4): 635–643.

Boyce BF, Aufdmorte TB, Garrett IR, Yates AJ, Mundy GR **1989**, Effects of interleukin-1 on bone turnover in normal mice, *Endocrinology* 125: 1142–1150.

Butz S, Wuester C, Scheidt-Nave C, Goetz M, Ziegler R **1994**, Forearm BMD as measured by peripheral quantitative computed tomography in a German reference population, *Osteoporosis Int.* 4: 179–184.

Carter DR, Bouxsein ML, Marcus R **1992**, New approaches for interpreting projected bone densitometry data, *J. Bone Miner. Res.* 7: 137–145.

Chalhoub N, Benachenhou N, Rajapurohitam V, Pata M, Ferron M, Frattini A, Villa A, Vacher J **2003**, Grey-lethal mutation induces severe malignant autosomal recessive osteopetrosis in mouse and human, *Nat. Med.* 9(4): 399–406.

Chung CR, Tsuji K, Nifuji A, Komori T, Soma K, Noda M **2004**, Micro-CT evaluation of tooth, calvaria and mechanical stress-induced tooth movement in adult Runx2/Cbfa1 heterozygous knock-out mice, *J. Med. Dent. Sci.* 51(1): 105–113.

Coleman RJ, Settergren D, Garber AK, Binkley N **1999**, Comparison of the peripheral X-ray imagery with standard DXA measurement of bone mass in rats and rhesus monkeys, *J. Bone Miner. Res.* 14(Suppl. 1): S491.

Compston JE **1995**, Bone density: BMC, BMD, or corrected BMD? *Bone* 16: 5–7.

Croce MA, Silvestri C, Guerra D, Carnevali E, Doraldi F, Tiozzo R **2004**, Adhesion and proliferation of human dermal fibroblasts on collagen matrix, *J. Biomaterials Applications* 18: 209–222.

Cummings SR, Black DM, Nevitt MC, Browner W, Cauley J, Ensrud K, Genant HK, Palermo L, Scott J, Vogt TM **1993**, Bone density at various sites for prediction of hip fractures, *Lancet* 341: 72–75.

Dambacher MA, Ito M, Mueller R, Neff M, Qin L, Schacht E, Schmitt S, Zhao YL **2004**, Bone structure *in vitro* and *in vivo* in animals and in men – a view into the future, *J. Mineralstoffwechsel.* 11(3): 13–19.

Dazai J, Bock NA, Nieman BJ, Davidson LM, Henkelman RM, Chen XJ **2004**, Multiple mouse biological loading and monitoring system for MRI, *Magn. Reson. Med.* 52: 709–715.

de la Pena LS, Billings PC, Fiori JL, Ahn J, Kaplan FS, Shore EM **2005**, Fibrodysplasia ossificans progressiva (FOP), a disorder of ectopic osteogenesis, misregulates cell surface expression and trafficking of BMPRIA, *J. Bone Miner. Res.* 20(7): 1168–1176.

Dequeker J, Nijs J, Verstraeten A, Geusens P, Gevers G **1987**, Genetic determinants of bone mineral content at the spine and radius: a twin study, *Bone* 8: 207–209.

Di Chiro G, Brooks RA **1979**, The 1979 Nobel prize in physiology or medicine, *Science* 206(4422): 1060–1062.

Eberhardt AW, Yeager-Jones A, Blair HC **2001**, Regional trabecular bone matrix degeneration and osteocyte death in femora of glucocorticoid-treated rabbits, *Endocrinology* 142(3): 1333–1340.

Erben RG **1997**, Embedding of bone samples in methylmethacrylate: An improved method suitable for bone histomorphometry, histochemistry, and immunohistochemistry, *J. Histochem. Cytochem.* 45(2): 307–313.

Fuchs H, Schughart K, Wolf E, Balling R, Hrabé de Angelis M **2000**, Screening for dysmorphological abnormalities – a powerful tool to isolate new mouse mutants, *Mamm. Genome* 11(7): 528–530.

Gailus-Durner V, Fuchs H, Becker L, Bolle I, Brielmeier M, Calzada-Wack J, Elvert R, Ehrhardt N, Dalke C, Franz TJ, Grundner-Culemann E, Hammelbacher S, Holter SM, Holzlwimmer G, Horsch M, Javaheri A, Kalaydjiev SV, Klempt M, Kling E, Kunder S, Lengger C, Lisse T, Mijalski T, Naton B, Pedersen V, Prehn C, Przemeck G, Racz I, Reinhard C, Reitmeir P, Schneider I, Schrewe A, Steinkamp R, Zybill C, Adamski J, Beckers J, Behrendt H, Favor J, Graw J, Heldmaier G, Hofler H, Ivandic B, Katus H, Kirchhof P, Klingenspor M, Klopstock T, Lengeling A, Muller W, Ohl F, Ollert M, Quintanilla-Martinez L, Schmidt J, Schulz H, Wolf E, Wurst W, Zimmer A, Busch DH, Hrabé de Angelis M **2005**, Introducing the German Mouse Clinic: open access platform for standardized phenotyping, *Nat. Methods* 2(6): 403–404.

Gardner JR, Hess CP, Webb AG, Tsika RW, Dawson MJ, Gulani V **2001**, Magnetic resonance microscopy of morphological alterations in mouse trabecular bone structure under conditions of simulated microgravity, *Magn. Reson. Med.* 45(6): 1122–1125.

Gasser JA **1995**, Assessing bone quantity by pQCT, *Bone* 17(S): 145–154.

Giunta C, Nuytinck L, Raghunath M, Hausser I, De Paepe A, Steinmann B **2002**, Homozygous Gly530Ser substitution in COL-5A1 cause mild classical Ehlers–Danlos Syndrome, *Amer. J. Med. Genet.* 109: 284–290.

Giwa OG, Anetor JI, Alonge TO, Agbedana EO **2004**, Biochemical observations in Blount's Disease (infantile tibia vara), *J. Natl Med. Assoc.* 96(9)9: 1203–1207.

Glüer CC, Faulkner KG, Estilo MJ, Engelke K, Rosin J, Genant HK **1993**, Quality assurance for bone densitometry research studies: concept and impact, *Osteoporosis Int.* 3: 227–235.

Grampp S, Jergas M, Glüer CC, Lang P, Brastow P, Genant HK **1993**, Radiological diagnosis of osteoporosis: current methods and perspectives. *Radiol. Clin. North Am.* 31: 1133–1145.

Gu W, Li X, Lau KH, Edderkaoui B, Donahae LR, Rosen CJ, Beamer WG, Shultz KL, Srivastava A, Mohan S, Baylink DJ **2002**, Gene expression between a congenic strain that contains a quantitative trait locus of high bone density from CAST/EiJ and its wild-type strain C57BL/6J, *Funct. Integr. Genomics* 1: 375–386.

Guldberg RE, Lin AS, Coleman R, Robertson G, Duvall C **2004**, Microcomputed tomography imaging of skeletal development and growth, *Birth Defects Res. C Embryo Today* 72(3): 250–259.

Hamrick MW **2003**, Increased bone mineral density in the femora of GDF8 knockout mice, *Anat. Rec. A Discov. Mol. Cell Evol. Biol.* 272(1): 388–391.

Han KO, Choi JT, Choi HA, Moon IG, Yim CH, Park WK, Yoon HK, Han IK **2005**, The changes in circulating osteoprotegerin after hormone therapy in postmenopausal women and their relationship with oestrogen responsiveness on bone, *Clin. Endocrinol. (Oxf.)* 62(3): 349–353.

Harada S, Rodan GA **2003**, Control of osteoblast function and regulation of bone mass, *Nature* 423: 349–355.

Harrison RG **1979**, A mummified foetus from the tomb of Tutankhamen, *Antiquity* 53: 19 ff.

Hayes RP, Nagy TR, Rahemtulla F, Bounelis P, Prince CW **2000**, Bone mineral density in a rat model of type 2 diabetes, *J. Dent. Res.* 79: 565.

Helfrich MH, Ralston SH **2003**, Methods in molecular medicine, *Bone Res. Protocols* 80.

Hernandez C, Beaupré G, Marcus R, Carter D **2001**, A theoretical analysis of the contributions of remodeling space, mineralization and bone balance to changes in bone mineral density during alendronate treatment, *Bone* 29(6): 511–516.

Ho AM, Johnson MD, Kingsley DM **2000**, Role of the mouse *ank* gene in control of tissue calcification and arthritis, *Science* 289: 265–270.

Hrabé de Angelis M, Flaswinkel H, Fuchs H, Rathkolb B, Soewarto D, Marschall S, Heffner S, Pargent W, Wuensch K, Jung M, Reis A, Richter T, Alessandrini F, Jakob T, Fuchs E, Kolb H, Kremmer E, Schaeble K, Rollinski B, Roscher A, Peters C, Meitinger T, Strom T, Steckler T, Holsboer F, Klopstock T, Gekeler F, Schindewolf C, Jung T, Avraham K, Behrendt H, Ring J, Zimmer A, Schughart K, Pfeffer K, Wolf E, Balling R **2000**, Genome-wide, large-scale production of mutant mice by ENU mutagenesis, *Nat. Genet.* 25(4):444–447.

Hui SL, Slemenda CW, Johnston CC **1989**, Baseline measurements of bone mass predicts fracture in white women, *Ann. Intern. Med.* 111: 355–361.

Huang YC, Kaigler D, Rice KG, Krebsbach PH, Mooney DJ **2005**, Combined angiogenic and osteogenic factor delivery enhances bone marrow stromal cell-driven bone regeneration, *J. Bone Miner. Res.* 20(5): 848–857.

Ichikawa Y, Sumi M, Ohwatari N, Komori T, Sumi T, Shibata H, Furuichi T, Yamaguchi A, Nakamura T **2004**, Evaluation of 9.4-T MR microimaging in assessing normal and defective fetal bone development: comparison of MR imaging and histological findings, *Bone* 34(4): 619–628.

Irwin S **1968**, Comprehensive observational assessment: Ia. A systematic, quantitative procedure for assessing the behavioral and physiologic state of the mouse, *Psychopharmacologia* 13: 222–257.

Ito M, Azuma Y, Takagi H, Kamimura T, Komoriya K, Ohta T, Kawaguchi H **2003**, Preventive effects of sequential treatment with alendronate and 1 alpha-hydroxyvitamin D3 on bone mass and strength in ovariectomized rats, *Bone* 33(1): 90–99.

Jackson IJ, Abbott CM **2000**, *Mouse Genetics and Transgenics*, Oxford University Press.

Jiang Y, Zhao J, White DL, Genant HK **2000**, Micro CT and Micro MR imaging of 3D architecture of animal skeleton, *J Musculoskelet. Neuronal Interact.* 1(1): 45–51.

Kaplan FS, Smith RM **1997**, Fibrodysplasia ossificans progressiva (FOP), *J. Bone Miner. Res.* 12(5): 855.

Karsenty G **2001**, *The Molecular Basis of Skeletogenesis*, Wiley, Chichester, UK; *Novartis Foundation Symposium* 232: 6–22.

Kasukawa Y, Baylink DJ, Guo R, Mohan S **2003**, Evidence that sensitivity to growth hormone (GH) is growth period and tissue type dependent: studies in GH-deficient lit/lit mice, *Endocrinology* 144(9): 3950–3957.

Kawamoto T **2003**, Use of a new adhesive film for the preparation of multi-purpose fresh-frozen sections from hard tissues, whole-animals, insects and plants, *Arch. Histol. Cytol.* 66(2): 123–143.

Kawano H, Sato T, Yamada T, Matsumoto T, Sekine K, Watanabe T, Nakamura T, Fukuda T, Yoshimura K, Yoshizawa T, Aihara K, Yamamoto Y, Nakamichi Y, Metzger D, Chambon P, Nakamura K, Kawaguchi H, Kato S **2003**, Suppressive function of androgen receptor in bone resorption, *Proc. Natl Acad. Sci. USA* 100(16): 9416–9421.

Keller H, Kneissel M **2005**, SOST is a target gene for PTH in bone, *Bone* June 6; [Epub ahead of print].

Kingsley DM **2001**, The molecular basis of skeletogenesis, Wiley, Chichester, UK; *Novartis Foundation Symposium* 232: 213–234.

Klein RF, Mitchell SR, Phillips TJ, Belknap JK, Orwoll ES **1998**, Quantitative trait loci affecting peak bone mineral density in mice, *J. Bone Miner. Res.* 13: 1648–1656.

Klein RF, Shea M, Gunness ME, Pelz GB, Belknap JK, Orwoll ES **2001**, Phenotypic characterization of mice bred for high and low peak bone mass, *J. Bone Miner. Res.* 16(1): 63–71.

Komminoth P, Long A, Ray R, Wolfe H **1992**, *In situ* polymerase chain reaction detection of viral DNA, single copy gene and gene rearrangements in cell suspensions and cytospins, *Diagn. Mol. Pathol.* 1: 85–97.

Kroger H, Vainio P, Nieminen J, Kotaniemi A **1995**, Comparison of different models for interpreting bone mineral density measurements using DXA and MRI technology, *Bone* 17: 157–159.

Li M, Healy DR, Li Y, Simmons HA, Crawford DT, Ke HZ, Pan LC, Brown TA, Thompson DD **2005**, Osteopenia and impaired fracture healing in aged EP4 receptor knockout mice, *Bone* 37(1): 46–54.

Li W, Zhao GF, Xu SW, Wang JW, Zhou YF, Liu J **2005**, Role of dual energy X-ray absorptiometry in monitoring fracture healing in a rat femoral fracture model, *Chin. J. Traumatol.* 8(2): 121–125.

Linden L **2003**, The bone and joint decade 2000–2010. *Bull. World Health Organization* 81(9): 629.

Link TM, Majumdar S, Lin JC, Newitt D, Augat P, Ouyang X, Mathur A, Genant HK **1998**, A comparative study of trabecular bone properties in the spine and femur using high resolution MRI and CT, *J. Bone Miner. Res.* 13(1): 122–131.

Lund AM, Nicholls AC, Schwartz M, Skovby F **1997**, Parental mosaicism and autosomal dominant mutations causing structural abnormalities of collagen I are frequent in families with osteogenesis imperfecta type III/IV, *Acta Paediatr.* 86(7): 711–718.

Lyon MF, Rastan S, Brown SDM **1996**, *Genetic Variants and Strains of the Laboratory Mouse*, Oxford University Press.

Majumdar S, Kothari M, Augat P, Newitt DC, Link TM, Lin JC, Lang T, Lu Y, Genant HK **1998**, High-resolution magnetic resonance imaging: three-dimensional trabecular bone architecture and biomechanical properties, *Bone* 22(5): 445–454.

Malkin I, Dahm S, Suk A, Kobyliansky E, Toliat M, Ruf N, Livshits G, Nurnberg P **2005**, Association of ANKH gene polymorphisms with radiographic hand bone size and geometry in a Chuvasha population, *Bone* 36(2): 365–373.

Mariani FV, Martin GR **2003**, Deciphering skeletal pattering: clues from the limb, *Nature* 423: 319–325.

Mayer-Kuckuk P, Gade TP, M Buchanan I, Doubrovin M, Ageyeva L, Bertino JR, Boskey AL, Blasberg RG, Koutcher JA, Banerjee D **2005**, High-resolution imaging of bone precursor cells within the intact bone marrow cavity of living mice, *Mol. Ther.* 12(1): 33–41.

McWalter EJ, Wirth W, Siebert M, von Eisenhart-Rothe RM, Hudelmaier M, Wilson DR, Eckstein F **2005**, Use of novel interactive input devices for segmentation of articular cartilage from magnetic resonance images, *Osteoarthritis Cartilage* 13(1): 48–53.

Mee AP, Hoyland JA, Braidman IP, Freemont AJ, Davies M, Mawer EB **1996**, Demonstration of vitamin D receptor transcripts in actively resorbing osteoclasts in bone sections. *Bone* 18: 295–299.

Melton LJ 3rd, Looker AC, Shepherd JA, O'Connor MK, Achenbach SJ, Riggs BL, Khosla S **2005**, Osteoporosis assessment by whole body region vs. site-specific DXA, *Osteoporosis Int.* 2005 [Epub ahead of print].

Meunier P and Boivin G **1997**, Bone mineral density reflects bone mass but also the degree of mineralization of bone: therapeutic implications, *Bone* 21(5): 373–377.

Misof BM, Roschger P, Baldini T, Raggio CL, Zraick V, Root L, Boskey AL, Klaushofer K, Fratzl P, Camacho NP **2005**, Differential effects of alendronate treatment on bone from growing osteogenesis imperfecta and wild-type mouse, *Bone* 36(1): 150–158.

Moller M, Horsman A, Harvald B, Hauge M, Henningsen K, Norbin BEC **1978**, Metacarpal morphology in monozygotic and dizygotic elderly twins, *Calcif. Tissue Res.* 25(2):197–201.

Morrison NA, Qi JC, Tokita A, Kelly PJ, Crofts L, Nguyen TV, Sambrook PN, Eisman JA **1994**, Prediction of bone density from vitamin D receptor alleles, *Nature* 367: 284–287.

Morriss-Kay GM, Iseki S, Johnson D **2001**, *The Molecular Basis of Skeletogenesis*, Wiley, Chichester, UK; *Novartis Foundation Symposium* 232: 102–121.

Muller R, Hahn M, Vogel M, Delling G, Ruegsegger P **1996**, Morphometric analysis of noninvasively assessed bone biopsies: comparison of high-resolution computed tomography and histologic sections, *Bone* 18(3): 215–220.

Mundlos S **2001**, *The Molecular Basis of Skeletogenesis*, Wiley, Chichester, UK;*Novartis Foundation Symposium* 232: 81–101.

Nagy TR, Clair AL **1999**, Validation of body composition measurements of mice using DXA, *Obesity Res.* 7(Suppl. 1): 27S.

Nagy TR, Wharton D **1999**, Precision and accuracy of in vivo bone mineral measurements of mouse femurs using DXA, *J. Bone Miner. Res.* 14(Suppl. 1): S493.

Nagy TR, Wharton D, Blaylock M, Powell S **1999**, Precision and accuracy of in vivo bone mineral measurements of mice using dual-energy X-ray absorptiometry, *FASEB J.* 13(5): A912.

Nesbitt S, Charras G, Lehenkari P, Horton M **2000**, Three-dimensional imaging of bone-resorbing osteoclasts: spatial analysis of matrix collagen, cathepsin K, MMP-9 and TRAP by confocal microscopy, *J. Bone Miner. Res.* 15: 1219.

Nielsen SP, Hermansen F, Bärenholdt O 1993, Interpretation of lumbar spine osteodensitometry in women with fractures, *Osteoporosis Int.* 3: 276–282.

Nielsen SP, Kolthoff N, Bärenholdt O, Kristensen B, Abrahamsen B, Hermann AP, Brot C **1998**, Diagnosis of osteoporosis by planar bone densitometry: can body size be disregarded? *Brit. J. Radiol.* 71: 934–943.

Nielsen SP, Slosman D, Sorensen OH, Basse-Cathalinat B, DeCassin P, Roux C, Meunier PJ **1999**, Influence of strontium on bone mineral density and bone mineral content measurements by dual X-ray absorptiometry, *J. Clin. Densitometry* 2(4): 371–380.

Nielsen SP **2000**, The fallacy of BMD: a critical review of the diagnostic use of dual X-ray absorptiometry, *Clin. Rheumatol.* 19: 174–183.

Nunez C, Tan YX, Zingaretti G, Punyanitya M, Rubiano F, Wang ZM, Heymsfield SB **2000**, The best predictive model for estimating fat-free mass, *Ann. NY Acad. Sci.* 904: 333–334.

Ornitz DM **2001**, *The Molecular Basis of Skeletogenesis*, Wiley, Chichester, UK; *Novartis Foundation Symposium* 232: 63–80.

Parfitt AM, Drezner MK, Glorieux FH, Kanis JA, Malluche H, Meunier PJ, Ott SM, Recker RR **1987**, Bone histomorphometry : standardization of nomenclature, symbols and units, *J. Bone Miner. Res.* 2(6): 595–610.

Patton JT, Kaufman MH **1995**, The timing of ossification of the limb bones, and growth rates of various long bones of the fore and hind limbs of the prenatal and early postnatal laboratory mouse, *J. Anat.* 186(Pt 1): 175–185.

Peacock M, Turner CH, Econs MJ, Foroud T **2002**, Genetics of osteoporosis, *Endocrine Rev.* 23(3): 303–326.

Phillips CL, Bradley DA, Schlotzhauer CL, Bergfeld M, Libreros-Minotta C, Gawenis LR, Morris JS, Clarke LL, Hillman LS **2000**, Oim mice exhibit altered femur and incisor mineral composition and decreased bone mineral density, *Bone* 27(2): 219–226.

Pizette S, Niswander L **2001**, *The Molecular Basis of Skeletogenesis*, Wiley, Chichester, UK; *Novartis Foundation Symposium* 232: 23–43.

Prentice A, Parsons TJ, Cole TJ **1994**, Uncritical use of bone mineral density in absorptiometry may lead to size-related artifacts in the identification of bone mineral determinants, *Am. J. Clin. Nutr.* 60: 837–842.

Pressel T, Bouguecha A, Vogt U, Meyer-Lindenberg A, Behrens BA, Nolte I, Windhagen H **2005**, Mechanical properties of femoral trabecular bone in dogs, *Biomed. Eng. Online* 4(1): 17.

Raouf A, Seth A **2002**, Discovery of osteoblast-associated genes using cDNA microarrays, *Bone* 30(3): 463–471.

Rauch F, Glorieux FH **2004**, Osteogenesis imperfecta, *Lancet* 363: 1377–1385.

Recinos RF, Hanger CC, Schaefer RB, Dawson CA, Gosain AK **2004**, Microfocal CT: a method for evaluating murine cranial sutures *in situ*, *J. Surg. Res.* 116(2): 322–329.

Ritman EL **2004**, Micro-computed tomography-current status and developments, *Annu. Rev. Biomed. Eng.* 6: 185–208.

Rittenberg B, Partridge E, Baker G, Clokie C, Zohar R, Dennis JW, Tenenbaum HCJ **2005**, Regulation of BMP-induced ectopic bone formation by Ahsg, *Orthop. Res.* 23(3): 653–662.

Rogers DC, Fisher EM, Brown SD, Peters J, Hunter AJ, Martin JE **1997**, Behavioral and functional analysis of mouse phenotype: SHIRPA, a proposed protocol for comprehensive phenotype assessment, *Mamm. Genome* 8: 711–713.

Rothschild BM, Turner KR, DeLuca MA 1988, Symmetrical erosive peripheral polyarthritis in the Late Archaic Period of Alabama, *Science* 241(4872): 1498–501.

Roux S, Lambert-Comeau P, Saint-Pierre C, Lepine M, Sawan B, Parent JL **2005**, Death receptors, Fas and TRAIL receptors, are involved in human osteoclast apoptosis, *Biochem. Biophys. Res. Commun.* Jun 3; [Epub ahead of print].

Roy D, Swarbrick C, King Y, Pye S, Adams J, Berry J, Silman A, O'Neill T **2005**, Differences in peak bone mass in women of European and South Asian origin can be explained by differences in body size, *Osteoporosis Int.* [Epub ahead of print].

Russell G, Mueller G, Shipman C, Croucher P **2001**, *The Molecular Basis of Skeletogenesis*, Wiley, Chichester, UK; *Novartis Foundation Symposium* 232: 251–257.

Russell WL, Bangham JW, Russell LB **1998**, Differential response of mouse male germ-cell stages to radiation-induced specific-locus and dominant mutations, *Genetics* 148: 1567–1578.

Samuels A, Perry MJ, Gibson R, Tobias JH **2001**, Effects of combination therapy with PTH and 17beta-estradiol on long bones of female mice, *Calcif. Tissue Int.* 69(3): 164–170.

Seeliger F, Leeb T, Peters M, Bru M, Brugmann M, Fehr M, Hewicker-Trautwein M **2003**, Osteogenesis Imperfecta in two litters of dachshunds, *Vet. Pathol.* 40: 530–539.

Seibel MJ, Robins SP, Bilezikian JP **1999**, *Dynamics of Bone and Cartilage Metabolism*, Academic Press Inc.

Shapiro JR, Primorac D, Rowe DW **1996**, Osteogenesis imperfecta: current concepts.

In *Principles of Bone Biology*, Academic Press Inc.

Shimizu M, Higuchi K, Bennett B, Xia C, Tsuboyama T, Kasai S, Chiba T, Fujisawa H, Kogishi K, Kitado H, Kimoto M, Takeda N, Matsushita M, Okumura H, Serikawa T, Nakamura T, Johnson TE, Hosokawa M **1999**, Identification of peak bone mass QTL in a spontaneously osteoporotic mouse strain, *Mamm. Genome* 10: 81–87.

Shum L, Coleman CM, Hatakeyama Y, Tuan RS **2003**, Morphogenesis and dysmorphogenesis of the appendicular skeleton, *Birth Defects Res. C Embryo Today* 69(2): 102–122.

Slemenda CW, Christian JC, Reed T, Reister TK, Williams CJ, Johnston CC Jr **1992**, Long-term bone loss in men: effects of genetic and environmental factors, *Ann. Intern. Med.* 117: 286–291.

Sorenson JA, Duke PR, Smith SW **1989**, Simulation studies of dual-energy X-ray absorptiometry, *Med. Phys.* 16: 75–80.

Srivastava AK, Mohan S, Wergedal JE, Baylink DJ **2003**, A genomewide screening of N-ethyl-N-nitrosourea-mutagenized mice for musculoskeletal phenotypes, *Bone* 33: 179–191.

Suda T, Kobayashi K, Jimi E, Udagawa N, Takahashi N **2001**, *The Molecular Basis of Skeletogenesis*, Wiley, Chichester, UK; *Novartis Foundation Symposium* 232: 235–250.

Sun D, Krishnan A, Zaman K, Lawrence R, Bhattacharya A, Fernandes G **2003**, Dietary n-3 fatty acids decrease osteoclastogenesis and loss of bone mass in ovariectomized mice, *J Bone Miner. Res.* 18(7): 1206–1216.

Superti-Furga A, Bonafe L, Rimoin DL **2002**, Molecular-pathogenetic classification of genetic disorders of the skeleton, *Am. J. Med. Genet.* 106: 282–293.

Tablante NL, Estevez I, Russek-Cohen E **2003**, Effect of perches and stocking density on tibial dyschondroplasia and bone mineralization as measured by bone ash in broiler chickens, *J. Appl. Poult. Res.* 12: 53–59.

Taicher GZ, Tinsley FC, Reiderman A, Heiman ML **2003**, Quantitative magnetic resonance (QMR) method for bone and whole-body-composition analysis, *Anal. Bioanal. Chem.* 377(6): 990–1002.

Van't Hof RJ, Armour KJ, Smith LM **2000**, Requirement of the inducible nitric oxide synthase pathway for IL-1-induced osteoclastic bone resorption, *Proc. Natl Acad. Sci. USA* 97: 7993–7998.

Villanueva AR, Mehr LA **1977**, Modifications of the Goldner and Gomori one-step trichrome stains for plastic-embedded thin sections of bone, *Am. J. Med. Technol.* 43(6): 536–538.

Vis M, Bultink IE, Dijkmans BA, Lems WF **2005**, The effect of intravenous pamidronate versus oral alendronate on bone mineral density in patients with osteoporosis, *Osteoporosis Int.* [Epub ahead of print].

Wachsmuth L, Engelke K **2004**, High-resolution imaging of osteoarthritis using microcomputed tomography, *Methods Mol. Med.* 101: 231–248.

Watts NB **1999**, Clinical utility of biochemical markers of bone remodeling, *Clin. Chem.* 45(8B): 1359–1368.

Weber MH, Sharp JC, Latta P, Sramek M, Hassard HT, Orr FW **2005**, Magnetic resonance imaging of trabecular and cortical bone in mice: comparison of high resolution *in vivo* and *ex vivo* MR images with corresponding histology, *Eur. J. Radiol.* 53(1): 96–102.

Weinstein RS **2001**, Glucocorticoid-induced osteoporosis, *Rev. Endocr. Metab. Disord.* 2(1): 65–73.

WHO Technical Report Series; 843, **1994**, Assessment of fracture risk and its application to screens for postmenopausal osteoporosis.

Whyte MP **1999**, Rare bone diseases. In *Dynamics of Bone and Cartilage Metabolism*, Academic Press, Chapter 43: 605–621.

Won YY, Chung YS, Park YK, Yoo VY **2003**, Correlations between microcomputed tomography and bone histomorphometry in Korean young females, *Yonsei Med. J.* 44(5): 811–815.

Woolf AD, Zeidler H, Haglund U, Carr AJ, Chaussade S, Cucinotta D, Veale DJ, Martin-Mola E **2004**, Musculoskeletal pain in Europe: its impact and a comparison of population and medical perceptions of treatment in eight European countries, *Ann. Rheum. Dis.* 63(4): 342–347.

Wu J, Wang X, Chiba H, Higuchi M, Nakatani T, Ezaki O, Cui H, Yamada K, Ishimi Y **2004**, Combined intervention of soy isoflavone and moderate exercise prevents body fat elevation and bone loss in ovariectomized mice, *Metabolism* 53(7): 942–948.

Zelzer E, Olsen BR **2003**, The genetic basis for skeletal diseases, *Nature* 423: 343–334.

Appendix

μCT Volumetric Data Processing

Certain procedures should be followed when adopting a method for volumetric data processing in the identification and geometric extraction of single and multiple trabecular elements. Usually the first step is to segment the volume dataset for the removal of bone, and then a ROI is encompassed the trabecular region. Within the slice, the contour of the trabecula is regionalized and then traced in the subsequent adjacent slices. Geometric trabecular criteria are defined to ascertain whether the trabecula connects to or traces a node, this will then determine whether or not the particular trabecula is included in the following slices. This node-extension procedure continues until the algorithm stops at the point where the selection criteria are not met. Ultimately, the slices identified are packed together to produce the trabecular VOI from which the morphometric parameters can be estimated. For the acquisition of CT images, it is generally necessary to determine the desired resolution before starting the scan procedure. The chosen resolution is always a compromise between time and quality (i. e. it is related to the number of projections and samples, and the radiation level). A higher resolution will result in longer acquisition times.

MRI Principles

In MRI images are generated from the magnetic resonance signal (i. e. the relaxation properties) of excited ^1H atoms. The object/tissue of interest is placed in a powerful, uniform magnetic field where the spins (i. e. angular momentum) of the atomic nuclei with non-zero spin numbers all align in either opposite parallel or anti-parallel directions to the magnetic field offset at an angle from the direction of the static magnetic field. Summation of the effect of the vast quantity of altered-spin nuclei in a small volume produces a detectable change in field. The magnetic dipole moment of the nuclei then move around the axial field. While the proportion of opposite parallel and anti-parallel nuclei is nearly equal, slightly more are oriented at the low energy angle. The frequency with which the dipole moments move is called the *Larmor frequency*. The tissue is then temporarily exposed to pulses of electromagnetic energy (RF pulse) in a plane perpendicular to the magnetic field, causing some of the magnetically aligned hydrogen nuclei to assume a temporary non-aligned high-energy state. The frequency of the pulses is dictated by the Larmor Equation $\omega = \gamma B$, where ω is the angular frequency of a processing proton, B is the strength of the magnetic field, and γ is the gyromagnetic ratio, a constant unique to the nucleus of each element. During the high-energy relaxation and realignment phase, the nuclei emit energy that is collected in order to provide information about their local environment. This realignment with the magnetic field is called *longitudinal relaxation* and the time taken in milliseconds for a certain percentage of the tissue nuclei to realign is termed *T1*. In addition, the *transverse relaxation* time known as *T2* is calculated for T2-weighted imaging which depends on the local dephasing of spins following the application of a transverse energy pulse. In medicine, a paramagnetic contrast agent is usually administered, and both pre-contrast T1-weighted images and post-contrast T1-weighted images are compared. In boney tissue, since only the marrow contains ^1H protons in high concentrations in the fluid form, T2 times become measurable. On the other hand, bone has a low mobile proton density and therefore has a low T2. The consequence is a low intensity resonance signal from bone, and higher intensity signal from the marrow.

Eventually, in order to image the different voxels (3-D pixels) of bone different orthogonal magnetic gradients are applied. An in-depth explanation of the procedures can be found in manufacturers' reports or textbooks on medical imaging techniques. It should be borne in mind that MRI is useful because various combinations of the gradients can be interrelated during the process so that slices can be taken in any orientation. In order to create the image, spatial information must be recorded along with the information regarding tissue relaxation. For this reason, magnetic fields with an intensity gradient are applied in addition to the strong alignment field to allow encoding of the position of the nuclei. The information is subsequently subjected by computer to inverse Fourier transformation into real space to obtain the targeted image and detailed anatomical information. Typical medical resolution is about 1 mm^3, while research models can exceed 1 μm^3, depending on the gradient strength and other technical and physical features.

4
Clinical Chemical Screen

Martina Klempt, Birgit Rathkolb, Bernhard Aigner, and Eckhard Wolf

4.1
Introduction

The clinical chemical screen consists of laboratory diagnostic procedures suitable for detecting defects in various organ systems, changes in metabolic pathways and hematological disorders. The methods used are automated routine procedures which permit the high-throughput screening of a large number of mice for a broad spectrum of clinical chemical parameters in blood and urine including substrates, electrolytes and enzymes, as well as hematological parameters such as red and white blood cell counts.

4.1.1
Relevance of the Screen

Variance in clinical chemistry and hematological characteristics of the laboratory mouse occurs as a result of multiple genetic and environmental factors. In humans, most inherited metabolic disorders lead directly or indirectly via altered organ functions to changes in laboratory diagnostic parameters. The identification of mutant mice showing similar clinical chemical deviations in either phenotype- and/or gene-driven procedures provides appropriate novel animal models for the identification of causative mutations and the pathological consequences thereof [1]. As the homeostasis of clinical chemistry parameters is regulated by poly-genic factors, quantitative trait loci (QTL) affecting chosen parameters and eventually influencing disease susceptibility are revealed using mouse strains which show considerable differences in the parameters in question [2]. Further phenotypic and genetic analyses promote the integrated evaluation of the function of the mammalian genome and represent an important contribution to the development of the systems biology of complex organisms.

Standards of Mouse Model Phenotyping. Edited by Martin Hrabé de Angelis, Pierre Chambon, and Steve Brown

ISBN: 3-527-31031-2

4.1.2
Biology and Medical Application

4.1.2.1 Biology of Clinical Chemical Parameters

Physiological values of clinical chemical and hematological parameters show substantial differences in the large number of different mouse strains available. Many factors including individual features, mouse husbandry and experimental procedures additionally affect these values. The biology of the clinical chemistry of the mouse has been comprehensively reviewed [3]. Clinical chemical mouse data from various projects have been published [3, 4] (http://www.jax.org/phenome; http://www. eumorphia.org).

Interpretation of the results of the clinical chemical screen requires the preceding determination of the physiological range of the respective parameters in a sufficient number of control mice. The 95 % range of the values is defined to be the reference range thereby eliminating outlier data. The 95 % range covers the data range including two standard deviations above and below the mean if the data for the parameter in question follows a Gaussian distribution [3].

4.1.2.2 Medical Application

Mouse models for human diseases leading to deviations of clinical chemical parameters have been established in both phenotype-driven and gene-driven projects.

Phenotype-driven Methods In phenotype-driven projects, the clinical chemistry of spontaneous and induced mutants is analyzed to identify similarities in the pathologic phenotype of mutant mice and human patients. ENU (N-ethyl-N-nitrosourea) has been used in various mouse mutagenesis programs to produce random mutations. Specific pathological states have been identified by appropriate routine procedures allowing the screening of large numbers of mice for a broad spectrum of parameters. Forward genetics techniques result in the detection of the chromosomal site and the subsequent identification of the causative mutation in the established lines [5–7].

In the Munich ENU Mouse Mutagenesis Project, a screening profile of clinical chemical parameters was established for the analysis of offspring of chemically-mutagenized mice of the inbred C3HeB/FeJ (C3H) genetic background. Breeding of the affected mice and screening of the offspring confirmed the transmission of the altered phenotype to subsequent generations, thereby revealing a mutation as the cause of the aberrant phenotype [1, 8]. Screening of more than 15,000 G1 animals and of G3 mice from more than 230 pedigrees for dominant and recessive mutations, respectively, revealed over 100 mutants with deviations in plasma levels of various plasma substrates and electrolytes as well as plasma enzyme activities, and/or hematological parameters (see Section 4.2.2).

Gene-driven Methods The clinical chemistry profile covering a large range of organ functions is also an essential component in the analysis of mouse models generated by gene-driven approaches including non-homologous additive gene transfer as well as targeted knockout/knock-in procedures. Thus, the clinical chemical screen not only permits the accurate and efficient examination of the expected effects but also the discovery of additional, more subtle consequences of particular genetic modifications, for example in knockout mice without obvious phenotypic alterations [9].

Furthermore, in-depth clinical chemical examinations contribute to the understanding of the pathomechanisms in transgenic mouse models with interesting disease phenotypes. For example, qualitative urinary protein analysis (see Section 4.4.2.3) is important for monitoring the progression of changes in the kidney in growth hormone (GH) transgenic mice, a model widely used to study chronic renal failure [10].

4.2 Diseases in Mouse and Humans

4.2.1 Diagnostic Impact of Clinical Chemistry

Many spontaneous and induced mouse mutants show remarkable deviations in clinical chemical and hematological values. The diagnostic impact of clinical chemical parameters for specific pathologic alterations has been described previously [3, 8] and is summarized for selected blood parameters (substrates, proteins, electrolytes, enzymes) in Tab. 4.1.

Table 4.1 Diagnostic impact of clinical chemical blood values in human diseases.

Parameter	Elevated values	Reduced values
Substrates		
Bilirubin	Icterus	
Cholesterol	Hypothyroidism, nephrotic syndrome, diabetes mellitus, myeloma, hepatic disease	Hyperthyroidism
Creatinine	Heart insufficiency, kidney disease, dehydration, urinary tract disorder	Cachexia
Glucose	Diabetes mellitus, stress, pancreatitis, spasms	Glycogen synthesis/storage disorder, gluconeogenesis defect, glucagon deficiency, ketone metabolism defect, malignant tumor, liver disease, kidney disease
Triglycerides	Hyperlipoproteinemia, pancreatitis, diabetes mellitus	Hyperthyroidism
Urea	Heart insufficiency, kidney disease, gastrointestinal bleeding	Urea cycle disorder, protein catabolism defect, hepatic disease
Uric acid	Gout, massive cell death	Congenital molybdenum cofactor deficiency, kidney disease

Table 4.1 Diagnostic impact of clinical chemical blood values in human diseases. (Continued)

Parameter	Elevated values	Reduced values
Proteins		
Ferritin	Anemia (aplastic, chronic hemolytic, megaloblastic, sideroblastic), hemochromatosis	Iron deficiency
Total protein	Plasmocytoma, chronic inflammation, hepatic cirrhosis, dehydration, hemolysis	Protein synthesis defect, liver disease, gastrointestinal tumor, malabsorption syndrome, protein loss, hyperthyroidism
Transferrin	Iron deficiency	Infection, neoplasm, liver disease
Unsaturated iron binding capacity	Iron deficiency	Hemochromatosis
Electrolytes		
Calcium	Hyperparathyroidism, bone tumor	Hypoparathyroidism, pseudo-h., vitamin D deficiency, hyperphosphatemia, Mg deficiency, pancreatitis, kidney disease
Chloride	Hypoaldosteronism	Kidney disease
Iron	Iron deficiency	Hemochromatosis, anemia (megaloblastic, sideroblastic)
Phosphorus, inorganic	Kidney disease	Malabsorption, vitamin D deficiency, rachitis
Potassium	Massive cell death, kidney disease, K homeostasis/ transport disorder	Diarrhoea, Cushing-Syndrome, kidney disease, chronic hepatic disease
Sodium	Dehydration, Morbus Conn	
Enzymes		
Alanine aminotransferase	Liver disease	
Alkaline phosphatase	Skeletal disease, hepatobiliary disease, cholestasis	Hypophosphatasia
α-Amylase	Pancreatitis, peritonitis, kidney disease, diabetic acidosis	Cystic fibrosis, pancreatic disease
Aspartate aminotransferase	Liver disease, myocardial infarction, muscular dystrophy	
Creatine kinase	Polymyositis, muscular dystrophy	
Lactate dehydrogenase	Liver disease, muscle disease, anemia (hemolytic, megaloblastic)	
Lipase	Acute pancreatitis	

4.2.2
Clinical Chemistry in Selected Disorders

The significant impact of clinical chemical and hematological analysis of mouse mutants in establishing disease models is demonstrated in selected disorders which were part of the focus in the Munich ENU Mouse Mutagenesis Program, by comparing mouse and human disease phenotypes. The three disorders which were chosen as examples and discussed in detail below are hypercholesterolemia as detected by the clinical chemical analysis of plasma; albuminuria as detected by the investigation of urine and acute myeloid leukemia (AML) detected in the hematology profile screening.

4.2.2.1 **Hypercholesterolemia**

Hypercholesterolemia characterized by an increased proportion of low density lipoprotein cholesterol (LDL-C) compared to high density lipoprotein cholesterol (HDL-C), is known as a main factor contributing to the development of human diseases, e. g. cardiovascular disease and Alzheimer's disease. Except for a subset of rare monogenic forms, most cases of hypercholesterolemia occur as a consequence of or associated with polygenic and multifactorial disorders. LDL-C predominates in humans and is sensitive to diet-induced elevations. In contrast, mice have high concentrations of HDL-C and low levels of LDL-C.

In gene-driven projects, many transgenic mouse lines showing hypercholesterolemia have been produced and these mimic the alterations seen in human diseases [3]. In the phenotype-driven project involving clinical chemistry analysis of ENU-mutagenized C3H mice, we established nine mutant mouse lines showing hypercholesterolemia. A single line showed deviations in additional blood chemistry parameters. Thus, the lines produced will contribute to the search for alleles which selectively cause increased primary plasma total cholesterol levels [11]. The clinical chemistry screen of another ENU mouse mutagenesis program revealed dislipidemic mouse lines of a different genetic background (BALB/c × C3H) in which additional mutations causing hypercholesterolemia may be detected [12]. In addition, QTL analysis for increased HDL-C levels using inbred mouse lines showing different values for this parameter resulted in the identification of several candidate genes [13].

4.2.2.2 **Albuminuria**

Nephropathies include various multifactorial disorders caused by genetic/inherited and/or environmental/acquired factors. Several monogenic disorders are known to result in progressive renal insufficiency. Once chronic renal insufficiency is established, it tends to progress to end-stage kidney failure irrespective of the initiating nephropathy.

In glomerular diseases, proteinuria appears as a consequence of the pathologic transglomerular passage of high-molecular weight (HMW) proteins due to the increased permeability of the glomerular capillary wall and the impaired re-absorption by the epithelial cells in the proximal tubuli. The load of these proteins in the tubular lumen leads to the saturation of the re-absorptive mechanism of the tubular cells. This results in the urinary excretion of proteins including the low-molecular-weight (LMW) proteins, which are reabsorbed under physiological conditions. Thus, proteinuria is used as marker for the course of glomerular diseases and is also used to monitor the patient's response to treatment. In addition, proteinuria represents an independent risk factor for the progression to renal failure [14].

In humans, quantitative as well as qualitative examination of urinary protein excretion is used to detect early kidney lesions. Physiologically, mice excrete large amounts of LMW proteins in their urine (major urinary proteins, MUPs), however, qualitative deviations in the urinary protein excretion pattern have been shown to occur in samples containing the same total concentration of urinary proteins. Therefore, the determination of albumin, an intermediate-molecular-weight (IMW) protein (~ 70 kDa), in the qualitative urinary protein analysis was chosen to identify mutant mice which might represent novel nephropathy models [10, 15, 16].

The search for ENU-induced mutants exhibiting albuminuria included more than 2000 G1 animals and nearly 50 G3 pedigrees which were screened using qualitative SDS-polyacrylamide gel electrophoresis (SDS-PAGE) of spot urine samples. Two mutant lines showing a low phenotypic penetrance of albuminuria were established. In addition, the albuminuria screen was used to analyze ENU mutant lines showing increased plasma urea levels to clarify whether severe kidney lesions are involved in the abnormal phenotype. This analysis revealed severe albuminuria in mice which are affected by a recessive mutation leading to increased plasma urea and cholesterol levels. Both the mapping and subsequent identification of the causative mutation and the precise pathological analysis of the mutant phenotype will establish unique models for nephropathies.

4.2.2.3 **Acute Myeloid Leukemia (AML)**

Acute myeloid leukemia (AML) is the most common form of human leukemia where specific genetic alterations lead to profound disturbances of early hematopoiesis. An increase in proliferation and a decrease in differentiation result in the appearance of a clonal myeloid blast population in the bone marrow and peripheral blood. As a consequence, myeloid cells predominate and lymphoid cells are diminished in the peripheral blood. AML can develop from transformed cells within the different hematopoietic cell lineages. Regardless of the origin of the cell, AML pathogenesis was observed to be a multi-step process involving alterations of both proto-oncogenes and tumor suppressor genes. Most patients diagnosed with AML finally die of the disease. Therefore, there is an urgent need to establish novel therapeutic strategies for this disease entity; this can be accomplished by the identification and functional analysis of the crucial players in the pathogenesis of leukemia [17].

Examination of the pathobiology of human leukopoiesis has been facilitated by the establishment of murine leukemia models using bone marrow transplantation and gene transfer strategies. However, these techniques are usually limited to the analysis of the consequences of already identified leukemia-specific genetic alterations. The generation of animal models for the subsequent identification of as yet unknown critical alterations in human leukemias is carried out by various phenotype-driven methods including random mutagenesis of mice by chemicals, radiation or retroviruses ([17] and references therein).

In the Munich ENU Mouse Mutagenesis Program, we screened for hematopoietic disturbances typically found in leukemias or myeloproliferative syndromes by analyzing differential white blood cell counts and immune-phenotyping of peripheral blood. We used myeloid predominance in the blood circulation to identify an attractive new model for myeloid leukemia. These mice harbor a semi-dominant mutation. Homozygous mutant mice develop a myeloproliferative syndrome and die before reaching adulthood. The T-cells and B-cells are completely displaced at 5 weeks of age. The hematological alterations in heterozygous mutant animals progress more slowly. Thus, the phenotypic alterations of the mutants are similar to those described for AML which makes this line a promising model for the functional analysis of AML.

4.3 Clinical Chemistry as Diagnostic Tool

4.3.1 History

The use of mice as a research tool started in the 15 th century [18], and the first investigations into human clinical chemistry were published in the 18 th century. Examination and collection of clinical chemical data from mice dramatically increased in the late 1970 s [19].

Clinical chemistry analysis started with the chemical and microscopical analysis of urine in the 18 th century. Investigations in this field were published by a small number of pioneers and the usefulness of these analyses was discussed controversialy. In the middle of the 19 th century, several findings in urine and blood, such as proteinuria, glucosuria as well as glucose and bile pigment in blood, became known as "diagnostic signs". At the beginning of the 20 th century, chemical analyses of body fluids were widely accepted as useful diagnostic tools among the medical community. Most of the tests routinely used today were developed during the first decades of the 20 th century after the introduction of colorimetric methods in the field, and were optimized and standardized during the 1960 s and 1970 s. The first clinical chemistry autoanalyzers were constructed in the 1950 s and 1960 s [19]. In mouse strains, clinical chemical markers were used for genetic control and monitoring before polymorphic genetic markers were introduced [20].

Mouse hematology using microscopic techniques was already established in the first half of the 20 th century, and hematological analysis in general was simplified by the development of automated blood cell counters in the mid-20 th century. Analyzers function either by the principle of electrical impedance or by the reflection of laser light. In both cases, every single cell induces an electric signal with the voltage directly correlated to the cell volume [21].

4.3.2
State of the Art

Technical improvements led to a drastic reduction in the volume of sample needed for the determination of a single parameter from 500 µl required for manual analysis to 2–30 µl for current clinical chemistry autoanalyzers. This significantly reduced the limiting factor of the blood sample volume and – together with technical improvements in other phenotyping methods – produced efficient tools for phenotyping the physiology and patho-physiology of normal and mutant mice. Current veterinary clinical chemical screening procedures in livestock and laboratory animals for clinic and research purposes are often carried out using the high-throughput techniques employed for human samples including the same equipment and reagents [3].

The clinical chemistry profile plays a major role in established mouse phenome projects where extensive data are collected with standardized protocols for the valid and reproducible clinical chemical characterization of a large number of different mouse strains [4] (http://www.jax.org/phenome; http://www.eumorphia.org).

4.4
Technical Requirements and Screening Protocols

4.4.1
Technical Requirements

Analogous to the high-throughput examination of human samples, the automated mouse clinical chemical screen allows the efficient analysis of a large number of samples. Blood collection, sample preparation and sample analysis are standardized to produce valid and reproducible results [3, 22]. Although dependent on the hygiene status of the colony, all animal body fluids including blood and urine of laboratory mice are potentially infectious to the personnel handling them and, therefore, need to be treated carefully according to the general laboratory rules.

4.4.1.1 Blood Collection

Blood collection in mice is an important issue in the refinement of methods for animal handling. The respective care, skill and experience are necessary for this procedure, and it should be carried out according to the respective animal welfare legislation.

The limiting factor for the clinical chemical analysis in mice is the volume of the blood sample. The total blood volume accounts for 7.5 % of the body weight of the mouse. Of the total blood volume 10 % can be removed once without causing significant alterations, and up to 15 % of the blood volume can be collected if fluid replacement is carried out. For repeated sampling, 10, 7.5 and 1 % of the total blood volume can be removed every fortnight, every week and each day, respectively [23, 24].

Different blood collection procedures have been described for the mouse. Heart puncture and decapitation are terminal bleeding procedures with the plasma normally being hemolytic. Retroorbital venous sinus puncture, tail vein sampling and saphenous vein puncture are routinely carried out in mice. The advantages of puncturing the retroorbital sinus under general anesthesia are the collection of large sample volumes, the short manipulation time of the procedure, the absence of marked hemolysis, the achievement of reproducible analytical results and the possibility of repeated collection at the same site. However, choice of the most suitable technique for blood collection is highly dependent on the skills and the experience of the personnel [25].

Thus, blood collection by retroorbital sinus puncture is recommended for the clinical chemical screen and carried out as follows: for a general clinical chemical screen fasting the mice overnight is not recommended. The mouse is anesthetized with ether and puncture of the retroorbital sinus is carried out with a 0.8-mm non-heparinized microhematocrit capillary (Laborteam K&K, Munich, Germany). From 3-month-old mice, 300 µl blood can be collected in Li-heparin-treated tubes (KABE Labortechnik, Elsenroth, Germany) to produce 130 µl plasma and additional 50 µl blood is collected in EDTA-coated tubes (KABE Labortechnik, Elsenroth, Germany). The tubes are immediately inverted several times to obtain optimal mixing with the anticoagulants. The sampling time is recorded. The time interval for a repeated blood sampling of the same mouse is at least 2 weeks.

4.4.1.2 Sample Preparation

The preparation of blood and urine samples for use in clinical chemistry analysis, and blood samples for use in the hematology analysis is described below.

Clinical Chemistry of Blood For clinical chemistry analysis 300 µl of blood is collected in Li-heparin-coated tubes and processed after incubation for 2 h at room temperature. A total of 130 µl of plasma obtained by centrifugation (4500 × g, 10 min) is transferred to a fresh 1.5-ml Eppendorf tube (Eppendorf, Hamburg, Germany). For the primary screen (see Section 4.4.2.1), the plasma is diluted with the same volume of H_2O_{dest} and thoroughly mixed for 5 s, whereas undiluted plasma is used for the secondary screen (see Section 4.4.2.2). Before putting the tubes into the autoanalyzer, the samples are centrifuged again at 4500 × g for 10 min. Undiluted plasma samples can be stored at +4 °C for a few days only and should be kept at – 20 °C for longer periods. However, the activity of lactate dehydrogenase must be determined immediately after blood collection in order to obtain valid results.

Clinical Chemistry of Urine Urine excretion can be induced during the handling and/or fixation of the mice or after the application of gentle pressure to the lower abdomen. Spot urine is collected using a Petri dish and is then transferred to a tube. Before putting the tubes in the autoanalyzer, the samples are centrifuged at 4500 × g for 10 min. Samples can be stored at +4 °C for 3 days but should be kept at – 20 °C for long-term storage. However, the activity of α-amylase should be analyzed immediately after urine collection in order to obtain valid results.

Hematology Blood for hematology analysis is collected in EDTA-coated tubes, thoroughly mixed and used immediately for the automated determination of the red and white blood cell counts. Blood smears are prepared from fresh EDTA-treated blood. Using a microhematocrit capillary, a small volume (< 3 µl) is placed on one end of a glass slide on the same side as the label. A second glass slide is placed on the first at an angle of 45° and used to spread the blood along the edge of the second slide. The smear is prepared by moving the second slide along the first to give an elliptical distal edge to the blood smear known as the "feathered edge", where the cells are sufficiently spread out to allow subsequent microscopic examination.

4.4.1.3 Sample Analysis

The technical requirements for sample analysis include autoanalyzer equipment for the clinical chemistry and hematology assessments.

Clinical Chemistry of Blood and Urine The clinical chemical parameters of blood plasma and urine samples including substrates, proteins, electrolytes and enzyme activities are automatically analyzed using appropriate routine procedures in accordance with the techniques employed for human samples. In our screening procedure we employ an Olympus AU400 autoanalyzer (Olympus, Hamburg, Germany) and reagents adapted from those used with human samples from Olympus (Hamburg, Germany) and Roche (Mannheim, Germany). However, determination of creatinine values using the Jaffe method has been shown to overestimate the plasma values in mice by two to six times (Meyer et al. 1985). This method therefore can be used as a convenient assay for the detection of severe disturbances of creatinine metabolism or kidney failure, but should be replaced by HPLC-determination for exact analysis (Yuen et al. 2004). Calibration and quality control are performed according to the manufacturer's protocols. Daily quality control is carried out prior to the analysis of the mouse samples with calibration samples obtained from the manufacturer using a physiologic and an abnormally high-value human sample for each parameter within the respective linear measurement range of the analyzer. A technical service is carried out at quarterly intervals. The procedure is adjusted for the analysis of mouse samples and thus 1.5 ml Eppendorf tubes (Eppendorf, Hamburg, Germany) are employed thereby minimizing the dead volume. The assays and principles of detection together with the linear measurement ranges of the autoanalyzer are listed in Table 4.2 for the blood and urine parameters established in our mouse screen.

Hematology Analysis of hematologic parameters is carried out using an Animal Blood Counter (Scil, Viernheim, Germany) validated by the manufacturer for the analysis of mouse blood. Calibration and quality control precede sample analysis using a dog blood standard (ABX Minotrol 16) provided by the manufacturer (Scil, Viernheim, Germany). A technical service is undertaken at biannual intervals.

The number of red (RBC) and white blood cells (WBC) and platelets (PLT) as well as the cell size are measured by electrical impedance, and hemoglobin (HGB) is quantified by spectrophotometry. Mean corpuscular volume (MCV), mean platelet

Table 4.2 Measurements of substrates, proteins, electrolytes and enzyme activities in blood and urine established for the mouse.

Parameter	Assay	Detection test	Unit[a]	Linear range[b] Blood	Urine
Substrates					
Bilirubin, direct	3,5-Dichlorophenyldiazonium tetrafluoroborate	Photometric color	µmol/l mg/dl	0–171 0–10	
Bilirubin, total	3,5-Dichlorophenyldiazonium tetrafluoroborate	Photometric color	µmol/l mg/dl	0–513 0–30	
Cholesterol	Cholesteroloxidase peroxidase (CHOD-PAP)	Enzymatic color	mmol/l mg/dl	0.64–18 25–700	
Cholesterol, high density lipoprotein (HDL-C)	Immunoinhibition	Enzymatic color	mmol/l mg/dl	0.05–4.65 2–180	
Cholesterol, low density lipoprotein (LDL-C)	Selective protection	Enzymatic color	mmol/l mg/dl	0.26–10.3 10–400	
Creatinine	Jaffé	Kinetic color	µmol/l mg/dl	18–2200 0.2–25	88.4–53040 1–600
Glucose	Hexokinase	Enzymatic UV	mmol/l mg/dl	0.6–45 10–800	0.05–39 1–700
Triglycerides	Glycerolphosphateoxidase peroxidase (GPO-PAP)	Enzymatic color	mmol/l mg/dl	0.11–11.4 10–1000	
Urea	Glutamate dehydrogenase (GLDH)	Kinetic UV	mmol/l mg/dl	0.8–50 5–300	2.5–1000 15–6000
Uric acid	Uricase, Peroxidase (POD), N-ethyl-N-(2-hydroxy-3-sulfo-propyl)-3-methylaniline (TOOS), 4-aminophenazone	Enzymatic color	µmol/l mg/dl	11.9–1487 0.2–25	11.9–16362 0.2–275
Proteins					
Ferritin	Latex agglutination	Immuno-turbidimetric	µg/l	8–450	
Microalbumin	Turbidimetric end point	Immuno-turbidimetric	mg/dl		0.5–30
Total protein	Biuret	Photometric color	g/l	30–120	
Transferrin	Turbidimetric end point	Immuno-turbidimetric	g/l	0.75–7.5	
Unsaturated iron binding capacity	Nitroso-PSAP	Photometric color	µmol/l µg/dl	9.8–71.4 55–400	
Electrolytes					
Calcium	Arsenazo III	Photometric color	mmol/l	1–4	

Table 4.2 Measurements of substrates, proteins, electrolytes and enzyme activities in blood and urine established for the mouse. (Continued)

Parameter	Assay	Detection test	Unit[a]	Linear range[b] Blood	Urine
Chloride	Target value ISE indirect	Ion-selective electrode	mmol/l	50–200	15–400
Iron	2,4,6-Tri[2-pyridyl]-5-triazine	Photometric color	µmol/l µg/dl	1.8–178.6 10–1000	
Phosphorus, inorganic	Complex formation with molybdate	Photometric UV	mmol/l mg/dl	0.32–6.4 1–20	3.2–112 10–350
Potassium	Target value ISE indirect	Ion-selective electrode	mmol/l	1–10	2–200
Sodium	Target value ISE indirect	Ion-selective electrode	mmol/l	50–200	10–400
Enzymes					
Alanine aminotransferase (EC 2.6.1.2)	IFCC, GSCC[c]	Kinetic UV	µkat/l U/l	0.05–8.33 3–500	
Alkaline phosphatase (EC 3.1.3.1)	Orthophosphoric-monoester-phosphorhydrolase	Kinetic color	µkat/l U/l	0.08–25 5–1500	
α-Amylase (EC 3.2.1.1)	α-1,4-glucan-4-glucanohydrolase	Kinetic color	µkat/l U/l	0.08–25 5–1500	0.08–25 5–1500
Aspartate aminotransferase (EC 2.6.1.1)	IFCC, GSCC[c]	Kinetic UV	µkat/l U/l	0.05–16.7 3–1000	
Creatine kinase (EC 2.7.3.2)	IFCC, GSCC[c]	Kinetic UV	µkat/l U/l	0.17–33.4 10–2000	
Lactate dehydrogenase (EC 1.1.1.27)	IFCC, GSCC[c]	Kinetic UV	µkat/l U/l	0.4–20 25–1200	
Lipase (EC 3.1.1.3)	Colorimetric	Kinetic color	µkat/l U/l	0.08–10 3–600	

a The SI unit of the parameter is given in the first line in cases where two units are listed.

b Linear measurement range for the Olympus AU400 autoanalyzer (Olympus, Hamburg, Germany) and the reagents for human samples adapted according to the manufacturer's instructions. Minor variations in the ranges may occur for different production units of the standardization reagents.

c Assay based on the recommendations of the International Federation for Clinical Chemistry and the German Society for Clinical Chemistry.

volume (MPV) and red blood cell distribution width (RDW) are calculated from the cell volume measurements and the hematocrit (HCT) is calculated as MCV × RBC. Mean corpuscular hemoglobin (MCH) and mean corpuscular hemoglobin concentration (MCHC) are calculated as HGB/RBC and HGB/HCT, respectively. Automatic white blood cell differentiation according to the cell volume results in preliminary values for granulocytes, lymphocytes and monocytes. However, due to the specifics of mouse hematology, these last three values do not match the results of the precise manual differential white blood cell count (see Section 4.4.2.3) and therefore only detect severe deviations.

4.4.2 Screening Protocols

The clinical chemical screen described here was established for the Munich ENU Mouse Mutagenesis Program and is based on three steps. The automated high-throughput primary screen allows the efficient analysis of a large number of mice for a broad range of alterations in various organ systems and metabolic pathways. Automated high-throughput secondary screens were established for the confirmation of defined organ defects and the comprehensive analysis of tissue-specific profiles. The tertiary screen is not carried out as a high-throughput screen but consists of more specific and thus time consuming in-depth examinations of a relatively small number of mice for the detailed characterization of particularly interesting disease phenotypes.

Thus, the screen is suitable for the detection of a plethora of diverse defects in organ systems and metabolic pathways. Parameter selection may be adapted according to the aim of the proposed project.

4.4.2.1 Primary Screen

The primary screen is arranged to efficiently detect alterations in a broad range of organ systems, metabolic pathways and hematologic values with a reliable number of parameters. It consists of three steps: the automated screens of clinical chemical parameters in both blood and urine and the automatic screening of hematological parameters. The limiting factor in mice is the volume of the blood sample.

Blood The clinical chemistry screen of the blood is carried out in 130 µl of plasma diluted with the same volume of H_2O_{dest} to the total volume of 260 µl and includes the following 20 parameters, (1) substrates: cholesterol, creatinine, glucose, triglycerides, urea, uric acid; (2) proteins: ferritin, total protein, transferrin; (3) electrolytes: calcium, chloride, phosphorus, potassium, sodium; (4) enzyme activities: alanine aminotransferase (EC 2.6.1.2), alkaline phosphatase (EC 3.1.3.1), α-amylase (EC 3.2.1.1), aspartate aminotransferase (EC 2.6.1.1), creatine kinase (EC 2.7.3.2), lipase (EC 3.1.1.3).

Urine The clinical chemistry screen of the urine is carried out in 130 µl of sample and includes the following 10 parameters: creatinine, glucose, urea, uric acid, microalbumin, chloride, phosphorus, potassium, sodium, α-amylase.

Table 4.3 Organ-profile analysis of mouse blood.

Tissue	Parameter
Bone/muscle	Calcium, phosphorus, alkaline phosphatase, creatine kinase, lactate dehydrogenase
Iron pathway	Ferritin, transferrin, iron, lactate dehydrogenase, unsaturated iron binding capacity
Kidney	Creatinine, total protein, urea, uric acid, chloride, potassium, sodium
Liver	Direct bilirubin, total bilirubin, cholesterol, high density lipoprotein cholesterol, low density lipoprotein cholesterol, alkaline phosphatase, alanine aminotransferase, aspartate aminotransferase
Pancreas	Cholesterol, high density lipoprotein cholesterol, low density lipoprotein cholesterol, glucose, triglycerides, α-amylase, lipase

Hematology The hematology screen analyzes the following 13 parameters in 50 µl of EDTA-treated blood, (1) red blood cells: hematocrit, hemoglobin, mean corpuscular hemoglobin, mean corpuscular hemoglobin concentration, mean corpuscular volume, red blood cell count, red blood cell distribution width; (2) white blood cells: granulocytes, lymphocytes, mean platelet volume, monocytes, platelet count, white blood cell count.

4.4.2.2 Secondary Screen

Secondary screens have been established for the comprehensive clinical chemical analysis of specific organ defects. In cases where the primary screen results reveal deviations from normal suggesting the functional alteration of defined organs, the respective secondary screen is carried out on a subsequent blood sample for confirmation of the primary test result and examination of the tissue-specific profile.

The tissue-specific profiles established in our screen for bone/muscle, iron pathway, kidney, liver and pancreas are listed in Table 4.3; 130 µl of undiluted plasma is used for each profile. Additional secondary screens may be arranged to further analyze organ profiles with the mouse blood sample volume and the time intervals between the examinations as the limiting factors for the number of parameters that can be screened (see Section 4.4.1.1).

4.4.2.3 Tertiary Screen

The tertiary screen is used for an in-depth analysis of pathogenic pathways underlying particularly interesting disease phenotypes. The following four tests are routinely established in our screening profile: (1) glucose tolerance test, (2) blood gas measurement in the blood clinical chemistry analysis, (3) urinary protein electrophoresis in the urine clinical chemistry analysis, and (4) differential white blood cell count in the hematological analysis.

Glucose Tolerance Test The glucose tolerance test is carried out to detect the presence and extent of disturbances of glucose metabolism in mice ([3] and references therein). Overnight fasted mice receive 20% glucose in a sterile 0.9% NaCl solution per os at a dose of 1.5 mg/g body weight. A small amount of blood (2–3 μl) collected from the tail tip at 0, 30, 60, 90, 120 and 180 min after glucose administration is applied directly onto a test stripe and blood glucose is measured with a Precision XTRA™ blood glucose meter (Abbott, Wiesbaden, Germany). Insulin is measured in parallel at 0, 30 and 180 min after glucose administration in 5 μl plasma or serum using the ultrasensitive mouse insulin ELISA system (Mercodia, Uppsala, Sweden).

Measurement of Blood Gas Blood gas analysis is carried out to identify any changes in the acid–base balance which may indicate a respiratory or metabolic disorder. This analysis is applied in the examination of metabolic pathways as several mouse strains react to food restriction with hypometabolic adaptation followed by hypothermia, which alters the blood gas values. The main problems associated with the valid determination of blood gas concentrations in mice are the small sample volumes and the instability of the gases after blood collection ([3] and references therein).

Blood collection is carried out under general anesthesia by puncturing the retro-orbital venous plexus with a non-heparinized 0.8 mm microhematocrit capillary (Laborteam K&K, Munich, Germany) which results in the removal of mixed venous blood; a sample size of 85 μl is required for the analysis. Blood gas parameters including pH, pO_2, pCO_2 and bicarbonate concentration are measured using the ABL5 blood gas analyzer system (Radiometer, Copenhagen, Denmark). The analysis should be carried out immediately after blood collection in order to achieve reproducible results. The animal's body temperature is measured for the automatic temperature correction of the blood gas values.

Electrophoresis of Urinary Protein Mice physiologically excrete large amounts of low molecular weight (LMW) proteins in their urine (major urinary proteins, MUPs). Therefore, qualitative analysis of urinary proteins is carried out using SDS-polyacrylamide gel electrophoresis (SDS-PAGE) which may indicate early glomerular lesions (albumin, an intermediate-molecular-weight (IMW) protein (~ 70 kDa)), advanced glomerular lesions (high molecular weight (HMW) proteins) or tubular lesions (proteins ranging between MUPs and albumin) in a screen for kidney diseases [15].

Spot urine samples of 20 μl are taken from the mice and stored at – 20 °C. Urine samples are boiled for 10 min after 1 : 2 dilution with sample buffer (62.5 mM Tris-HCl pH 6.8, 2% SDS, 25% glycerol, 0.01% bromophenol blue, 5% 2-mercaptoethanol). A 10-μl aliquot of the samples is electrophoresed (25 mM Tris, 200 mM glycine, 0.1% SDS) in a Tris-HCl-polyacrylamide gel (4–20% gradient) using the Bio-Rad Mini Protean II system (Bio-Rad Laboratories, Hercules, CA, USA). Protein bands are visualized by staining with Coomassie Brilliant Blue. The broad range SDS-PAGE standard (Bio-Rad Laboratories, Hercules, CA, USA) is used as the molecular weight standard for the detected bands. Wild-type mice are used as controls. Stained gels are photographed and dried for documentation.

This method of protein detection in urine was successfully evaluated using growth hormone (GH) transgenic mice which were reported to inevitably develop progressive kidney lesions including glomerulosclerosis with secondary tubulo-interstitial lesions. Albumin was identified by comparison with the molecular weight standards. In addition, albuminuria was confirmed by Western blot analysis using rabbit anti-mouse albumin antibody (Biotrend, Cologne, Germany) and horseradish peroxidase-conjugated swine anti-rabbit antibody (DAKO Diagnostika, Hamburg, Germany) [10]. The albumin/creatinine ratio is determined to correct for variations in the concentration of urine samples.

Differential White Blood Cell Count The differential white blood cell count is carried out by microscopic examination of blood smears. Air-dried blood smears are stained with May–Grünwald reagents using the Hemacolor® staining set (Merck, Darmstadt, Germany) and are then analyzed microscopically under oil immersion at a magnification of 100 × by trained technicians. To determine the quantitative distribution of leukocyte subpopulations, 100 white blood cells are differentiated. In addition, the appearance of pathologic cells and changes of red blood cell morphology is also analyzed. The white blood cell count is divided into lymphocytes, neutrophils, monocytes, eosinophils, basophils and additional cell types ([28] and references therein).

4.5 Logistics of the Screen

The logistics of the screen are designed to optimize execution and outcome of the analysis by implementing comprehensive automation, standardization and quality control protocols. Unrecognized laboratory environment-specific factors may lead to significant deviations in the results of highly standardized experiments [29].

4.5.1 General Considerations

Effective standardization of mouse husbandry includes the evaluation of abiotic and biotic factors relating to the environment in order to produce valid and reproducible results from the clinical chemical screen. The standardization of mouse husbandry has been extensively reviewed ([30] and the references therein). Mouse husbandry is undertaken according to the specific-pathogen-free (SPF) hygiene standards of the Federation of European Laboratory Animal Science Associations (FELASA) protocols (http://www.felasa.org) and is continuously under review.

Implementation of environmental enrichment for the improvement of mouse husbandry is desirable from the point of view of animal welfare, but may negatively affect efforts to establish standardization thus leading to changes in the values of some clinical chemical parameters as well as increasing the variability of the results. Further research will promote progress in this field [31].

Choosing a mouse strain that will be appropriate to the aims of the proposed project can be simplified by consulting the published data relevant to the clinical chemistry of the mouse [3] (http://www.jax.org/phenome; http://www.eumorphia.org). Both sexes should be evaluated separately. In addition, the aim of the project determines the logistics of sample collection and analysis by taking into account the volume of the blood sample as the limiting factor (see Section 4.4.1.1).

Interpretation of the results of the clinical chemical screen requires the preceding determination of the reference range of the respective parameters in control mice. For this purpose, the published datasets of clinical chemical values of mouse strains are of limited use because of the large number of critical variables which cannot be completely standardized between different laboratories. Therefore, analysis of a sufficient number of controls is essential for each individual project [3].

4.5.2
Lessons from ENU Mutants

Having used the inbred C3H genetic background in the Munich ENU Mouse Mutagenesis Project, a defined breeding scheme was established to detect dominant and recessive mutations leading to clinical chemical alterations. Three-month-old mice were used for the primary screen (see Section 4.4.2.1). In cases where the clinical chemistry appeared to be abnormal, the parameters of interest were re-examined in a second analysis after a period of 3 weeks. Mice which also showed abnormalities in their clinical chemistry in the second analysis were used for subsequent breeding to establish mutant lines [1, 8].

The clinical chemistry screen of more than 15,000 G1 animals and G3 mice from more than 230 pedigrees identified animals with deviations for only some of the parameters listed in Section 4.4.2.1. In addition, more than half of all mutants found with plasma substrate deviations showed an abnormal total cholesterol level (Rathkolb et al., unpublished data). The markedly variable frequency of detection of mutations in the parameters under investigation suggests that deviations in some parameters are lethal and/or different numbers of genes are involved in the various physiological processes which are monitored by the clinical chemistry screen.

In the search for mutants showing deviations in their clinical chemistry (see Section 4.2.2), we identified more than 100 mice consistently showing increased plasma total cholesterol levels in analyses carried out both before and after the 3-week time interval. Transmission of the altered phenotype to subsequent generations led to the production of only nine hypercholesterolemic lines. The breeding methods used for the establishment of mutant lines detected nearly 80% of the fertile hypercholesterolemic mice which did not transmit hypercholesterolemia to the offspring. The extent of the abnormal plasma total cholesterol values in these mice was not indicative of the successful establishment of a mutant line. A total of 10 newly-bred hypercholesterolemic mice was necessary to establish one mutant line with hypercholesterolemia. A high ratio in the failure of the transmission of the defect to the offspring was also found for other parameters in our clinical chemistry screen. One reason for this may be the occurrence of non-genetic hypercholesterolemia due to the inherent impossibility of carrying out complete standardization of husbandry and experimental methods. In addition, loss of the additive phenotypic effect by segregation of multiple mutations in the subsequent generations may also lead to failure in the transmission of the aberrant phenotype.

Nearly half of the established nine lines showed varying degrees of incomplete penetrance of the mutant phenotype in the subsequent offspring. Appearance of hypercholesterolemia due to the interaction of multiple non-linked mutations leads to lower numbers of affected offspring. Future linkage experiments in these lines will reveal the number and chromosomal positions of the mutations which are involved in the development of hypercholesterolemia. It has yet to be established whether deviations in the number of offspring per litter and/or in the sex ratio of the offspring indicating early loss of mutants provide an alternative explanation for the observed decrease in frequency of mice showing hypercholesterolemia.

Thus, improvement in the characterization of the clinical chemistry of mutant lines may be achieved by analyzing the animals more frequently and/or by specific challenge tests, for example feeding the animals with experimental diets in the hypercholesterolemia screen. The early examination time-point in our screen was chosen to avoid loss of mutant animals due to severe health disorders. Screening mice at an older age might lead to a higher level of phenotypical penetrance of the mutation as well as to additional deviations in the clinical chemistry due to other ENU-induced mutations [11].

4.6 Trouble Shooting

Clinical chemistry analysis is standardized to produce valid and reproducible results [3, 22] (http://www.jax.org/phenome; http://www.eumorphia.org). Unrecognized laboratory environment-specific factors may lead to significant deviations in the results of highly standardized experiments [29].

Evaluation of the results of the clinical chemistry analyses includes both the technical and biological features of the experiment. The technical evaluation examines the calibration and quality control of the equipment and analytic procedures together with the effects of any *in vitro* factors. The biological evaluation analyzes the plausibility of the results and their correspondence to control and/or published values together with the effects of any *in vivo* factors.

A negative evaluation may be due to the occurrence of incorrect or unexpected results when compared to published data as well as to the lack of reproducible results. This may be a result of systematic and/or random failures in the screen [22].

4.6.1
Factors Interfering *In Vivo*

Factors which may interfere *in vivo* include all those which affect the sample up to the point of sample collection. They are classified as constant (e. g. genetics, sex, age) and variable (e. g. individual physiology, body weight, activity, performance, time of day, environment, social interference, health status, nutrition, interference with humans, experimental procedures) factors.

We observed that an increase in temperature of 2–3 °C in the mouse facility caused by a defect in the air-conditioning system, led to a drastic increase of up to 50 % and to a high variability in the activity of alanine aminotransferase and aspartate aminotransferase thus highlighting the need for comprehensive and controlled standardization.

4.6.2
Factors Interfering *In Vitro*

In vitro factors which may lead to anomalies in the analyses include all those associated with treatment of the sample starting with the sampling procedure. Sample collection (e. g. method, personnel) and preparation (e. g. storage) procedures as well as the assay and method of the sample analysis (e. g. calibration, measurement range, and contamination) should be examined.

Blood sampling from the retroorbital plexus by inexperienced technicians frequently leads to hemolysis of the samples which is associated with aberrant results in the clinical analysis.

With regard to the analytic method used, the validity of the results may be low in immunoassays using reagents specific for substrates from other species, for example human. The reagents should be checked for their specificity and relevance to mouse samples.

4.7
Short-term Outlook

The clinical chemical screen described above consists of three steps. The primary screen gives a broad overview of the clinical chemistry of the mice, whereas – based on the results of the primary screen – detailed analyses are carried out in the secondary and tertiary screens. Evaluation of the data from large-scale ENU mouse mutagenesis projects may facilitate the optimization of parameter selection for the primary screen by excluding and/or changing parameters where physiologic values are vital factors. Tissue-specific profiles in the secondary screen may be newly established, expanded and/or optimized. In addition, further improvement of the technical equipment will lead to the establishment of new protocols in the tertiary screen for the in-depth analysis of selected disorders which are apparent from deviations observed in the clinical chemistry.

Comparability of results in clinical chemistry screens of different projects will increase in the future with the use of standardized protocols (http://www.jax.org/phenome; http://www.eumorphia.org).

Improvement in mouse husbandry and refinement of animal manipulation in the phenotyping screens are the continued aims of current research to improve animal welfare and will be promoted within the standardization requirements of such experiments.

References

1 B. Rathkolb, E. Fuchs, H. J. Kolb, I. Renner-Muller, O. Krebs, R. Balling, M. Hrabé de Angelis, E. Wolf *Exp. Physiol.* **2000**, 85, 635–644.

2 Members of the complex trait consortium *Nat. Rev. Genet.* **2003**, 4, 911–916.

3 W. F. Loeb, F. W. Quimby *The Clinical Chemistry of Laboratory Animals*, Taylor & Francis, Philadelphia, **1999**.

4 K. Paigen, J. T. Eppig *Mamm. Genome* **2000**, 11, 715–717.

5 M. H. Hrabé de Angelis, H. Flaswinkel, H. Fuchs, B. Rathkolb, D. Soewarto, S. Marschall, S. Heffner, W. Pargent, K. Wuensch, M. Jung, A. Reis, T. Richter, F. Alessandrini, T. Jakob, E. Fuchs, H. Kolb, E. Kremmer, K. Schaeble, B. Rollinski, A. Roscher, C. Peters, T. Meitinger, T. Strom, T. Steckler, F. Holsboer, T. Klopstock, F. Gekeler, C. Schindewolf, T. Jung, K. Avraham, H. Behrendt, J. Ring, A. Zimmer, K. Schughart, K. Pfeffer, E. Wolf, R. Balling *Nat. Genet.* **2000**, 25, 444–447.

6 P. M. Nolan, J. Peters, M. Strivens, D. Rogers, J. Hagan, N. Spurr, I. C. Gray, L. Vizor, D. Brooker, E. Whitehill, R. Washbourne, T. Hough, S. Greenaway, M. Hewitt, X. Liu, S. McCormack, K. Pickford, R. Selley, C. Wells, Z. Tymowska-Lalanne, P. Roby, P. Glenister, C. Thornton, C. Thaung, J. A. Stevenson, R. Arkell, P. Mburu, R. Hardisty, A. Kiernan, A. Erven, K. P. Steel, S. Voegeling, J. L. Guenet, C. Nickols, R. Sadri, M. Nasse, A. Isaacs, K. Davies, M. Browne, E. M. Fisher, J. Martin, S. Rastan, S. D. Brown, J. Hunter *Nat. Genet.* **2000**, 25, 440–443.

7 J. Beckers, M. Hrabé de Angelis *Curr. Opin. Chem. Biol.* **2002**, 6, 17–23.

8 B. Rathkolb, T. Decker, E. Fuchs, D. Soewarto, C. Fella, S. Heffner, W. Pargent, R. Wanke, R. Balling, M. Hrabé de Angelis, H. J. Kolb, E. Wolf *Mamm. Genome* **2000**, 11, 543–546.

9 C. A. Pinkert *Comp. Med.* **2003**, 53, 126–139.

10 E. Wolf, R. Wanke, in L. F. M. van Zutphen, M. van der Meer *Welfare of Transgenic Animals*, Springer-Verlag, Heidelberg, **1997**, pp. 26–47.

11 M. Mohr, M. Klempt, B. Rathkolb, M. Hrabé de Angelis, E. Wolf, B. Aigner *J. Lipid Res.* **2004**, 45, 2132–2137.

12 T. A. Hough, P. M. Nolan, V. Tsipouri, A. A. Toye, I. C. Gray, M. Goldsworthy, L. Moir, R. D. Cox, S. Clements, P. H. Glenister, J. Wood, R. L. Selley, M. A. Strivens, L. Vizor, S. L. McCormack, J. Peters, E. M. Fisher, N. Spurr, S. Rastan, J. E. Martin, S. D. Brown, A. J. Hunter *Mamm. Genome* **2002**, 13, 595–602.

13 X. Wang, B. Paigen *Arterioscler. Thromb. Vasc. Biol.* **2002**, 22, 1390–1401.

14 P. Saborio, J. Scheinman *Curr. Opin. Pediatr.* **1998**, 10, 174–183.

15 T. Doi, L. J. Striker, C. C. Gibson, L. Y. Agodoa, R. L. Brinster, G. E. Striker *Am. J. Pathol.* **1990**, 137, 541–552.

16 R. J. Beynon, J. L. Hurst *Biochem. Soc. Trans.* **2003**, 31, 142–146.

17 M. Smith, M. Barnett, R. Bassan, G. Gatta, C. Tondini, W. Kern *Crit. Rev. Oncol. Hematol.* **2004**, 50, 197–222.

18 H. Morse, in H. L. Foster, J. D. Small, J. G. Fox *The Mouse in Biomedical Research*, Academic Press, New York, **1981**, pp. 1–16.

19 L. Rosenfeld *Clin. Chem.* **2002**, 48, 186–197.

20 L. M. Silver *Mouse Genetics: Concepts and Applications*, Oxford University Press, New York, **1995**.

21 L. Thomas *Labor und Diagnose*, TH-Books, Frankfurt/Main, **2000**.

22 H. Greiling, A. M. Gressner *Lehrbuch der Klinischen Chemie und Pathobiochemie*, Schattauer, Stuttgart, **1995**.

23 D. B. Morton, D. Abbot, R. Barclay, B. S. Close, R. Ewbank, D. Gask, M. Heath, S. Mattic, T. Poole, J. Seamer, J. Southee, A. Thompson, B. Trussell, C. West, M. Jennings *Lab. Anim.* **1993**, 27, 1–22.

24 M. A. Suckow, P. Danneman, C. Brayton *The Laboratory Mouse*, CRC Press, Boca Raton, **2001**.

25 K. H. Diehl, R. Hull, D. Morton, R. Pfister, Y. Rabemampianina, D. Smith, J. M. Vidal, C. van de Vorstenbosch *J. Appl. Toxicol.* **2001**, 21, 15–23.

26 M. H. Meyer, R. A. Meyer Jr, R. W. Gray, R. L. Irwin *Anal Biochem.* **1985**, 144, 285–290.

27 P. S. Yuen, S. R. Dunn, T. Miyaji, H. Yasuda, K. Sharma, R. A. Star *Am. J. Physiol. Renal Physiol*, **2004**, 286, F1116–F1119.

28 N. Everds, in H. J. Hedrich, G. Bullock *The Laboratory Mouse*, Elsevier Academic Press, London, **2004**, pp. 271–286.

29 J. C. Crabbe, D. Wahlsten, B. C. Dudek *Science* **1999**, 284, 1670–1672.

30 H. J. Hedrich, G. Bullock *The Laboratory Mouse*, Elsevier Academic Press, London, **2004**.

31 I. A. Olsson, K. Dahlborn *Lab. Anim.* **2002**, 36, 243–270.

5
Exploration of Metabolic and Endocrine Function in the Mouse

Marie-France Champy, Carmen A. Argmann, Pierre Chambon, and Johan Auwerx

5.1
General Introduction

Endocrine and metabolic dysfunctions are amongst the most common diseases in developed societies [1]. Examples of such diseases include obesity, insulin resistance, type 2 diabetes, hyperlipidemia and atherosclerosis. Mouse models are becoming increasingly popular tools in the study and characterization of molecular and physiological aspects of these diseases. Therefore the characterization of endocrine and metabolic dysfunction in mice has become of utmost importance for both basic and clinical scientists. Endocrine and metabolic disturbances are, however, often subtle in their presentation so that their detection is often a challenge to even the most experienced mouse physiologists. In this chapter, we will discuss in detail the rationale and technical aspects of the current metabolic and endocrine tests used in mice.

5.1.1
Investigating a Mouse with Endocrine and Metabolic Dysfunction

In humans, clinical history and physical examination constitute critical elements in the general process of a diagnosis. Although obtaining a clinical history in mice is not possible, physical examination, as in humans, can provide important indications of endocrine or metabolic dysfunction. Numerous examples illustrate this principle neatly. For example, diseases of hormone deficiency or excess are physiologic determinants of physical traits including stature, weight, complexion, hairiness and behavior. Body characteristics such as thinness or obesity can result from abnormalities in lipid or glucose metabolism. Loss of body weight in conjunction with an increase in food consumption can be the consequence of malabsorption or hypermetabolism (thyroid disease). Close observation of mouse behavior also reveals important information, for example nervousness may be associated with hyperthyroidism. Therefore all endocrine or metabolic investigations should start with an in-depth analysis of home cage behaviors using such tests as SHIRPA or the dysmorphological screen [2]. Screening of standard blood parameters e. g. glucose, lipids, calcium and electrolytes also provides indications for the diagnosis of

Standards of Mouse Model Phenotyping. Edited by Martin Hrabé de Angelis, Pierre Chambon, and Steve Brown

ISBN: 3-527-31031-2

endocrine or metabolic dysfunction which can be further investigated with a wide array of more specialized and detailed tests covering the whole endocrine and metabolic systems.

5.1.2
Principles of Endocrine and Metabolic Testing

Familiarity with certain principles and rules facilitates the skilful diagnosis of endocrine and metabolic dysfunctions. In particular, endocrine and metabolic disorders are most often indicated by either an excess or deficiency of a certain biologically-active hormone or metabolite. Alternatively these disorders can be the consequence of inadequate cellular responses to a particular hormone or metabolite. These principles help guide the phenotyping paradigms which we will propose.

Hormones and metabolites can be characterized by the pathway and origin of their production, their storage and release into the serum, and their degradation and metabolism. Although hormone and metabolite levels can be regulated in all of these aspects, the production rate is often the primordial factor determining their levels. Interestingly production of most metabolites and hormones is regulated directly or indirectly by the metabolite or hormone in question through a positive and interconnected negative feedback loop. This control has important implications since plasma levels of a metabolite or hormone often only make sense if the appropriate regulatory factor is taken into account (e. g. glucose and insulin). Elevations in levels of metabolites/hormones and their regulatory factors in the absence of any evidence of hormone excess often reflect a state of resistance (e. g. glucose and insulin elevation in insulin resistance). A final interesting correlate of regulated hormone secretion is that they underpin various dynamic endocrine or metabolic tests for the evaluation of discrete disturbances in metabolic and endocrine function (e. g. intraperitoneal insulin tolerance test).

Another general point that merits some discussion concerns the standardization of these tests. Animal housing and handling conditions will have a major impact on the outcome of phenotypic analysis, including endocrine and metabolic testing. Notorious examples include the impact of fasting on plasma glucose or free fatty acids levels (Fig. 5.1). Additional factors include: the number of animals per cage (housing density); the type of diet; diurnal rhythm; the blood collection procedure (retro-orbital or tail puncture, with or without anesthesia); age and gender of the mice. The establishment of the effects of these factors has prompted the European Union consortium, Eumorphia, to propose a set of guidelines for studying metabolism and endocrine function in mice (http://www.eumorphia.org/servlet/ECFLP.Frameset). An in-depth discussion of these variables can also be found in a recent review [3].

5.1.3
Strain in Relation to Mouse Models of Metabolic Disease

Strain background, strain type, and substrain of mice must be carefully chosen for metabolic/endocrine studies as they have an impact on almost every variable (see

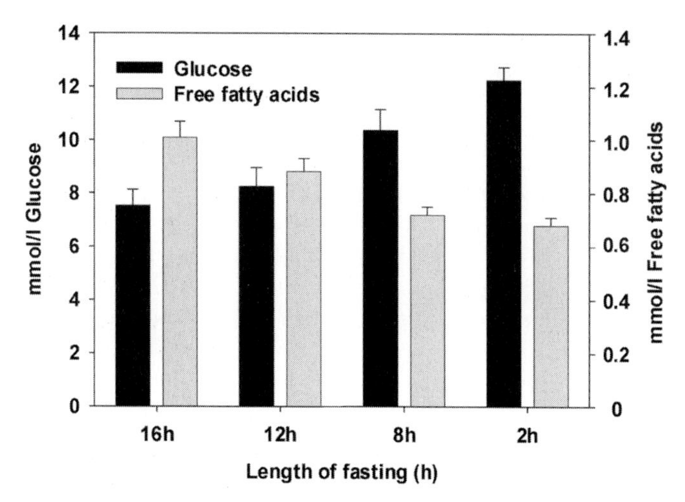

Fig. 5.1 Correlation between the duration of fast and the effect on plasma glucose and free fatty acid levels. Male C57Bl/6 mice were fasted for 2, 8, 12 or 16 h prior to blood sampling and the measurement of glucose and free fatty acid levels. The levels of glucose and free fatty acids in the blood are inversely related to the duration of the fast [3].

also Fig. 5.3 B). In general C57BL/6 mice have a good breeding performance and are one of the preferred genetic backgrounds for a large number of congenic strains, covering both polymorphic and mutant loci. However, this strain of mouse has low plasma cholesterol and low plasma triglyceride (TG) but high plasma glucose levels. C3H mice have low blood pressure, high plasma cholesterol, TG and glucose. In contrast, 129 inbred strains have high plasma cholesterol and low TG levels, but are poor breeders and exhibit severe behavioral abnormalities. Because of these strain-specific differences it is essential to select the most appropriate mouse strain for a particular study. Furthermore, it is clear that results obtained from one genetic background do not necessarily translate into another background. This becomes particularly important if one considers the huge number of studies performed in gene-targeted mice on a non-homogeneous genetic background due to the frequent use of 129 ES cells for targeting. If the background strain of the phenotyped mouse, differs from that used to generate the knockout mice, we recommend that mice be backcrossed for at least nine generations to avoid the confounding factors contributed by flanking donor chromosomal DNA. Alternatively, flanking donor chromosomal DNA can be monitored by using a marker-assisted selection protocol, leading to a more rapid production of congenic strains compared to traditional backcrossing [4].

Some naturally-occurring mutations have provided us with valuable mouse models for metabolic/endocrine investigation [5]. In view of their widespread use we will briefly mention some of the features of these models. In the context of diabetes and obesity several useful models, including the Lep^{ob}, $Lepr^{db}$, Ay, tubby, and KK/Ay mice, have increased our understanding of the molecular pathogenesis of obesity and diabetes. *Lep*^{ob} mice do not express leptin, increase in weight rapidly and develop severe obesity [6]. In addition to obesity, these mice exhibit hyperphagia, a diabetes-like syndrome of hyperglycemia, glucose intolerance, elevated plasma insulin, subfertility and an increase in hormone production from both the pituitary and adrenal glands. *Lepr*^{db} mice, on the other hand, carry a point mutation in the leptin receptor gene, *Lepr* [7]. These mice recapitulate the symptoms of human diabetes quickly in that they become obese at around 3 to 4 weeks of age and have elevated plasma insulin (by day 10 to 14) and glucose (by 4 to 8 weeks). Furthermore, the homozygote mice are polyphagic, polydipsic, and polyuric. The yellow Ay mouse is also very hyperinsulinemic and obese but often presents without the concomitant vasculopathy that may be present in other strains [8]. The KK/Ay mouse is yet another model for obesity and insulin-dependent diabetes mellitus [9]. These mice also develop hyperglycemia, hyperinsulinemia and glucose intolerance, but unlike the *Lepr*^{db} and *Lep*^{ob} mice, KK/Ay mice also show a remarkable diet-induced hypertriglyceridemia, a common feature of the metabolic syndrome. In our opinion this makes them one of the most valuable models for the study of the metabolic syndrome (www.jax.org/jaxmice). In addition to the above-mentioned models which have been useful in the area of metabolic disease, other natural mutations have proved to be of value in the study of endocrine and metabolic function. A good example is the Ames mouse, which is characterized by a primary pituitary deficiency causing the absence of or extreme reduction in the production of growth hormone, prolactin and thyroid stimulating hormone [10]. This mouse model has greatly advanced our understanding of pituitary function.

5.2 Evaluation of Energy Homeostasis

Increases or decreases in body weight can result from changes in food consumption (energy intake) or from altered energy expenditure [11]. The energy required to maintain body functions both at rest and during stress and exercise is provided in the form of the daily consumption of carbohydrates, fat and protein. This equilibrium can be disturbed as a result of changes in calorie intake or altered physical or metabolic energy expenditure (the burning of muscle tissue versus brown adipose tissue), parameters that are greatly influenced by endocrine and metabolic function (Fig. 5.2 A).

5.2.1 Body Weight and Food Intake

Monitoring body weight and food intake is a relatively simple test and provides the initial assessment of metabolic homeostasis. We recommend as an initial screening the evaluation of body weight and food intake over a 6-week period, between 6 and 12 weeks of age. Body weight should be recorded by weighing the mice twice a week, at the same time of day. Care should be taken to avoid stressing the animals since this can impact on body weight gain. In the case of longer studies (e. g. aging)

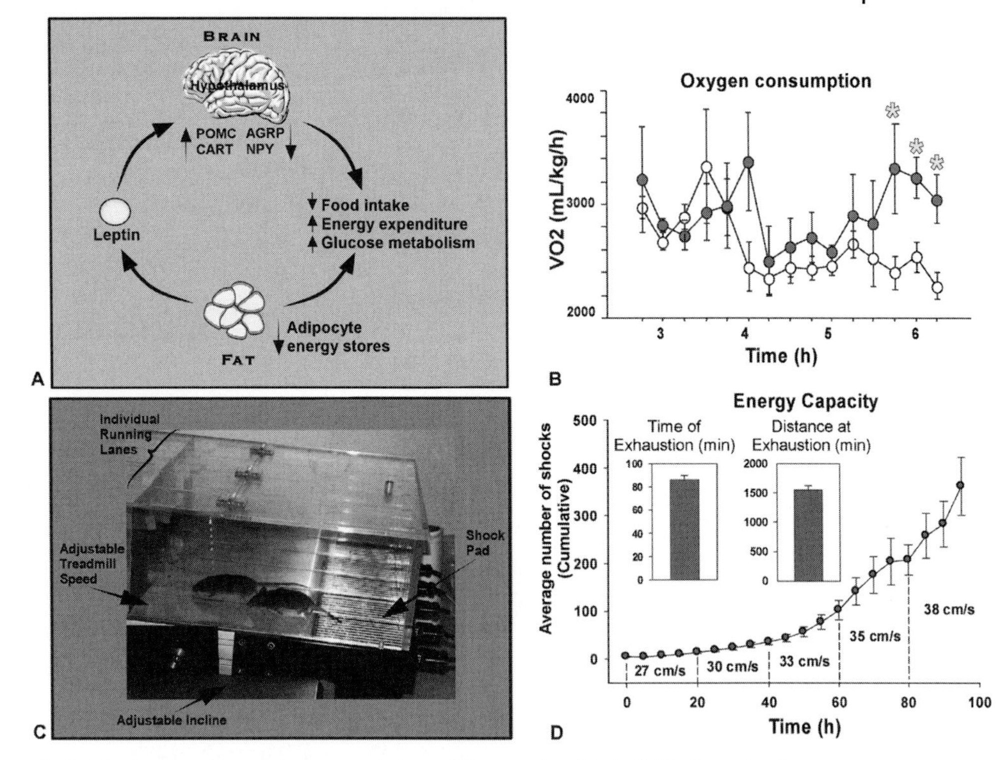

Fig. 5.2 Evaluation of energy homeostasis in the mouse. (A) Energy intake (food intake) and energy expenditure are the two important components of the total energy balance. The equilibrium can be disturbed when there are changes in endocrine function, which can alter calorie intake or energy expenditure [13]. (B) Oxygen consumption in wild-type (white dots) and TIF2 knockout mice (red dots) was measured over 6 h. Mutant mice have an elevated VO_2 which is indicative of increased energy expenditure [13]. (C) Exercise capacity can be evaluated using an exercise treadmill. Mice are placed in individual running lanes and the speed, incline and degree of shock are altered. (D) Energy capacity in C57Bl/6 mice was measured using an incremental exercise protocol starting with a speed of 27 cm/s with 2–3 cm/s increases every 20 min. During the experiment, there was an increase in the number of shocks with exhaustion being defined as 100 shocks within a 5-min interval. Time and distance run at exhaustion indicates energy capacity (unpublished observations).

the frequency of weighing can be reduced to a bimonthly recording. Body mass index is calculated as the body weight divided by the square of the length.

Weighing the food given and the food remaining in the grid of the cages on a twice-weekly basis is a good and reliable method for evaluating food consumption. We often measure food intake when mice are housed at a density of four mice per cage. As an alternative and more precise measurement of food intake, mice can be housed individually in metabolic cages and their food intake calculated on a daily basis over 4 days.

5.2.2
Energy Expenditure by Indirect Calorimetry

Energy expenditure is evaluated through indirect calorimetry by measuring oxygen consumption with specifically-designed equipment [12]. The system monitors gas volume concentrations of oxygen (O_2) and carbon dioxide (CO_2) at the inlet and outlet ports of a partially sealed chamber which is forcibly ventilated by a stream of ambient air at a known flow rate, either by positive or negative pressure. The concentration differences measured between the ports together with the flow rate and pressure are used to compute oxygen consumption (VO_2), carbon dioxide production (VCO_2) and the respiratory exchange ratio (RER) calculated as VO_2/VCO_2 (Fig. 5.2 B). The chambers are equipped with a third port which is used for temperature measurement. Heat is calculated from the VO_2 and RER using the standard formula: heat (H) = calorific value × VO_2 × 0.001; calorific value = 3.815 + 1.232 × RER. In practice, mice are placed into the metabolic chambers individually while they have *ad libitum* access to water. Ideally indirect calorimetry should be carried out at thermoneutrality (~ 30–34 °C in the mouse). If the temperature drops below 30 °C, mice suffer a huge increase in metabolic rate, which can confound the outcome of the tests. Indirect calorimetry is commonly performed under two different experimental conditions. First, indirect calorimetry can be carried out under a short period (6 h) of food deprivation. Alternatively the test can be performed over 24 h while the mice have free access to food. Obviously the experimental outcome needs to be interpreted with this in mind. O_2 and CO_2 measurements are taken at regular time points during the whole experiment and VO_2 and VCO_2 values are expressed as ml/kg/h. As energy expenditure varies with body weight it is important to normalize the values obtained for body mass. Based on allometric scaling methods metabolic body mass is estimated as the body weight to the power of 0.75 or 0.67 ($ml/kg^{0.75}/h$ or $ml/kg^{0.67}/h$) [12]. Ideally however, VO_2 should be corrected to lean body mass and not total body mass. For groups of animals with roughly the same body composition and fat mass, normalization is not required. However, in obese animals adipose tissue consists mainly of "metabolically inactive" storage tissue and normalization of body weight to the power of 0.75 or 0.67 also gives an incorrect estimate of metabolic body mass. In such cases lean body mass (which reflects "metabolically active tissue") is best obtained from the analysis of body composition as determined by densitometry or an equivalent method (see below). Taking all this into account we prefer to perform indirect calorimetry in young animals, when body weight differences are usually less pronounced.

5.2.3
Cold Test

Adaptive thermogenesis is another important component of energy expenditure since it enables the body to adapt its metabolism to changes in environmental conditions, such as a sudden drop in temperature. Because of the relatively higher surface area of small homeothermic animals such as the mice, heat loss occurs much faster than in larger animals. Consequently, small animals will burn more calories once the temperature has dropped below thermoneutrality in order to maintain their core body temperature. The cold test is a simple test by which such adaptive thermoregulation can be evaluated [13]. In practice, mice are individually placed into a cage without food but with free access to water. The initial body temperature is recorded by inserting a small thermoprobe into the rectum of the animal. Mice are then placed in a cold room at 4 °C and the rectal body temperature is recorded every hour for 6 h. Results are presented graphically as a curve showing the decrease in body temperature over the duration of the experiment.

5.2.4 Exercise Test

One of the most severe, yet physiologically relevant, stresses to the cardiovascular system that relies on the utilization of metabolic substrates is exercise. The whole body adapts to exercise as the result of a coordinated response of multiple organ systems, including the cardiovascular, pulmonary, endocrine-metabolic, immunologic and skeleto-muscular systems. Consequently, alterations in the functioning of any one of these organ systems can significantly affect the response to exercise. Furthermore, like the cold test, exercise can often elicit subtle physiologic phenotypes that may not be evident at baseline but are only manifested during the stress of exercise testing.

Acute low intensity exercise increases muscle workload and thereby the energy requirements of working muscles. This increase in energy demand is directly related to oxygen consumption by the muscle through mitochondrial oxidative metabolism and ADP phosphorylation. In order to meet the increased oxygen demand of the muscle, oxygen delivery must be increased to the working muscle tissue, which in turn stresses the cardiovascular system. When these demands are not met, compensatory mechanisms which include increased anaerobic metabolism, accumulation of lactate and exercise intolerance, or a quicker onset of fatigue are brought into play. One standard method for assessing exercise performance is the treadmill exercise. In general treadmill systems are composed of a belt, which is enclosed in a plexiglass chamber, with a stimulus device consisting of a metal shock grid attached to the rear of the belt (Fig. 5.2 C). The speed and slope of the belt are adjusted electronically. The devices can be confined to a metabolic chamber allowing for the concurrent measurement of O_2 consumption and CO_2 production. Furthermore the combination of exercise testing with telemetry implant devices or catheterization, allows blood pressure and heart rate to be recorded during exercise [14].

In general we suggest an incremental exercise protocol whereby mice begin with a 20-min acclimatization period at 27 cm/s and a 5 ° incline followed by 3 cm/s increases in running speed until exhaustion. A mouse is considered to be exhausted when it receives approximately 100 shocks (at 2 mA each shock) in a period of 5 min. Mice in general tolerate speeds of up to 83 cm/s and the level of shock is specific for each apparatus. The performance of the mice is evaluated by recording the duration of running and the total distance covered (Fig. 5.2 D). When changes in exercise capacity are detected further investigation is required both in cardiovascular, sensory motor, and skeletal muscle function. If a treadmill is not available, forced swimming has also been commonly used to score exercise capacity. This low-cost approach is particularly useful when there are difficulties in inducing the mice to run. The assessment of cardiovascular parameters is, however, limited due to the technical limitations of the aqueous environment.

There are several confounding factors that must be discussed with regard to exercise testing. The first issue is acclimatization of the animal to the exercise instrument. In general we acclimatize animals by using a habituation protocol on the day preceding the actual running test. With this procedure the mice are placed in the chamber where they run at 27 cm/s for 10 min, this generally lessens the "shock" of the experiment the following day and the mice perform better during the actual test. Longer periods of habituation may result in some degree of conditioning, since it has been shown that chronic treadmill exercise (12 weeks) leads to an increase in exercise capacity. Genetic background is also a confounder as there are substantial differences in exercise capacity between different inbred mouse strains [15, 16]. For example, C57Bl/6J have one of the worst records for exercise capacities on the treadmill but one of the best on the voluntary wheel; whereas DBA mice have the worst capacity for exercise on the wheel but a much better capacity on the treadmill [16].

5.2.5
Lean and Fat Composition of the Body

Bone densitometry and body composition (lean and fat content) are currently often measured using dual-energy X-ray absorption (DEXA) analysis [17, 18]. This method eliminates the need for destructive chemical analysis and allows multiple consecutive measurements throughout the lifetime of the animal. The mice are briefly anesthetized (by intraperitoneal injection of a ketamine–xylazine solution) and placed on the disposable positioner of the apparatus. The densitometer provides bone mineral and body composition results from body imaging in less than 5 min. Analysis of the results is based on the total body region of interest (ROI), the area from which the bone density is estimated. The software inserts an oval-shaped exclusion area to exclude the head and its soft tissue. The analysis provides bone mineral density in g/cm^2, lean, fat and total tissue in grams, and percent fat. Recently other methods of determining body fat mass have become available. Most notably, EchoMRI quantitative magnetic resonance, which is capable of analyzing the body composition of mice *in vivo* in less then 1 min. EchoMRI counts hydrogen molecules of different bonding types to determine the mass of different parameters. This technique is very precise and advantageous in that it does not require anesthesia or sedation of the mouse, but its major drawback is that it does not provide information relating to bone mineral density.

5.3
Evaluation of Standard Clinical Chemistry Blood Parameters

Analysis of a subset of standard serum parameters is a good screening procedure for asymptomatic mice and can yield a wealth of information concerning metabolic and endocrine parameters. For screening purposes and to reduce variability of the results we recommend that initial serum or plasma blood parameters be measured following an overnight fast in young adult mice (10–12 weeks of age). After overnight fasting, blood can be collected by retro-orbital or tail puncture from anesthetized mice. The volume of blood obtained must not exceed more than 10% of the total body weight of the animal. Approximately 100 µl of plasma is used for the determination of plasma biochemical parameters. We often perform a second serum analysis after a shorter (4 h) fast or even in absence of fasting, especially when we suspect abnormalities of lipid and glucose homeostasis. Furthermore, retesting in older mice rapidly yields information concerning the evolution of biochemical and metabolic parameters. In cases where abnormal blood parameters are detected more extensive exploration is carried out as described below. The measurement of ions, urea, creatinine, hepatic and cardiac enzymes, total protein, albumin, glucose, TG, total cholesterol, total bilirubin and total bile acids can be undertaken manually or by using an automated laboratory workstation using specific colorimetric or enzymatic methods. The suitability of using an automated system requires careful consideration, especially since around 15 parameters will need to be investigated in a relatively small sample of plasma or serum. The amount of serum necessary for testing is thus a predetermining factor in the choice of the equipment. Clinical chemistry analysis is extremely useful as an initial screen, yielding information about multiple organ systems. Hence, we consider it to be a primary test which indicates the direction that further investigations should take.

5.4
Evaluation of Glucose Homeostasis

When an initial blood screen indicates altered levels of insulin and/or glucose, the metabolic tests described in this section can be used to explore the animal's glucose homeostasis and insulin sensitivity in more detail. Diabetes or glucose intolerance can either be due to a primary deficiency in insulin production/secretion (mimicking type 1) or be a consequence of insulin resistance (type 2). Insulin resistance develops when higher than normal concentrations of insulin are needed to exert its normal physiological effects, which include inhibiting hepatic glucose production (hepatic gluconeogenesis) and increasing muscle and fat cell uptake of glucose [19]. As a consequence the pancreas compensates by increasing the amount of insulin secreted in order to maintain normal blood sugar levels. However, this compensation is temporary and is usually followed by pancreatic dysfunction and the development of type 2 diabetes (Fig. 5.3 A). There are several indicators of insulin resistance including elevated fasting insulin levels, increased fasting glucose and impaired glucose tolerance. The tests used to explore glucose homeostasis are described in order of invasiveness and ease, with the least invasive being described first.

5.4.1 HOMA (Homeostasis Assessment Model)

The homeostasis assessment model is a mathematical model for predicting values for insulin sensitivity and β-cell function when fasting plasma glucose and fasting insulin concentrations are known [20]. The calculation is as follows: (insulin μU/ml)/22.5 × (glucose mg/dl)/18. When two values are compared, for example, a wild-type and control, a higher score may indicate insulin resistance. This test gives a static measure of insulin resistance, whereas the tests described next are dynamic tests for insulin resistance. The HOMA is widely used in epidemiological studies in humans, or for monitoring changes in insulin resistance with time in individual patients [21]. In the mouse this test also provides a rapid comparison between the control and mutants. Furthermore, in humans, the estimates of insulin resistance by HOMA have been shown to correlate well with estimates from the euglycemic clamp [20].

5.4.2 Meal Tolerance Test (MTT)

One of the most natural ways to evaluate glucose tolerance is to measure serum glucose levels both before and after a standard meal. The meal tolerance test shows the effect of a standard meal on glucose metabolism and insulin secretion. Animals are therefore fasted for approximately 16 h. A standardized meal (chow diet) is then administered in the form of a food pellet of a specific weight which the hungry animal will then eat. Plasma glucose and insulin are measured in blood collected 1 h after feeding [22].

5.4.3 Intra-Peritoneal or Oral Glucose Tolerance Test (IPGTT or OGTT)

Insulin resistance is assessed in a glucose tolerance test by following the metabolic clearance of a standard glucose load. Essentially, blood glucose measurements are taken several times over a period of 3 h. In a non-diabetic state the glucose levels rise and then fall quickly, whereas, in a diabetic state, glucose levels rise higher and fail to fall as rapidly as they do in the non-diabetic state. Initially, the basal glucose level is measured by placing a drop of blood collected from the tail (time 0) on a glucometer, a solution of 20% glucose in sterile saline (0.9% NaCl) at a dose of 2 g glucose/kg body weight is then administered to the mice by intra-peritoneal injection (IPGTT) or by oral gavage (OGTT). Blood is sampled from the tail vein for glucose determination at 15, 30, 45, 60, 90, 120, 150, and 180 min post-glucose injection. For data analysis, the plasma glucose concentrations are plotted against time, producing a curve which represents the increase in blood glucose and its return to

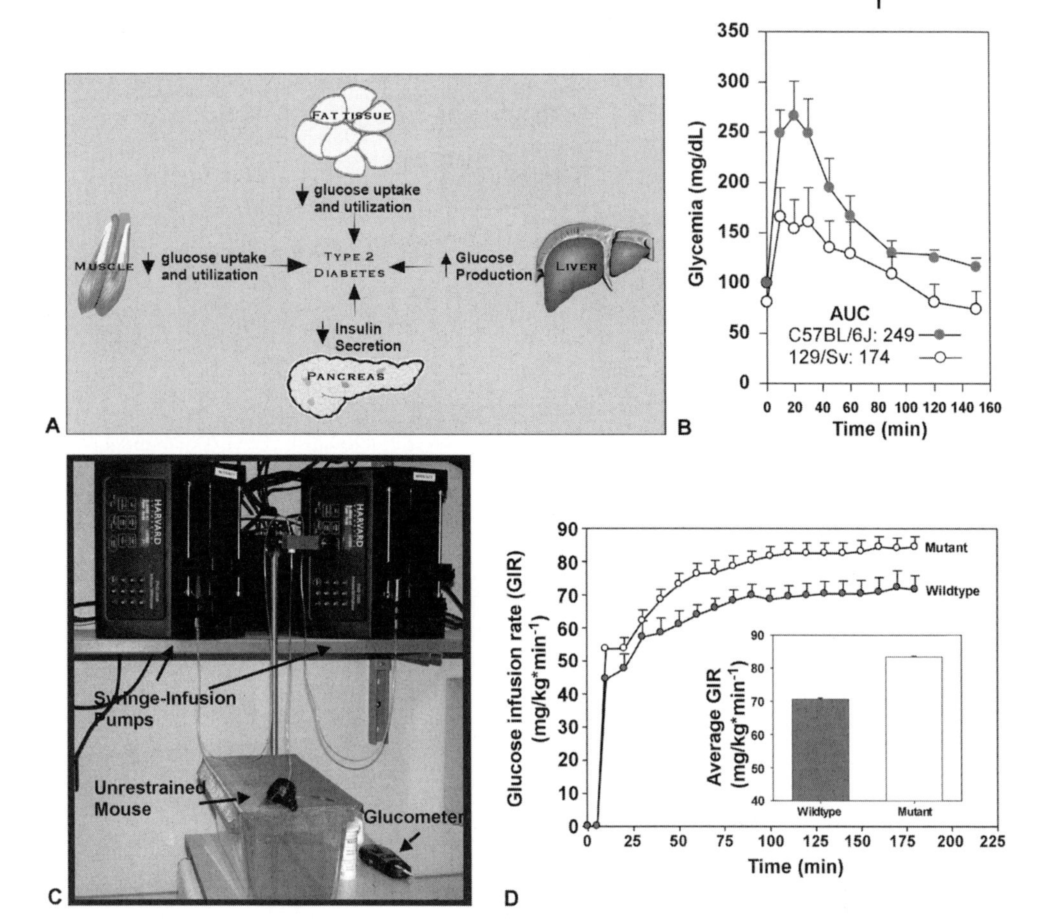

Fig. 5.3 Evaluation of glucose homeostasis. (A) Insulin resistance develops when higher than normal concentrations of insulin, are needed to exert its normal physiologic effects. As a consequence the pancreas compensates by increasing the amount of insulin secreted in order to maintain normal blood sugar levels. However, this compensation is temporary and is usually followed by pancreatic dysfunction and the development of type 2 diabetes. (B) The oral glucose tolerance test (OGTT) measures insulin resistance by examining the metabolic clearance of a standard oral glucose load. C57BL/6J mice, as compared to the 129Sv mice, are significantly more glucose intolerant as indicated by the increased area under the curve for glucose clearance (AUC) [3]. This study also highlights the effect of the genetic background of the mouse. (C) The hyperinsulinemic euglyemic clamp is the gold standard for assessing insulin resistance. Syringe-infusion pumps enable simultaneous insulin and glucose infusion at a variable or constant rate. Mice are awake and unrestrained. Blood taken from the tail vein is measured with a glucometer. (D) The degree of insulin sensitivity is indicated by the average glucose infusion rate (GIR) over the last 60 min of the clamp, when blood glucose is stabilized at euglycemia (unpublished observations). An increase in the GIR indicates efficient blood glucose clearance and insulin sensitivity.

the normal level [23]. The incremental area under the curve (AUC) is calculated and is a representative measure of insulin sensitivity such that an increase in the area under the curve indicates a decrease in the glucose tolerance, often caused by reduced insulin sensitivity. Like all metabolic and endocrine tests the influence of genetic background on these tests is important (Fig. 5.3 B). The glucose tolerance test indicates the efficiency of the liver in concentrating and storing glucose, the capacity of the pancreas to produce insulin, and the amount of active insulin it produces and the sensitivity of the cells in the body to the action of insulin. The advantage of the OGTT over the IPGTT or the clamp is that this test is performed under normal physiological conditions, as glucose is given orally thus preserving the effects of entero-hormonal responses and the physiological kinetics of glucose absorption and insulin secretion.

The outcome of all these tests may indicate normal or abnormal glucose tolerance. Decreased glucose tolerance, in which the blood glucose curve shows a sharp peak before declining more slowly than normal is usually suggestive of insulin resistance and type 2 diabetes. However, other disorders such as hemochromatosis, increased levels of circulating glucocorticoids or epinephrine, need to be excluded. Increased glucose tolerance, where the blood glucose levels peak at lower than normal levels, are indicative of increased insulin secretion or improved insulin sensitivity. This can be seen in conditions associated with stimulated metabolism, such as overproduction of insulin, deficiencies in glucocorticoid production or in cases where the supply of energy-producing substrate is limited (e. g. liver deficiency, hepatic enzymatic deficiency, malabsorption, post-exercise).

5.4.4
Intra-Peritoneal Insulin Sensitivity Test (IPIST)

A relatively quick and effective test to estimate insulin sensitivity in the animal is an IPIST. This test estimates insulin resistance from the rate of decline in glucose levels following an intravenous bolus injection of insulin. Overnight fasted animals are given an i. p. injection of regular insulin (0.5 IU/kg) and blood is sampled from the tail vein at 15, 30, 60, and 90 min post-injection and glucose is analyzed as described above. Insulin sensitivity is determined by calculating the rate of decline of the log transformed glucose concentrations, estimated by linear regression. A rate constant can be derived which is expressed as the percentage decline in glucose per minute. In the case of insulin-resistance the fall in glucose is retarded and the recovery of normal glucose levels takes longer. Because this test can be completed rapidly, it is not subject to the problems of interference from any counter-regulatory hormones which may have been released and reflects the combination of suppression of hepatic glucose output and stimulation of peripheral glucose uptake by insulin.

5.4.5
Glucose Clamps

The current gold standard method for measuring insulin resistance is the euglycemic clamp. In this method glucose is "clamped" at a predetermined value

(5 mmol/l for euglycemia, 15 mmol/l for hyperglycemia and 2.5 mmol/l for hypoglycemia) by titrating a variable-rate of glucose (glucose infusion rate, GIR) against a fixed-infusion rate of insulin [20, 24]. In some instances, rendering a diabetic mouse normoglycemic is artificial and is not representative of their usual metabolic state and therefore, iso-glycemic clamps are an alternative. Since supraphysiological insulin concentrations are attained, hepatic glucose output should no longer be an issue in general. Glucose clamps are technically demanding, invasive and time consuming and their use should be considered carefully, especially given the relatively comparable data that can be acquired from glucose tolerance tests [20].

In practice, a catheter is established in the femoral vein under anesthesia (ketamine and xylazine) 2–3 days in advance of the study; the catheter is fed underneath the mouse‘s skin and fixed behind the head [25, 26] (Fig. 5.3 C). After surgery, mice are housed individually and allowed to recover for at least 48 h to give them sufficient time to regain their body weight. The clamp procedure is undertaken in awake, unrestrained, unstressed and light-cycle inverted mice (1 week acclimatization) following a 5-h fast. Mice are acclimatized to the tops of cages while the catheter is attached to a syringe-infusion pump for at least 1 h prior to glucose/insulin administration. It should be noted that the catheter from the mouse must be bifurcated to allow for simultaneous constant and variable injections of insulin and glucose, respectively. Baseline glucose values are measured by tail vein sampling prior to the injection of insulin. Catheter placement is assessed with a short priming dose (6 µl/min for 1 min) of insulin prior to the constant infusion of insulin at a flow rate of 2 µl/min, equivalent to 18 mU of insulin/kg/min. The infusion rate can also be calculated as a dose per unit of surface area rather than body weight, in order to avoid over-insulinization of overweight animals. Blood glucose values are monitored every 5 min throughout the test and glucose is infused (20 % solution in saline) at a variable rate until euglycemia (plus or minus 15 %) has been reached and maintained. At this point the animal is "clamped" and the degree of insulin resistance is inversely related to the amount of glucose necessary to maintain the required blood glucose concentrations. The effect of the clamp is visualized graphically by plotting plasma glucose concentration versus time. The GIR (mg glucose/kg animal × min) is then calculated as the average rate over the last 60 min of the clamp. When the average GIR of one animal is greater than that of another, this indicates greater insulin sensitivity or more rapid clearance of glucose from the plasma (Fig. 5.3 D).

5.4.6 Utilization of Glucose by Individual Tissues

This section outlines modifications that can be incorporated into the euglycemic clamp to further define insulin action on a whole body and individual tissue basis. To estimate insulin-stimulated whole-body glucose turnover, a prime-continuous infusion of [3-^{3}H]glucose can be infused throughout the clamp procedure (10 µCi bolus, 0.1 µCi/min). To estimate insulin-stimulated glucose uptake in individual tissues, 2-deoxy-d-[1-^{14}C] glucose (2-[^{14}C]DG) can be administered as a bolus

(10 µCi) 75 min after initiation of the clamps. Blood samples collected during and at the end of the clamp are then used to measure plasma [^{3}H]glucose, $^3H_20$, 2-[^{14}C]DG and insulin concentrations. The rate of insulin-stimulated whole-body glucose turnover is determined as the ratio of the [^{3}H]glucose infusion rate (dpm/min) to the specific activity of plasma glucose (dpm/µmol) during the final 30 min of the clamps. Hepatic glucose production during the clamp can be assessed by subtracting the steady-state glucose infusion rate from the whole-body glucose turnover rate. At the end of the clamp, mice are sacrificed and various tissues are collected including muscle (gastrocnemius, tibialis anterior and quadriceps), liver and adipose. Tissue 2-[^{14}C]DG-6-phosphate content can be used to estimate insulin-stimulated glucose transport activity since 2-DG is a glucose analog that is phosphorylated but not metabolized [25].

5.4.7
Insulin Secretion Test

The hyperglycemic clamp is used for measuring β-cell insulin secretion [27] and is basically a modified euglycemic clamp, but in general is easier to carry out. In this test, only glucose (20%) is infused intravenously at a variable rate until a steady state of hyperglycemia (10–15 mmol/l) is achieved. As in the euglycemic clamp, repeated blood glucose measurements enable the GIR to be adjusted accordingly. The algorithms used to establish the GIR have been summarized elsewhere [28, 29]. In this instance a greater GIR could imply increased β-cell insulin secretion. A measure of insulin sensitivity can also be obtained by dividing the GIR required for maintenance of hyperglycemia by the mean of the plasma insulin concentration measured over the same period (the last 20–30 min of the 120-min clamp). At this point the relationship between insulin and glucose is sigmoidal and insulin resistance is likely to prevail when this curve is shifted to the right. In addition to insulin sensitivity, non-insulin-dependent glucose uptake, glucose effectiveness, β-cell secretory capacity, and hepatic insulin clearance can also be estimated. Although the hyperglycemic clamp offers the added advantage of assessing both insulin resistance and β-cell function from a single test, the prevailing glucose concentration is higher than that found in the normal physiological condition, thus complicating its interpretation. As a consequence the glucose infusion is an integrated measure of insulin-mediated glucose disposal, glucose-mediated glucose disposal and often, renal glycosuria. With the euglycemic clamp, the latter two aspects are minimized [20].

It must be kept in mind that insulin resistance cannot be measured in isolation. The circumstances under which it is assessed as well as the method used must be specified. For example, basal insulin resistance (as measured following an overnight fast or HOMA) is not synonymous with stimulated insulin resistance. Time of day and stress may also influence insulin resistance measurements as increased glucocorticoids, FFA, and adrenalin levels can increase hepatic glucose output and impact on peripheral glucose clearance and confound the interpretation of the results of these tests.

5.5
Measurement of Serum Lipids and Lipoprotein Parameters

Lipoproteins are supramolecular complexes consisting of lipids and proteins that function to transport non-polar lipids from their sites of synthesis to their sites of utilization (Fig. 5.4 A) [30]. These lipoproteins in general assume a spherical structure whereby the polar phospholipids, proteins and non-esterified or free cholesterol

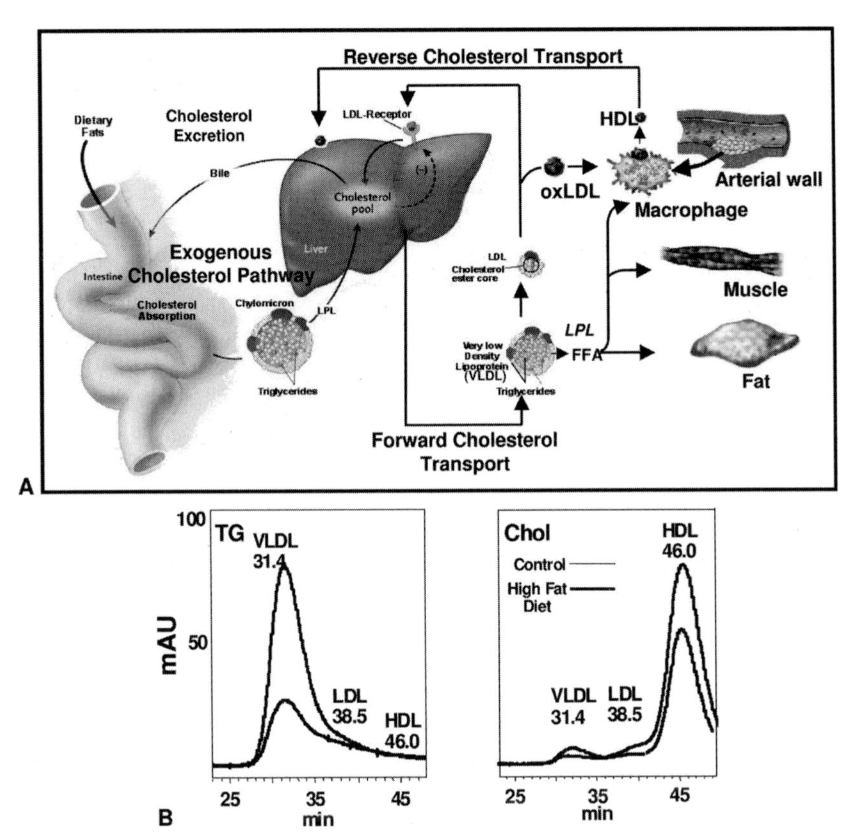

Fig. 5.4 Serum lipids and lipoproteins. (A) The transport of lipoproteins is achieved through three pathways. The exogenous pathway describes the absorption and packaging of dietary-derived lipids into large triglyceride (TG)-rich particles called chylomicrons. The forward cholesterol transport pathway involves the assembly and secretion of VLDL by the liver. The reverse cholesterol transport pathway involves the hepatic secretion of nascent or lipid-poor HDL that serves to remove and carry free cholesterol from the extra-hepatic tissues back to the liver where HDL is degraded and cholesterol is excreted (figure adapted from [45]). (B) Lipoproteins can be separated by gel filtration, which fractionates lipoproteins on the basis of particle size. Plasma VLDL, LDL and HDL are eluted from a Sepharose column with a NaCl buffer at a constant flow rate (35 µl/min), and cholesterol or TG levels are then measured. A typical elution profile for the different lipoproteins is shown (VLDL = 31.4 min; LDL = 38.5 min and HDL = 46.0 min). Animals on a high fat diet have a characteristic elevation in VLDL and LDL cholesterol as indicated by their respective enhanced lipoprotein peaks [46].

form a monolayer shell that facilitates the transport of core hydrophobic lipids, including cholesterol ester and TG, in the aqueous blood system. The non-polar lipids serve numerous biological functions including acting as a transportable metabolic energy pool, a structural component for cell membranes, and as the building blocks for hormones. The lipoprotein-associated proteins, or apolipoproteins, provide structural integrity to lipoproteins, act as ligands for cellular receptors and function as cofactors modulating enzymatic activity [31]. In humans, abnormalities of lipoprotein transport, synthesis and degradation, usually of genetic etiology, often leads to disease. Most notably, the metabolic syndrome has an important lipid component including: increased plasma levels of TG, apoB and small dense LDL particles, and reduced apoA-I and HDL cholesterol levels. Long-term dyslipoproteinemia associated with the metabolic syndrome increases the risk of atherosclerotic vascular disease and coronary artery disease [32].

Lipoproteins are transported in the circulation by three pathways: exogenous cholesterol transport, forward (endogenous) cholesterol transport and reverse cholesterol transport [33]. The exogenous pathway describes the absorption and packaging of dietary-derived lipids into large TG-rich particles called chylomicrons. The forward cholesterol transport pathway involves the assembly and secretion of VLDL by the liver which replaces chylomicrons as the main vehicle of TG transport in the circulation in the post-absorptive state. The reverse cholesterol transport pathway involves the hepatic secretion of nascent or lipid-poor HDL that serves to remove and carry free cholesterol from the extra-hepatic tissues back to the liver where HDL is degraded and cholesterol is excreted. The exogenous [34] and forward cholesterol [35] transport pathways are extensively reviewed elsewhere.

5.5.1 Serum Lipid Parameters

Investigation of lipid metabolism begins with the analysis of standard lipid serum parameters. To reduce variability of the results we recommend that serum or plasma blood parameters be recorded after an overnight fast. Approximately 50 μl of plasma is used for the determination of plasma lipid parameters. Measurements of TG, total cholesterol, LDL and HDL cholesterol and free fatty acids are carried out manually or with an automated laboratory workstation using specific colorimetric or enzymatic methods. HDL cholesterol is measured using an enzyme chromogen system following apoB precipitation with an anti-human beta-lipoprotein antibody, which removes apoB-containing lipoproteins (chylomicrons, VLDL, IDL, and LDL). LDL cholesterol is measured following a two-step reaction in which the first step eliminates non-LDL cholesterol and in the second step, LDL cholesterol is measured with the cholesterol oxidase reaction.

5.5.2 Isolation of Plasma Lipoprotein

When abnormalities are detected in total, HDL or LDL cholesterol levels it is useful to isolate and analyze the different lipoprotein fractions. Two major techniques are principally used for the separation of lipoprotein fractions, differential ultracentrifugation and chromatography. Differential ultracentrifugation is a preparative method for separating lipoproteins on the basis of their hydrated density. At solvent densities higher than that of the lipoproteins, the latter float at a rate dependent on their density, sizes and shape. The various classes of lipoprotein are sequentially isolated by stepwise increases in the solvent density through the addition of such salts as KBr, NaBr or NaCl [36] and represent a broad range of densities unequally distributed along a density gradient from 0.92 to 1.25 g/ml. Lipoprotein particles classified according to buoyant density include: chylomicrons ($d < 0.94$ g/ml); VLDL ($d = 0.94$–1.006 g/ml); IDL ($d = 1.006$–1.019 g/ml); LDL ($d = 1.019$–1.063 g/mL); and HDL ($d = 1.063$–1.21 g/ml). Their equivalent flotation rates established by analytical ultracentrifuge at a background density of 1.063 g/ml, are such that chylomicrons display flotation rates (Sf) greater than 400; while the flotation rates of VLDL, IDL and LDL are Sf 20–400, 12–20 and 0–12 respectively. At a background density of 1.2 g/ml, HDL_2 and HDL_3 display flotation rates of Sf 3.5–9 and 0–3.5, respectively. Once the lipoprotein fractions have been isolated the particle content of cholesterol, TG and different apolipoproteins can then be determined.

Lipoprotein fractions can also be separated by gel filtration. Using these techniques, fractionation is achieved on the basis of particle size. Briefly, plasma VLDL, LDL and HDL are separated on a Sepharose column by elution with a NaCl buffer at a constant flow rate (35 µl/min); cholesterol levels can be measured on-line by adding an enzymatic reagent to the eluted fractions and measuring the absorbance using a spectrophotometer (Fig. 5.4 B). In addition to serum lipid analysis this test shows how the cholesterol is partitioned between the different lipoprotein fractions. Chromatography is quicker than ultracentrifugation for the separation of lipoproteins, but there is sometimes a disparity between the fractions isolated by centrifugation and by chromatography. This disparity is at least partially due to the difficulty in deciding where the boundaries between the conventional fractions would lie on the chromatographic elution curve.

Identification of a lipoprotein disorder is best achieved by combining the plasma lipids levels with those in the isolated lipoprotein fractions. For example, when there is an elevation in plasma TG in the chylomicron fraction this may represent a disorder in chylomicron clearance including LPL-mediated hydrolysis, whereas elevated plasma cholesterol, TG and low HDL values may indicate the presence of the metabolic syndrome with insulin resistance. In addition, when analyzing cholesterol profiles it is important to take into account the differences between human and murine models; in mice HDL cholesterol represents between 65 and 75 % of the total cholesterol, whilst the LDL cholesterol fraction represents between 10 and 20 %. This underlies the difficulties experienced in mimicking human lipid disorders in mice.

5.5.3 Apolipoproteins

Alaupovic [37] introduced the so-called ABC nomenclature, which designated the apolipoproteins in alphabetical order by capital letters, their constitutive polypeptides by Roman numerals and the polymorphic forms of either apolipoproteins or polypeptides by Arabic numbers. Although not all the functions of apolipoproteins are known, several functions have been identified including: stabilizing lipid emulsions (apoA, apoB, apoC), acting as cofactors and modulators of enzymatic reactions (apoC, and apoA-I), managing the export of lipids out of cells (apoE and apoB), assisting in the clearance of lipids by target organs and cells through specific receptor interactions (apoB, apoE and apoA), and inhibiting with platelet aggregation (apoH) [38]. The apolipoprotein components of lipoprotein fractions prepared by ultracentrifugation procedures can be analyzed by electrophoresis on non-reducing SDS-PAGE gels. Protein samples are heat-denatured and loaded onto 4 to 20% gels, separated by electrophoresis and visualized with Coomassie blue stain. These techniques give profiles and semi-quantitative values for the apolipoprotein content in blood (apoB-100, B-48, apoA-I, A-II, apoE and apoC). The distribution of apoC and apoE subspecies can be analyzed by isoelectric focusing gel electrophoresis. Some apolipoproteins can be measured individually by immunoassays. All these assays are based on the principle of the antigen–antibody reaction; the main difficulty is obtaining specific antibodies against mouse apolipoproteins as at present very few are commercially available.

Apolipoprotein measurements on their own are not clinically useful and should be used in conjunction with plasma lipid profiles and lipoprotein fraction analysis. Taken as a single measurement, the apoB level may provide information about the number of potentially atherogenic lipoproteins present, as there is one apoB molecule per particle of chylomicron, VLDL, LDL or IDL. In humans the presence of small, dense LDL particles, has been shown to be associated with the metabolic syndrome and an elevated LDL apoB level may represent the presence of small dense LDL.

5.6 Measurement of Hormones

Hormones affect almost all aspects of human function to provide a homeostatic internal environment. They control body growth, development, reproduction, energy homeostasis and facilitate the body‘s response to physiological and psychological stresses. Hormones, which can be derived from steroid compounds, amino acids or lipids, accomplish these functions by altering the rate of enzyme activity, by modifying cell membrane transport, or by inducing secretory activity. Thus in addition to routine serum chemistry, the levels of a large panel of hormones should be measured in order to further explore different endocrine and metabolic systems. These tests may concern the hypothalamic and pituitary axis (follicle stimulating hormone (FSH), luteinizing hormone (LH), growth hormone (GH), adrenocorticotrophin hormone (ACTH)), the endocrine pancreas (insulin and glucagon), steroid hormones (estradiol, progesterone, testosterone, corticosterone), thyroid hormones (T3, T4), and adipose tissue hormones (adiponectin and leptin). Concentrations of most hormones in blood or urine are much lower than those of general chemistry analytes, and specialized techniques, often only available in specialized laboratories, are necessary to measure these low concentrations. Moreover these techniques require larger amounts of blood, which often poses a problem in the mice because it is only possible to collect a total volume of blood of between 200 and 300 µl at each sampling.

Generally, hormone levels are measured by radioimmunoassay or enzymatic immunoassay (both competitive or immunometric). In competitive radioimmunoassay there is competition between a radioactive and a non-radioactive antigen for a fixed number of antibody binding sites. The amount of labeled antigen bound to the antibody is inversely proportional to the concentration of unlabeled antigen in the sample. Separation of the free and bound antigen is achieved by centrifugation if the antibody is attached to a tube or beads or if a double antibody system is used. Most of the immunometric assays are sandwich assays, which use an antigen sufficiently large to allow the concurrent binding of two primary antibodies directed to different binding sites and two secondary antibodies, a capture antiserum directed to one of the antigenic sites of the antigen attached to a solid phase and a signal antiserum directed to a second antigenic site on the antigen attached to an assay signal system (radiolabeled or enzymatic system). As the antigen concentration increases, the signal increases progressively.

Kits specific for some hormonal measurements or specific antibodies against mouse hormones are now commercially available. Steroid hormones, testosterone, estradiol and progesterone can be measured in mice with a human radioimmunoassay procedure; the main problem is that there are species-specific effects of free fatty acid and binding proteins in human and mouse serum. To compensate for these effects steroid-depleted mouse plasma can be added to the standard reaction. For other hormones like insulin and leptin, immunoenzymatic kits specific for mouse hormones and requiring only small volumes of plasma are now available.

Successful new technologies include the recently developed "multiplex analysis" systems. The advantages of these systems are the simultaneous measurement of several hormones in a single small-volume plasma sample. This multi-analyte detection works by incorporating several technologies including bioassays, internally-dyed microspheres, fluidics, lasers and the latest in high-speed digital signal processors. Uniform, internally-dyed polystrene microspheres with 100 different color ratios allows for 100 multiplexed assays to be performed in one reaction with each uniquely colored microsphere containing a different assay. Each set of beads is conjugated to a different capture antibody; the conjugate beads are then mixed and incubated with the plasma samples and specific antigens (hormone) in the wells of a microtiter plate. To detect and measure each capture antigen, a second fluorescent-labeled antibody that specifically binds the antigen is added. A fluidics-based dual laser detector captures the fluorescent signals, identifies the assay carried on each bead and then quantifies the reaction using computer algorithms. Mouse endocrine multiplex assays measuring insulin, amylin, leptin, glucagon and glucagon-like peptide-1 are commercially available (see www.Linco.com).

It should be noted that for hormonal measurement, blood must be collected at a fixed time of the day so as to avoid diurnal variations (e. g. corticosterone secretion is higher at 17.00 hours than at 08.00 hours). The length of the fasting period can also affect the measurement of hormone levels, for example insulin levels increase after a meal when the glucose levels are rising. As many hormones are secreted in a pulse-like manner, for example GH and LH, it is sometimes necessary to check the hormone level after repeated sampling. All these factors must be taken into account when measuring hormone levels.

In addition to static hormone measurement, dynamic endocrinology tests can also be used. In general these tests involve administering a hormone (insulin) or metabolite (glucose) and following the metabolic response over time. For example, following a bolus injection of LHRH in the LH-releasing hormone (LHRH) test, blood is sampled at 0, 20 and 60 min for the measurement of LH and follicle stimulating hormone. Several such tests that are used to investigate glucose homeostasis, including the hyperinsulinemic euglycemic clamp, OGTT and IPGTT, have already been described. Another useful dynamic test is the dexamethasone suppression test used in humans to diagnose Cushing's syndrome, in which a single dose of dexamethasone is administered to suppress the normal release of ACTH, after which plasma glucocorticoid levels are measured over time.

5.7 Reproduction and Fertility

The reproductive traits of male and female mice are evaluated to identify defects linked to gametogenesis, placentation, the mammary glands, and nursing. Analysis of these reproductive traits provides information regarding reproductive endocrine function and should be complimented with hormonal measurements. This is carried out according to standardized procedures; three mutant mice of each gender (8 weeks of age) are mated with one wild-type C57BL/6J animal. The number of pups born from the litters is recorded at birth and 3 weeks later at weaning. Males and females are separated after they have produced two litters or after 6 months, whichever occurs first. In cases where the fertility of homozygous animals cannot be determined the reproductive abilities of heterozygotes should be analyzed.

Parental behavior is assessed using the pup retrieval test. The dam is sequestered on the nest using a cardboard barrier while the pups are scattered randomly around the home cage. Isolated mouse pups emit ultrasonic vocalizations which alert the parent to the fact that the pup is out of the nest, this in turn triggers retrieval behavior in the parent. The length of time before the first pup is retrieved, the total number of pups retrieved during the test period and the time taken to retrieve all the pups is recorded. The test is carried out on the third day after birth and the dams are given up to 10 min to return the pups to the nest.

5.8 Bile Acids

Bile consists of bile salts, cholesterol, phosphatidylcholine and bilirubin and is secreted from the hepatocytes into the bile canaliculi [39]. Bile salts are molecules that are synthesized from cholesterol. They are essential as the generators of bile flow in the hepatocyte and as physiological detergents that facilitate the absorption of dietary lipids and fat-soluble vitamins. In addition, they are crucial for the excretion of hepatic cholesterol into the bile. Therefore bile salt excretion and biosynthesis constitute the major pathways for cholesterol catabolism. In addition, bile formation facilitates the secretion of waste products such as bilirubin which is derived from heme metabolism.

Bile salts constitute a group of structurally variable molecules. When secreted from the hepatocytes primary bile salts, such as cholic acid (CA) and chenodeoxycholic acid (CDCA), are conjugated with either glycine or taurine. Mice preferentially synthesize tauroconjugates, whereas in humans glycoconjugates predominate. In the intestine bile salts are subjected to deconjugation and 7α-dehydroxylation by microbial enzymes, resulting in non-conjugated and secondary bile salts. The secondary bile salts of CA and CDCA are deoxycholic acid (DCA) and lithocholic acid (LCA) respectively.

Secreted bile is usually diverted to and stored in the gall bladder, however, when a meal is ingested, bile flows into the duodenum and intestine. The bile salts are efficiently (95 %) absorbed again by passive diffusion and active transport from the terminal ileum, and transported back to the liver via the portal vein, which completes their enterohepatic circulation. Each bile salt molecule may complete 4–12 cycles per day between the liver and intestine. Because of this efficient recirculation, only a small proportion of the bile salt pool is derived from *de novo* biosynthesis.

The balance between intestinal absorption and hepatic uptake of bile salts determines the serum bile salt concentration. Normally hepatic extraction of bile salts is constant and efficient. Because of the high levels of bile salt in the intestine during digestion, the postprandial serum bile salt levels are generally higher than the fasting levels. Fasting serum bile salt levels can be used in the diagnosis of liver disease in conjunction with standard liver function tests. Serum bile salt levels are a more sensitive tool for the diagnosis of minor hepatic derangements, since they reflect the functional activity of the hepatocyte and not the acute leakage of enzymes. Fasting bile salt levels are probably more reproducible and are therefore a better determinant of liver dysfunction. Total bile salts are measured in serum or plasma by a specific enzyme measurement or a radioimmunoassay. These methods do not discriminate between the different bile salt molecules. However, identification of individual bile salts can be achieved with a more detailed analysis using mass spectrometry [40].

The size of the bile salt pool, the kinetics of bile salt turnover and the excretion of neutral and acidic sterols in the faeces can all be determined using other methods. The bile salt pool is by definition the total amount of bile salts present within the enterohepatic circulation. This can be determined by dissecting the liver, gall bladder and the entire small intestine. After homogenization of these tissues, bile salts can be extracted and quantified by any method of choice: a specific enzyme measurement or radioimmunoassay for total bile salt levels or mass spectrometric analysis for the identification of individual bile salt molecules [39, 41]. In addition to this the kinetics of individual bile salts such as cholic acid in plasma, can be determined using a microscale stable isotope dilution technique. Using this technique, pool size, fractional turnover rate (a measure for the half-life of a bile salt) and the synthesis rate of the bile salt of interest can be determined. By extracting neutral and acidic sterols from faeces the daily bile salt loss (acidic sterols) and cholesterol excretion (neutral sterols) can be estimated.

5.9 Post-Mortem Analysis and Histology

Systematic necropsy examination and careful histological analyses are invaluable tools for the post-mortem diagnosis of endocrine and metabolic problems in the mouse. Thus, whenever possible individual tissue analysis should be combined with *in vivo* metabolic phenotyping. Tissue analysis can be as simple as measuring its weight, to taking note of its color and morphometric features. In addition to histological and immunological analyses of the tissues, more complex tissue analyses may include the assessment of the tissue lipid content, RNA and protein content, DNA synthesis, and intracellular organelle function (size of the mitochondria for example).

The age at which morphological analyses should be undertaken is dependent on when the phenotype is expressed. If prenatal death of animals is suspected to be due to a disproportionately low number of homozygous mutant animals then identifying the time of fetal death requires euthanasia of pregnant dams at successive stages of pregnancy. When metabolic abnormalities present after birth or are progressive, it is best to screen animals in the early, mid and late stages of the condition. Comprehensive protocols for performing gross necropsy and fixation have been previously described in detail and will only be summarized here (http://www.eulep.org/NecropsyoftheMouse and http://www.eumorphia.org /servlet/ECFLP.Frameset) [42].

Following sacrifice the mouse is weighed, placed in dorsal recumbence and a standard necropsy is performed (Fig. 5.5 A). When metabolic disorders are suspected the following organs should be inspected and removed: liver, intestine (ileum, colon and duodenum), white adipose tissue (inguinal and epididymal), brown adipose tissue, muscle (quadriceps), pancreas, and spleen. All tissues should be weighed and organ weights expressed as a percentage of total body weight. Grossly evident pathology should be recorded and photographed. Tissue should be divided into sections for histological and molecular analysis. For molecular analysis tissue should be snap-frozen in liquid nitrogen immediately and stored at – 80 °C. For histological analysis, fresh tissue should be placed in plastic histology cassettes immediately and immersed in an appropriate fixative to avoid tissue autolysis. After fixation the tissues should be transferred to 70% ethanol and stored at 4 °C until processed. There are several choices of fixative, with neutral buffered 10% formalin being the most common as it is suitable for paraffin sections and staining with hematoxylin and eosin (H&E). Formalin (10%) is a convenient fixative as specimens can be fixed overnight and left indefinitely which makes it possible to evaluate lipid levels retrospectively, as substances that are normally soluble in ethanol would be lost during tissue processing. Paraformaldehyde (4%) which can be prepared in a

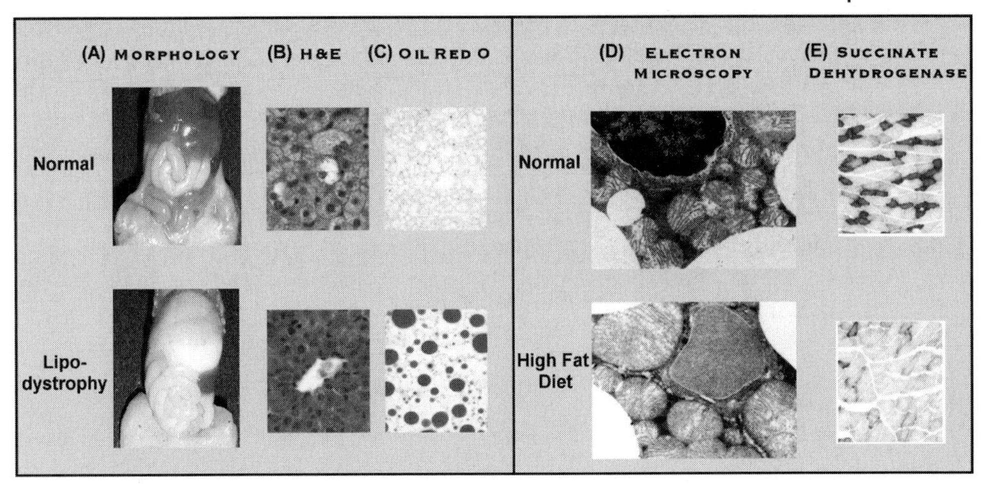

Fig. 5.5 Examples of post-mortem analysis and histology. A systematic necropsy examination and careful histological analyses are invaluable tools for the post-mortem diagnosis of endocrine and metabolic problems in the mouse. (A) Gross morphology can indicate tissue abnormalities including accumulation of fat in the liver [47]. (B) Hematoxylin and eosin (H&E) stain the proteins and DNA respectively, and highlight the cellular morphology. (C) Oil red O stains neutral lipid droplets a brilliant red and shows the accumulation of cholesteryl ester and triglycerides [47]. (D) Electron microscopy of tissue sections reveals mitochondrial size and cristae composition. Succinate dehydrogenase is a mitochondrial-specific enzyme and the product of the reaction which it catalyzes stains the mitochondria blue. In mice fed a high-fat diet, mitochondrial function is compromised (unpublished observations).

buffered solution at pH 7.0 and refrigerated until required, is mostly used as a fixative for electron microscopy and *in situ* hybridization and is slightly less harsh than formalin. Bouin's solution uses picric acid (permanently stains all tissues yellow), acetic acid and formalin, and although preferred by some researchers, the finer details of the sample are often lost with this staining method. An alternative to paraffin fixation is freezing the tissue in a clear jelly called O.C.T. compound (Tissue-Tek, Sakura Finetek, USA, Torrance, CA, USA). This is a thick, clear fluid used in conjunction with plastic base molds to bind and freeze fresh tissues which are floated in the mold and placed on dry ice. Frozen sections are ideal for immunostaining although it is difficult to cut thin frozen sections and from the point of view of the morphology of the sample, paraffin sections are superior.

The most conventional stain for formalin-fixed paraffin section is H&E [42]. Eosin stains the proteins pink and the hematoxylin stains negatively charged nucleic acids blue and gives a general picture of the tissue morphology (Fig. 5.5 B). For adipose tissue it is important to note the size of the adipocyte. In some instances no difference in tissue weight may be observed, however adipose tissue composed of more small cells compared to fewer big cells is a potentially pathologic finding. H&E also stains the pancreatic islets, which are responsible for glucose-stimulated insulin secretion. Oil red O, which stains lipids including TG and cholesteryl ester, is a commonly employed stain used in the assessment of neutral tissue lipid content (Fig. 5.5 C). Alternatively lipids can be extracted from frozen tissue sections with methanol–chloroform and analyzed enzymatically for cholesterol, TG and CE content as described for plasma lipid analysis. In addition to using electron microscopy for the assessment of mitochondrial activity (Fig. 5.5 D), mitochondrial density can be visualized by incubation with nitroblue tetrazolium (Fig. 5.5 E). This staining procedure is based on the visualization of diformazan, the product of the catabolism of nitroblue tetrazolium by the mitochondrial enzyme succinate dehydrogenase.

5.10 Molecular Imaging

Real-time imaging will be the next technique to provide information crucial to the understanding of the biological processes involved in endocrinology and metabolism. New imaging strategies have already been developed which provide fine temporal and spatial resolution enabling cellular and molecular changes to be detected in real time, *in vivo* [43]. Perhaps the most relevant for endocrine/metabolic investigation is the non-invasive imaging of the activity of reporter genes, usually an enzyme (luciferase or thymidine kinase), whose expression is placed under the control of a response element for a transcription factor (such as that of a nuclear receptor). The activity of the reporter is recorded with an optical camera (for luciferase reporters) or by positon emission tomography (for thymidine kinase reporter systems), which reflects the activity of the hormonal signaling pathway that activates the transcription factor in question. Such imaging strategies permit non-invasive visualization of the activity of hormones and endocrine signaling pathways. Although these strategies are still on the starting blocks, they will undoubtedly be in widespread use in the near future.

Other non-invasive imaging techniques with potential application to endocrinology include classical X-ray, single photon emission computer tomography (SPECT), magnetic resonance imaging (MRI) and ultrasound. Several of these techniques enable the study of cellular and molecular events. An in-depth review of the application of these techniques and their uses in small animals is provided elsewhere [43, 44].

Acknowledgements

We thank B. Thorens, L. El Fertak, S. Houten, M. Selloun, O. Wendling and members of the ICS and Auwerx laboratory for their helpful discussions and valuable input.

References

1 S. M. Grundy *J. Clin. Endocrinol. Metab.* **2004**, 89, 2595–2600.
2. D. C. Rogers, et al. *Mamm. Genome.* **1997**, 8, 711–713.
3. M. Champy, et al. *Mamm. Genome.* **2004**, 15, 768–783.
4. S. C. Collins, et al. *Mamm. Genome.* **2003**, 14, 350–356.
5. J. Lusis, et al. *The Genetic Basis of Common Diseases,* Oxford University Press, New York, **2002.**
6 Y. Zhang, et al. *Nature.* **1994**, 372, 425–432.
7 H. Chen, et al. *Cell.* **1996**, 84, 491–495.
8 W. C. Morgan *J. Natl. Cancer Inst.* **1950**, 11, 263–268.
9 H. Iwatsuka, et al. *Endocrinol. Jpn.* **1970**, 17, 23–35.
10 K. E. Borg, et al. *Proc. Soc. Exp. Biol. Med.* **1995**, 210, 126–133.
11 B. M. Spiegelman, et al. *Cell* **2001**, 104, 531–543.
12 R. K. Porter *Cell Mol. Life Sci.* **2001**, 58, 815–822.
13 F. Picard, et al. *Cell* **2002**, 111, 931–941.
14 B. Graham, et al. nAQ2 n*Methods* **2002**, 26, 364–370.
15 D. Bernstein *Physiol. Genomics* **2003**, 13, 217–226.
16 J. T. Lightfoot, et al. *Physiol. Genomics* **2004**, 19, 270–276.
17 T. Nagy *Obes. Res.* **2000**, 8, 392–398.
18 S. Grier, et al. *Invest. Radiol.* **1996**, 31, 50–62.
19 A. R. Saltiel, *Cell* **2001**, 104, 517–529.
20 T. M. Wallace, et al. *Diabet. Med.* **2002**, 19, 527–534.
21 T. M. Wallace, et al. *Diabet. Care* **2004**, 27, 1487–1495.
22 S. Rocchi, et al. *Mol. Cell.* **2001**, 8, 737–747.
23 K. Kaku, et al. *Diabetes* **1988**, 37, 707–713.
24 G. Pacini, et al. *Best Pract. Res. Clin. Endocrinol. Metab.* **2003**, 17, 305–322.
25 J. K. Kim, et al. *J. Clin. Invest.* **2001**, 108, 153–160.
26 R. Burcelin, et al. *Diabetes* **2001**, 50, 1282–1289.
27 D. Elahi *Diabet. Care.* **1996**, 19, 278–286.
28 R. A. DeFronzo, et al. *Am. J. Physiol.* **1979**, 237, E214–E223
29 D. R. Matthews, et al. *Diabet. Care.* **1989**, 12, 156–159.
30 G. L. Mills, et al. *Laboratory Techniques in Biochemistry and Molecular Biology,* Elsevier, Amsterdam, **1994**.
31 J. Shepherd, et al. *Am. J. Cardiol.* **1991**, 68, A5–A7.
32 J. D. Brunzell, et al. *Am. J. Med.* **2003**, 115 (Suppl. 8A), 24S–28S.
33 A. M. Gotto, et al. *Methods Enzymol.* **1986**, 128, 3–41.
34 P. H.R. Green, et al. *J. Lipid Res.* **1981**, 22, 1153–1173.
35 P. Libby *Am. J. Cardiol.* **2001**, 88, 3N–8N.
36 O. De Lalla, et al. *Methods in Biochemical Analysis,* Interscience, New York. **1954**.
37 P. Alaupovic *Prog. Lipid Res.* **1991**, 30, 105–138.
38 E. J. Schaefer, et al. *J. Lipid Res.* **1978**, 19, 667–687.
39 S. M. Houten, et al. *Ann. Med.* **2004**, 36, 482–491.
40 K. D. Setchell, et al. *J. Lipid Res* **1983**, 24, 1085–1100.
41 W. J. Griffiths, et al. *Mass Spectrom. Rev.* **2003**, 22, 81–152.
42 P. Wheater, et al. *Functional Histology,* Churchill Livingston, London, **1987**.
43 T. F. Massoud, et al. *Genes Dev.* **2003**, 17, 545–580.
44 C. H. Contag, et al. *J. Magn. Reson. Imaging* **2002**, 16, 378–387.
45 C. Lee, et al. *Endocrinology* **2003**, 144, 2201–2207.
46 M. Watanabe, et al. *J. Clin. Invest.* **2004**, 113, 1408–1418.
47 H. Koutnikova, et al. *Proc. Natl. Acad. Sci. USA* **2003**, 100, 14457–14462.

6
Behavioral and Neurological Phenotyping in the Mouse

Valter Tucci, Gonzalo Blanco and Patrick M. Nolan

6.1
Introduction

The elucidation of human and mouse genomes brings with it the promise of a comprehensive understanding of biological and disease mechanisms. The post-genomic era has been characterized by the development of a number of new approaches in the mouse to further this understanding. The study of gene function or functional genomics can be realized through the incorporation of systematic, hierarchical and high-throughput analyses. *In silico* and *in vitro* approaches are two such means by which gene function can be studied although, ultimately, the characterization of gene function in the living organism is essential. In the mouse two main approaches have been adopted to realize these goals: the gene-driven approach, where specific gene sequences are targeted and the phenotype of mice carrying the targeted mutation is systematically studied, and the phenotype-driven approach, where the phenotype of mice carrying randomly-generated mutations is studied and the mutant genes subsequently identified. Both approaches have been used to investigate the role of genes in human disease. In this chapter, we will introduce methods by which behavioral, neurological and neuromuscular traits in the mouse can be studied and comment on the application of these methods to the study of human neurological and psychiatric disease. Because of the complex nature of neurological and psychiatric disorders in humans, it is often difficult to identify appropriate models in mice. However upon careful consideration of phenotypic screening batteries and methods of data analysis we can begin to associate behavioral traits in mice with behavioral pathologies in human disease states.

The mouse is now a widely used laboratory species providing insights into the genetics of behavior and human disorders where behavior is affected. The use of laboratory species as models for behavioral and neurological disorders has several distinct rationales. Investigators working with humans are generally restricted by factors that cannot be controlled by the experimental design, such as missing observations and small family size; these problems lead to inferior data analysis compared to that achieved with experimental animals. Moreover gene hunting usually involves the collection of family pedigrees. This becomes impossible, for example,

Standards of Mouse Model Phenotyping. Edited by Martin Hrabé de Angelis, Pierre Chambon, and Steve Brown

ISBN: 3-527-31031-2

in humans affected by diseases that have a late onset. Importantly with animals it is possible to conduct a large number of experiments that would otherwise be unacceptable in human beings; for example, new therapies are regularly tested in animal models before being used in human populations.

Neurobehavioral genetics is the science responsible for revealing the intricate mechanisms leading to the expression of complex behavioral traits. Put in simple terms, this is a measure of how the nervous and neuromuscular systems perceive, act, learn, and remember processes that have been acquired from the environment through the sensory system. The key to neurological and behavioral phenotyping in the mouse has been the development of assays that measure these processes, or components thereof, the integration of data from multiple assays through data analysis and the correlation of behavioral phenotypes in mice with intermediate signatures in human neurological and psychiatric disorders (so-called endophenotypes; Fig. 6.1). In order to illustrate this, we have organized the chapter in the following manner: (1) an introduction to some of the achievements in furthering our understanding of specific neurological, neuromuscular and psychiatric disorders including the contribution of mouse genetics, (2) a consideration of methods of neurological/behavioral analysis and its application to mouse genetics and to the study of human disorders, (3) a description of some of the standard procedures used in neurological and behavioral phenotype analysis, (4) some practical approaches and considerations and (5) the outlook for future phenotypic analysis in neurological and behavioral fields. The field of neurobehavioral genetics is large and diverse, so this chapter cannot provide a complete and comprehensive coverage of all aspects of phenotyping. Instead, a range of topics has been selected, not only to give an informative and interesting account of mouse behavior but also to provide the background necessary to link mouse models to human diseases. One important feature we would like to emphasize is that behavioral analysis is not simply a matter of running a maze test. No test alone is informative, it is only after careful consideration of a subject's performance in multiple tests that a particular hypothesis can be developed and examined further.

6.2
Human Neurological and Psychiatric Disorders

Human neurological and psychiatric disorders are complex disorders that, more often than not, are heterogeneous, involve multiple genetic loci and are subject to any number of environmental influences. The aim of behavioral analysis in mouse mutants is not necessarily to model neurological and psychiatric disorders, this would be almost impossible as the complex multigenic components of the vast majority of these disorders could not be replicated in the mouse. Nevertheless, our hope is that the analysis of mouse mutants representing endophenotypes of these neurological and psychiatric disorders will help us to gain additional insight into human disease states. Many of these diseases/disorders exhibit overlapping trait anomalies which allow them to be grouped together as outlined below.

6.2.1 Neurodegenerative Disorders

These disorders are characterized by the fact that they include a pathology incorporating age-related degeneration of neurons. Members of this group include Alzheimer's disease, Parkinson's disease, Huntington's disease, frontotemporal dementias and spinocerebellar ataxias. Although similar mechanisms of degeneration are evident in many of these disorders, they are distinguishable by the fact that different regions of the central nervous system are targeted. Because of this, pathologies may be associated with a number of behavioral and neurological phenotypes including deficits in motor, cognitive and circadian function. Many of the major causative genes have been identified and characterized in mouse models with varying degrees of success, although an equal number of genetic and environmental factors remain to be identified. We will consider specific investigations in Alzheimer's disease (AD) and Parkinson's disease (PD) as examples.

In many cases human disease is genetically heterogeneous, which confounds attempts at gene hunting. This could be explained by the fact that the same disease pathology may arise from mutations in different genes (that commonly act within the same biochemical pathway). This is the case with AD, which can be caused by single mutations in any number of genes including presenilin 1 (PS1), presenilin 2 (PS2), and the amyloid plaque precursor protein (APP). The discovery of mutations in APP, PS1 and PS2 has led to the amyloid cascade hypothesis, which states that increased production of amyloid Aβ, or the more insoluble Aβ 42 species leads to amyloid deposition, neuronal dysfunction and clinical disease. The precise relationship and intermediary steps between amyloid deposition and clinical disease remains to be determined. However, there are a number of concurrent biochemical processes which may be of importance in the pathogenesis of AD, including the aggregation of tau into filaments and neurofibrillary tangles, the formation of oxygen free radicals in oxidative stress, mitochondrial dysfunction and the inflammatory response. Although there is a range of clinical and pathological evidence supporting the role of these pathways in AD, none has been implicated directly by the finding of mutations in affected pedigrees. Research into AD has included the development of transgenic murine models that can act as a test bed for the analysis of disease progression and the development of novel therapies. For example, in 1991, Quon and colleagues first described a transgenic APP mouse, using the human APP751 isoform transgene [1]. These authors concluded that the neuronal overexpression of human APP751 isoform leads to amyloid pathology. The behavioral phenotype of this mouse was initially reported in [2]. The authors reported a cognitive age-related deterioration in these mice; in particular they found deficits in spatial learning and spontaneous alternation (using the Y-maze). Recently this mouse line has been extensively characterized [3]. In a comparison of two lines of these APP mice, the authors reported a more severe cognitive impairment in one particular line. They concluded that the cognitive impairments may be independent of plaque formation as the two lines present similar levels of transgene expression. For many years there has been widespread debate as to whether amyloid plaques or neurofibrillary tangles represent the key issue in the disease [4]. Mouse models are being used to clarify this issue and support the hypothesis that APP mis-metabolism may be the trigger [5].

PD is another of the most common neurodegenerative diseases with a prevalence of 1 in 350, and a lifetime risk of 1 in 40. It is a slowly progressive disorder mainly characterized by degeneration of dopaminergic neurons in the substantia nigra and the ventral tegmental area leading to a multitude of motor and non-motor behavioral disturbances. Non-motor disturbances include visuospatial and attentional anomalies and cognitive abnormalities that are often associated with psychotic symptoms. Most cases are sporadic with unknown etiology, but about 1% of cases are familial. Polymeropoulos and co-workers [6] found a mis-sense mutation in the α-synuclein gene in a large kindred with autosomal dominant PD. A year later another mutation in α-synuclein was found in a German family [7]. Both these mutations were associated with more than 90% penetrance of the disease which implicated them in the etiology of PD in these families. α-Synuclein was then found to be the main component of the Lewy bodies [8]. These intracytoplasmic inclusions are the pathological hallmarks of PD, and this finding provided a direct link between α-synuclein and the etiology of sporadic PD. Since then, genetic variability in the α-synuclein gene has been found to be a risk factor for sporadic PD [9] and studies on α-synuclein in cell and mouse models have furthered our understanding of the pathogenesis. Thus, although inherited mutations in α-synuclein as a cause of PD are very rare, an understanding of the mechanisms by which these mutations cause PD in these families has provided important insights into the common sporadic form of the disease. Further genetic factors have been identified in the etiology of PD. To date nine genetic loci have been reported for PD and mutations in three additional genes, parkin, UCHL1 and DJ-1, have been identified in cases of genetically-inherited PD [10]. The identification of these genes has provided information about the molecular pathogenesis of both Mendelian PD and the more commonly occurring sporadic PD. No suitable mouse models of PD have been developed to date although these may arrive in the near future.

6.2.2 Mental Retardation Syndromes

Approximately 2–3% of the population are affected by syndromic disorders with associated mental retardation (MR). MR is more prevalent in males implicating genetic factors on the X-chromosome in a large number of cases [11]. These syndromic disorders are associated with a broad spectrum of behavioral and neurological deficits including those affecting cognition, attention and sensorimotor function. Currently, genetically-modified mice are the most common approach to the investigation of the role of genetic factors as the pathological basis of MR. Moreover, studies investigating genetically-modified mice with MR are also increasing our knowledge about the role that selected genes play in regulating aspects of normal cognitive and behavioral functions. Trisomy 21, or Down syndrome (DS), is a genetic MR that occurs in one of every 700 births [12]. The first mouse model for DS, a chromosomal translocation, was developed in 1975 [13]. Subsequently, many mouse models for DS which display phenotypic elements of the human disease have been generated [14].

6.2.3 Disorders Affecting Social Behavior

Society is responsible for controlling social behavior among children, adolescents and adults, in order to limit and treat a number of disorders such as anorexia, depression and autism. Biological and environmental factors both contribute to a social or anti-social behavior. Research provides support that biological and environmentally based anti-social behaviors interact with each other and may be overcome through social and educational programs as well as medical treatments. The tenth edition of the *International Classification of Diseases* [15] describes a group of disorders, called pervasive developmental disorders (PDD), which are characterized by qualitative abnormalities in reciprocal social interactions and in patterns of communication, and by a restricted, stereotyped, repetitive repertoire of interests and activities. Two examples of social behavioral disorders are considered below.

6.2.3.1 Anorexia

In modern society environmental and genetic factors contribute to the development of new disorders, such as eating-related disorders. An increasing number of adolescents are affected by anorexia, a debilitating disease that affects mostly girls but also a number of boys. The classical symptoms of anorexia are a lowered body weight, an obsessive concern about food, fear of gaining weight, depression, amenorrhea, bradycardia, abnormal blood pressure and abnormal temperature [16]. An increasing amount of information regarding the genetics of regulation of food intake has become available [17]. While a number of mouse models have been identified for obesity, only a few have been developed for the study of anorectic disease. A spontaneous mouse mutation, the autosomal recessive *anx* mutation, has been studied as a potential model of anorexia [18]. These mice show poor appetite, reduced body weight, body tremors, hyperactivity and uncoordinated gait. The phenotype in *anx/anx* mice appears at 5–8 days after birth and they die at 3–5 weeks. In addition, levels of leptin are reduced at postnatal day 8 [19]. Although a few more mouse models for anorexia have been developed, no model has so far proven to be ideal. Factors that confound the search for mouse models include the importance of environmental factors in the onset of this disease and the evolutionary divergence of social behavior in distinct mammalian species (i. e. mouse and man).

6.2.3.2 Autism

Autism is a developmental disorder characterized by impairment in social interaction or communication, and restricted or repetitive behavioral patterns [20]. Strong evidence for the heritability of autism comes from twin studies. A common idea among scientists in the field is that it is unlikely that animal behaviors will fully parallel the human disorder because of its complex nature and for the reasons described above. In general, studies concerning the genetic factors of autism show that the autistic phenotype results from an unfavorable combination of alleles of several genes that interplay with environmental factors. Such a multifactorial

Table 6.1 Autistic traits in humans and assays used to measure equivalent traits in mice.

Human autistic traits [21]	Mouse phenotypic traits	Phenotypic assay
Face processing	Social recognition	Social Recognition test [119]
Sensitivity to social reward Motor imitation	Social transfer of food preference	Social transfer of food preference test [163]
Memory (prefrontal-related)	Learning and memory	Timing behavior [164]
Planning and flexibility	Reaching and grasping	MoRaG (www.eumorphia.org)
Language	Vocalization	Vocalization analysis [165]

model of the disease may be the starting point from which to identify combinations of endophenotypes in mouse models. Following an interdisciplinary approach, Dawson and colleagues [21] proposed six broader phenotype autism traits in humans. Standardized behavioral assays that measure the preference of mice for initiating social interactions with novel conspecifics, has been suggested as a fruitful approach to the identification of mutant mouse models of autism [22]. A systematic investigation into the suitability of mouse models of autism could be conducted according to Tab. 6.1.

6.2.4
Mood Disorders: Depression, Manias and Schizophrenia

Depression and schizophrenia are two major psychiatric disorders causing suffering and disability in the patients affected. Three major factors contribute to depressive syndrome: genetic factors, prenatal and/or postnatal environmental factors, and life events (i. e. stress). Risk factors for schizophrenia appear, in part, to be genetically based, although external factors may contribute markedly to the emergence of the disorder. Evidence for the possible linkage of schizophrenia to many chromosomal regions, including 6 p24, 8 p, 13 q32, 18 p. 2, has been shown (for an overview see [23]). Recent reports implicate many genes, such as neuregulin1, dysbindin and RGS4, with the pathophysiology of schizophrenia in certain patients (for a review see [24]). Studies of neurodevelopmental mechanisms suggest that a combination of genetic susceptibility and environmental perturbations appears to be necessary for the expression of some psychotic disorders. Similarly, altered synthesis and/or release of neurotrophic factors during specific developmental periods may affect the pathological manifestation of these disorders (see [25]). Nerve growth factor (NGF) and brain-derived neurotrophic factor (BDNF) are proteins involved in neuronal survival and plasticity of dopaminergic, cholinergic and serotonergic neurons in the central nervous system (CNS). Changes in the modulation of these neurons in specific brain regions have been found in depression and schizophrenia.

Animal models, in particular rodents, have been used to characterize the role of neurotrophins in depression and schizophrenia under baseline conditions and following antidepressive and antipsychotic treatments. For example, administration of haloperidol, an antipsychotic, sedation-inducing drug, can result in lowered NGF levels in the bloodstream and hypothalamus of adult mice [26] and circulating NGF levels in schizophrenic patients [27]. Animal models of depression (Flinders sensitive line/Flinders resistant line; FSL/FRL rats) have been useful in the investigation of NGF and BDNF function in depressive syndrome. In general FSL animals show higher levels of both BDNF and NGF in the frontal cortex and occipital cortex, whereas in the hypothalamus only the levels of BDNF are affected [25]. However these results are not in accord with other reports showing that levels of neurotrophins are decreased in depression [28], although they are consistent with the observation that antidepressant drugs increase mRNA (BDNF) levels in the brain [29] and that BDNF itself has an antidepressant effect [30].

6.2.5
Anxiety

Human anxiety is generally considered to be a response of the subject to their environment. Anxiety is either a transitory response elicited in individuals when presented with a potentially threatening situation or an inherent trait of the individual. As many as 25% of the population may potentially, suffer from one of the many anxiety-related disorders including phobias, panic disorder and post-traumatic stress disorder. Various physiological and behavioral changes are associated with this phenomenon and pharmacological intervention has implicated several brain neurotransmitter systems in the expression of anxiety. Numerous studies have provided evidence for heritable components to anxiety disorders although evidence for linkage to single specific genetic loci has not been forthcoming (for a review see [31]).

Autonomic, neuroendocrine and behavioral abnormalities have been described in knockout mice and specific parametric differences have been observed in inbred strains. Anxiety in mice is commonly defined by increased avoidance of a novel environment (e. g. the center zone in an open field test) and/or increased fear reaction. Both these measurements rely on the motor activity of the mouse. Thus, in some cases, abnormal responses in animal models of anxiety may represent an activity rather than an emotional problem. Some of these issues may account for the difficulties in assessing the genetic basis of anxiety in mice. The use of more sophisticated and time-consuming tests, for example fear-potentiated startle, may help in eliminating false positive results. Studies looking for genetic variants of anxiety have, so far, revealed a complicated story. Genetic effects are smaller and more complex than expected. The elevated-plus maze, light–dark box and open field test have been used to map QTLs on chromosomes 1, 3, 4, 5, 6, 7, 10, 11, 12, 14, 15, 18, 19 and on the X chromosome [32–35]. These studies clearly indicate that anxiety-like behavior in mice is a multigenic trait with many loci of small effect. However, the use of functional genomics approaches to investigate the genetic variants underlying these small effect QTLs promises to provide novel functional insight. For example, the application of this strategy has recently demonstrated the regulator of G-protein signaling, *Rgs2*, to be a novel factor associated with anxiety in mice [36]. Similarly, a multitude of targeted mouse mutants show abnormalities in anxiety-like behaviors. Many of these specifically affect GABAergic and serotonergic neurotransmitter systems although targeted mutants affecting diverse processes such as cytoskeletal and extracellular matrix function show similar disturbances. Some genetic components have been the subject of more detailed study than others. The behavioral phenotype seen in 5-HT1A receptor knockout mice [37–39] suggests that it may be a suitable model for anxiety. These data are quite impressive considering that they were generated in three different laboratories, on three different genetic backgrounds and, presumably, under different experimental conditions. Moreover, anxiety was apparent not only in homozygote but also in heterozygote 5-HT1A receptor knockout mice. From a psychiatric point of view this may lead to the possibility that a partial receptor deficit represents a risk factor in affective disorders. Moreover, the 5-HT1A receptor knockout has been characterized for many additional anxiety-related symptoms, such as autonomic activation, increased stress responsiveness, and neuroendocrine abnormalities [40].

6.2.6 Neuromuscular Disorders, Myopathies and Neuropathies

Although the search for the genetic cause of inherited muscular dystrophies has been very successful, genetic causes for rare muscle diseases such as distal myopathies (MPDs) [41], hyaline body myopathies [42] and some uncharacterized forms of limb girdle dystrophies (LGMD) remain largely unknown. Nonetheless, a systematic method for identifying neuromuscular mouse mutants will ultimately have the greatest impact at the functional level, i. e. the elucidation of the cellular role of a number of proteins with a well-established causative role in human muscle disorders but whose molecular pathogenesis remains elusive. The mouse model has been shown to be crucial to the study of human disorders and this has been extensively exploited in the neuromuscular field. The availability of animal models for muscular dystrophies has not ceased to increase since the identification of dystrophin as the defective gene in Duchenne muscular dystrophy and the subsequent identification of the dystrophin–glycoprotein protein complex (DGC) [43–45]. Specific human disorders have been recreated by gene targeting of almost every component of the DGC and other sarcolemma-associated proteins [46]. There has been a synergistic effect in the search for the causes of inherited muscular dystrophies in humans and in further studies of mice which have led to the elucidation of complex molecular networks of disease-causing proteins. This network includes proteins at the sarcomere (titin, myotilin, telethonin), cytoskeleton (F-actin, filamin2, dystrophin), membrane associated or integral (dysferlin, caveolin3, integrinα7, sarcoglycans, dystroglycan), basal lamina (s-laminin) and extracellular matrix (collagen VI). These proteins share an essential role in preserving muscle integrity through the stresses of contraction and relaxation and any missing link in the network that compromises the physical stability of the sarcolemma will lead to a dystrophic process. Undoubtedly, these advances have led to molecular genetics playing a pivotal role in the diagnostic process of muscular dystrophies, providing a solid basis for early and accurate diagnosis [47]. In addition, various mouse models of muscular dystrophies have been exploited to explore the feasibility of alternative therapeutic targets. For example, overexpression of a mini gene of agrin, a protein involved in the formation of the neuromuscular junction, has been shown to significantly ameliorate the progression of the disease in a mouse model of congenital muscular dystrophy deficient in s-laminin [48]. Similarly, the ability of the insulin-like growth factor-1 (IGF-1) to counter the progression of muscle disease has been tested in mouse models. IGF-1 has been injected into muscles of laminin-deficient 129 ReJ *dy/dy* mice [49] and also upregulated in dystrophin-deficient *mdx* mice by means of genetic crosses with transgenic lines expressing muscle IGF-1 [50]. In both cases, significant amelioration of the muscle wasting or dystrophic process, respectively, was reported. Similarly, full-length Calpain 3 and alternative forms have been tested in the mouse as a potential target for LGMD2A therapy with the surprising result that these alternate forms of Calpain 3 presented impaired muscle development [51]. The use of forms of Calpain3 lacking exons 6 or 15 was clearly disadvantageous, again demonstrating the utility of the mouse in testing potential therapeutic approaches. Mouse models have also played an important role in the study of peripheral nerve disorders. For example, the recreation in the mouse of proximal spinal muscular atrophy (SMA), the second most common autosomal recessive inherited disorder in humans, has provided the most conclusive evidence of the role of SMN exon 7 and the SMN2 gene in the disease process (for a review, see [52]). The proof of concept in this field is illustrated by the identification of cytoplasmic dynein heavy chain (DNCH1) as the protein underlying the ENU-induced *loa* mutant. The *loa* gene was identified by positional cloning and further functional analysis established that this protein has a role in axonal retrograde transport and is associated with degenerative motoneuron disease in the mouse [53]. Although no human condition has been found to be linked to the *Dnchc1* homolog, the *loa* mutant will continue to play a pivotal role in providing mechanistic insights into this devastating disease.

6.3 Behavioral and Neurological Phenotyping in the Mouse

Behavior is studied in a variety of scientific disciplines; of these the most important are ethology, psychology, physiology and genetics. Obviously each of them serves a specific research plan, a set of questions (about behavior) that they are seeking to answer. Traditionally, behavioral biologists did not concentrate on the mouse for behavioral studies, concluding that mouse behavioral traits were not as stable as those seen, for example, in the rat. Thus the majority of rodent behavioral tests that are carried out in modern research laboratories, and subsequently the principles that guide them, have been developed using the rat, or some other species. It is rare to find that a behavioral paradigm has been designed specifically with the mouse in mind. In redesigning or redeveloping behavioral tests for the mouse as the subject, it is important to consider first, the principles that have evolved through decades of behavioral testing in other species and second, once behavioral models have been selected, it is important to validate them in carefully-considered mouse studies.

Behavior is a fundamental property of living things, including plants. By definition behavior involves movement of the arms, legs, fingers, eyes, the response of the heart and its effects on the circulation etc. Movement, moreover, involves the body skeleton, the muscles that move the skeleton, and the nervous system that controls and coordinates the movements of these muscles. Yet movement also involves actions directed towards environmental stimuli. Such movements/actions require a certain level of attention to the stimulus, assessment of environmental conditions and, finally, a goal-directed response, all of which are coordinated by the central nervous system. Moreover, factors such as prior experience and social interaction, which can modify this response, involve high levels of reasoning and cognition. Failure of one or many of these processes can lead to the debilitating features of all neurological and psychological disorders. Most neurological disorders involve compromised movement, and lead to a number of symptoms which disturb the daily interaction with the surrounding environment. Most psychiatric diseases, on the other hand, affect reasoning and cognition and result in disabilities affecting normal responses to the environment.

6.3.1
Neurological and Neuromuscular Function

One of the major challenges in identifying new mouse models of neuromuscular defects is effective diagnostics. With the exception of extreme phenotypes showing rapid disease progression such as the classical muscular dystrophy mutant *dy* (laminin2 deficient [54]) or the dysferlin-deficient mouse SJL-Dysf [55], in general, it has only been possible to classify spontaneous mouse mutants and transgenic lines as suffering from primary neuromuscular disorders after extensive histological analysis, e. g. the kyphoscoliosis-*ky* mutant [56], transgenic lines expressing expanded forms of the human poly(A)-binding protein nuclear 1 [57], transgenic lines overexpressing alpha-tropomyosin(slow) [58], caveolin 3-null mice [59] and others [60]. Interestingly, a knockout of the dysferlin gene has also been reported to be subsymptomatic under normal conditions [61]. Assessment of locomotor function in mice has not been carried out as extensively as in other model organisms. A systematic and objective attempt to phenotype mice resulted in the development of the SHIRPA protocol [62]. However, visible features such as abnormal gait or even tremors are not necessarily linked to a peripheral neuromuscular defect and it is noticeable that very few neuromuscular mutants have been reported in genome-wide phenotype-driven ENU screens primarily based on non-invasive tests [63, 64]. This is partially due to the fact that neuromuscular problems can often be subsymptomatic under normal conditions and a more thorough analysis may be required to expose any obvious impairment. For example, transgenic mice with elevated levels of Cu/Zn superoxide dismutase (Tg-CuZnSOD), the gene underlying familial ALS in humans, walked normally and showed no obvious signs of paralytic muscle dysfunction. This is despite the reported abnormal EMG recording and defects in the neuromuscular junctions [65] and motor axons [66] of these mice. However, after broad assessments significant differences were found between Tg-CuZnSOD mice and controls in rope grip and footprint ink tests [67]. Therefore, as also shown in the *mdx* mutant [68], once the neuromuscular nature of the problem has been demonstrated, it is often possible to distinguish the mutant from wild-type controls provided that enough mice are included in the tests to allow for appropriate statistical analysis. However, such experimental design cannot be readily extrapolated to high-throughput first-line screens in which a large number of mice need to be assessed over a short period of time.

Classically, ascertaining the neuromuscular phenotype would require extensive histological and electromyography analysis. Gross morphological changes in the muscle such as fibrosis, mononuclear cell infiltration or centrally nucleated fibres, can be exposed using standard H&E staining while exposure of myelination problems in peripheral nerve or spinal cord abnormalities often requires more specific stains such as Luxol fast blue, Weil's myelin technique or the Kultschitsky's Method (http://www.ncl.ac.uk/nnp/industry/histology/myelin.htm). Electromyography has been extensively applied in the mouse to refine the neuromuscular phenotype [69], attempt to discern the nature of the disorder [70], as an indicator of disease progression and evaluation of therapies [71, 72] or to characterize nerve conduction changes in mice with myelin protein gene defects [73]. Both methodologies have been applied in downstream characterizations of known mutants rather than as primary screens mainly because extensive histopathology is usually a terminal method and electromyography is technically challenging.

6.3.2
Learning and Cognition

Cognition refers to the ways in which animals acquire information through the senses, process this information, retain it and decide to act upon it [74]. Generally in animal laboratory forms of learning, the experimenter attempts to make use of the innate behavioral properties of a particular species. As an example, spatial learning is essential among many species for the determination and recognition of territoriality [75]. In the laboratory, investigating spatial learning traditionally means putting rats or mice into mazes and testing cognitive mapping versus simpler mechanisms such as learning fixed routes or responses. One of the simplest forms of this test would be the spontaneous alternation test where animals navigate through the arms of a T-maze or Y-maze [76]. However, the most utilized of spatial learning paradigms in mice is probably the water maze [77] which is known to depend on the integrity of the hippocampus. Hippocampus-related learning is concerned mostly with the mapping of spatial coordinates. The water maze was first developed using rats, and then adapted for use in mice. There are many reports of mouse mutants being deficient in this test. Spatial learning has been widely studied in mice because it involves the organization of a number of cognitive processes based on natural spatial orientation behavior. External and internal cues help the animal to find the way out of a maze (or onto a platform) during testing. As well as spatial navigation, a number of external variables may play a crucial role in the water maze test; these variables include stress caused by the water and the motor capabilities of the animal. These variables may be further investigated by secondary screening, although the level at which they affect the spatial learning parameters as recorded by the test can never be truly evaluated. Some alternative spatial learning assays, for example the Barnes test [78] or radial arm maze [79], may help in discriminating these behaviors.

Another feature that is utilized to study learning in rodents is negative reinforcement, for example mice tend to avoid negative stimuli such as a foot-shock. There are many so-called avoidance paradigms. Passive avoidance is a relatively easy test to carry out and can be incorporated into a large-scale screening program. However the data from this type of test can be difficult to interpret. In this test animals are placed in the light compartment of a two-compartment box. The other compartment is dark and contains a grid floor through which a brief electric current can be passed. The mouse passes into the dark compartment (as mice prefer the dark) where it receives a brief foot-shock. Memory is assessed by returning the mouse to the light compartment and recording the duration of the latent period before the mouse re-enters the dark compartment. However a number of factors, e. g. anxiety level, shock sensitivity etc., can alter performance on this test so these confounding variables should be ruled out before a direct effect on memory can be inferred. In a conditioned avoidance task the animal actively moves from one compartment to another to avoid receiving a shock. The onset of the shock follows a cue – the conditioned stimulus – and the animal has to learn the association between the cue and the shock. Again these tests have been used to assess function in transgenic and KO mice and there are wide variations in performance of these tasks between different mouse strains [80, 81]. Another paradigm, the fear-conditioning paradigm, measures the freezing response of mice when they expect to receive a foot-shock.

Further studies can be conducted in mice by determining whether particular drugs or drug combinations can affect the performance of animals in learning tests. The results may help in discriminating between the various behaviors that might play a role in the performance of an animal in a particular paradigm. Furthermore, this approach may help in defining which neural processes contribute towards a particular mutant phenotype.

6.3.3 Social Behavior

Mice are social animals as is demonstrated by the fact that social isolation among mice can lead to behaviors associated with depression and anxiety. The olfactory system in mice plays a critical role in social behavior. It is particularly sensitive to minute variations in odors, which in turn allow the mouse to explore its environment and its opponents. By using olfactory signals, mice can collect information regarding the age, status and intentions of other mice. Also, chemical information from odors serves to control mate choice and other aspects of sexual behavior. Mice and other species, create territorial patterns of urine deposition for marking individual and group spaces [82–84]. The odors of approaching unfamiliar male mice may trigger an alarm reaction followed by aggressive behavior. When laboratory mice are rendered surgically anosmic they show very different social behavior towards other intact mice. Aggressive behavior is diminished, with animals moving freely around the enclosure rather than confining themselves to particular areas and generally ignoring one another. When they do encounter another individual, they appear startled and move away from each other [85]. The use of olfaction by mice in mediating social encounters means that cage cleaning is an important source of variability [86]. It has been shown that standard methods of cage cleaning, in which only the substrate and parts of the cage are washed clean of scent marks can be detrimental to male mice by promoting aggression.

Aspects of social behavior in mice can be assessed in a systematic manner. Moy and colleagues [22] have recently developed a new automated procedure to assess sociability and the preference for social novelty among mice. To assess sociability, each mouse exploration is scored in a central habituated area, and in two side chambers, one containing an unfamiliar conspecific (stranger 1) in a wire cage, and the other containing an empty wire cage. Subsequently, preference for social novelty is assessed by presenting the test mouse with a choice between the first, now-familiar, conspecific (stranger 1) in one side chamber, and a second unfamiliar mouse (stranger 2) in the other side chamber.

In social terms, mice show a complex behavioral repertoire: searching for food, forming complex social organizations, building tunnels and constructing nests. Much information concerning the behavior of mice can be ascertained from simple observations of home cage activities. Conventional standard laboratory home cages may prevent much natural and motivated behavior (e. g. nesting, tunnelling, extensive locomotion). As a result, mice in laboratory conditions may frequently exhibit so-called abnormal behaviors, for example stereotypies [84]. Cage enrichment can increase the repertoire of behaviors that can be observed in the home cage environment.

6.3.4 Emotionality in Mice

A number of mouse paradigms have been developed to explore the biological basis of anxiety. The most widely used are the open field, elevated-plus maze and the light–dark box [87]. These devices all rely on the innate aversion of mice to open spaces and bright areas, which conflicts with their natural tendency to explore novel environments. An open field is a large square or round arena that is brightly lit and is normally white in color. When placed in this environment, mice preferentially spend time at the perimeter rather than at the center, and those mice exhibiting high anxiety freeze and defecate. A light–dark box consists of a two-compartment box, one compartment is white and brightly lit, while the other is darkened. Mice are free to move between compartments. Those with high anxiety tend to show an increased latency to emerge from, and a preference for, the dark compartment. The elevated-plus apparatus consists of an elevated platform in the shape of a "plus" sign. Two arms of the platform are open and two are enclosed. Mice with a low anxiety phenotype show a reduced latency to enter, and tend to spend more time in, the open arms of the apparatus. These experimental anxiety paradigms have been supported with the use of pharmacological compounds. Rodents, when administered anxiolytic drugs, spend more time in the center of the open field, on the open arms of the elevated plus maze and on the light side of the dark–light box [88, 89]. Further measures of anxiety can be determined using a holeboard test.

One of the features of depressed mood in humans is a feeling of helplessness. A similar state in rodents, termed behavioral despair, is observed when animals are placed in a stressful situation from which they cannot escape. Two such situations, being placed in a cylindrical tank of water and being suspended by the tail, result in characteristic immobile postures in mice. The duration of this immobility can be reduced by pre-treating animals with a range of antidepressant drugs although the response to these agents in animals is acute whereas chronic administration is required to treat human depression. Many mouse mutants have displayed either an increase or a decrease in immobility during these tests (for a review of animal models of depression see [90, 91]).

6.3.5 Processing Sensory Information in Mice

Deficiencies in attention and information processing have long been noted in patients with schizophrenia. It is believed that the inability to filter and process extraneous stimuli that are presented in rapid succession accounts for the tendency of schizophrenics to suffer from fragmentation of normal cognitive processes [92] and that these functions are vital to the maintenance of cognitive function [93]. A typical example of these attentional deficiencies is represented by impairments in sensorimotor gating. Sensorimotor gating is typically measured by the level of attenuation of a startle response upon presentation of a non-startle-inducing pre-pulse stimulus; this is commonly known as pre-pulse inhibition (PPI). PPI has been extensively investigated as a behavioral test that can be easily transferred between humans and mice. Different inbred strains of mice have shown a genetic component to PPI performance. Several single-gene mutations also show alterations in PPI. For example *Prodh* null mice are deficient for the gene that encodes proline dehydrogenase and these mice also show a significantly lower level of PPI in comparison to the wild-type controls [94]. *Prodh* is the mouse homolog of the Drosophila *SlgA* (sluggish-A) gene. The human homolog lies on chromosome 22 q11 in a region that has been associated with psychiatric disorders.

Mice as well as humans maintain a large number of physiological variables under the constant control of an internal clock. The predominant rhythm adopted is circadian (~24-h) but for mice the activity pattern is nocturnal. Humans and mice use circadian control to regulate diverse processes such as the sleep–wake cycle, locomotor activity, temperature regulation, metabolism, water/food intake and levels of circulating hormones. In mammals, the main circadian clock is located in the suprachiasmatic nuclei (SCN) of the hypothalamus. The SCN clock can autonomously regulate its body functions keeping all body systems constantly updated. However, under normal conditions it is reset by environmental cues such as sunrise/sunset, a process known as entrainment. Disturbances in circadian parameters have been associated with a number of psychiatric and neurological disorders in humans including seasonal affective disorder, depression, bipolar disorder and neurodegenerative disorders [95–99]. Moreover, polymorphisms in the pacemaker genes *clock* and period have already been associated with disorders such as familial advanced sleep phase syndrome [100] and seasonal affective disorder [95] and particular allelic forms can be associated with diurnal preference in humans [101].

Several important insights into circadian function have been established only from the analysis of mutant phenotypes in mice and other rodents. The cloning of the mouse *clock* gene [102, 103] is still a milestone in circadian studies. The *clock* mutant was identified in a behavioral screen of progeny of ENU-mutagenized animals. Cloning of the spontaneous hamster mutant, *tau*, has implicated post-translational mechanisms in clock function. The mutation, found in casein kinase I epsilon, results in less efficient phosphorylation of the period protein and a dramatic shortening of the biological clock [104]. Identification of the molecular components of the biological clock is far from complete and much work has concentrated on establishing how the pacemaker interacts with environmental cues (entrainment), how the molecular clock is translated into a synchronized neural signals and how output rhythms are generated and maintained.

6.3.6 Endophenotypes

A key issue in mouse phenotyping, particularly with respect to modeling human diseases, is the degree of similarity between mouse phenotypes and those in humans. Clearly, differences between the two species prevent the replication of the complex multigenic phenotypes. However gene-mapping for neurological and psychiatric diseases is only successful if the phenotypes that are adopted are informative for the particular disease under investigation. A progressively large consensus is spreading across laboratories that the identification of intermediate phenotypes, so-called endophenotypes, may represent a powerful approach to the mapping studies of neurological and psychiatric diseases. Endophenotypes are based on the assumption that a simpler genetic architecture is involved. Endophenotype characterization in studying diseases is evident in many psychiatric and behavioral disorders, for example impaired sensorimotor gating and working memory [105, 106] are associated with schizophrenia and locomotor hyperactivity is associated with attention-deficit hyperactivity disorder (ADHD) [107]. The systematic application of this approach to the analysis of mouse models can be visualized using a feed-forward network approach (Fig. 6.1). An initial unit (e. g. schizophrenia) might be associated with deficits in different associated traits (e. g. sensorimotor gating, attention, learning and memory and so on) which in turn might be investigated using different phenotypic assays (e. g. PPI as a measure of both sensorimotor gating and attention). A mouse mutant that is found to be deficient in some of these assays might represent a possible model for the human disease. The mouse model may never accurately reflect the human condition, but rather model some important characteristics of the human disease. For example *Prodh2* mice have a PPI deficiency and have been presented as a model for schizophrenia in humans although no other schizophrenia-associated traits have been investigated. To confound matters, traits associated with the human patient population are also variable, with different patients showing different sets of symptoms although the diagnosis is the

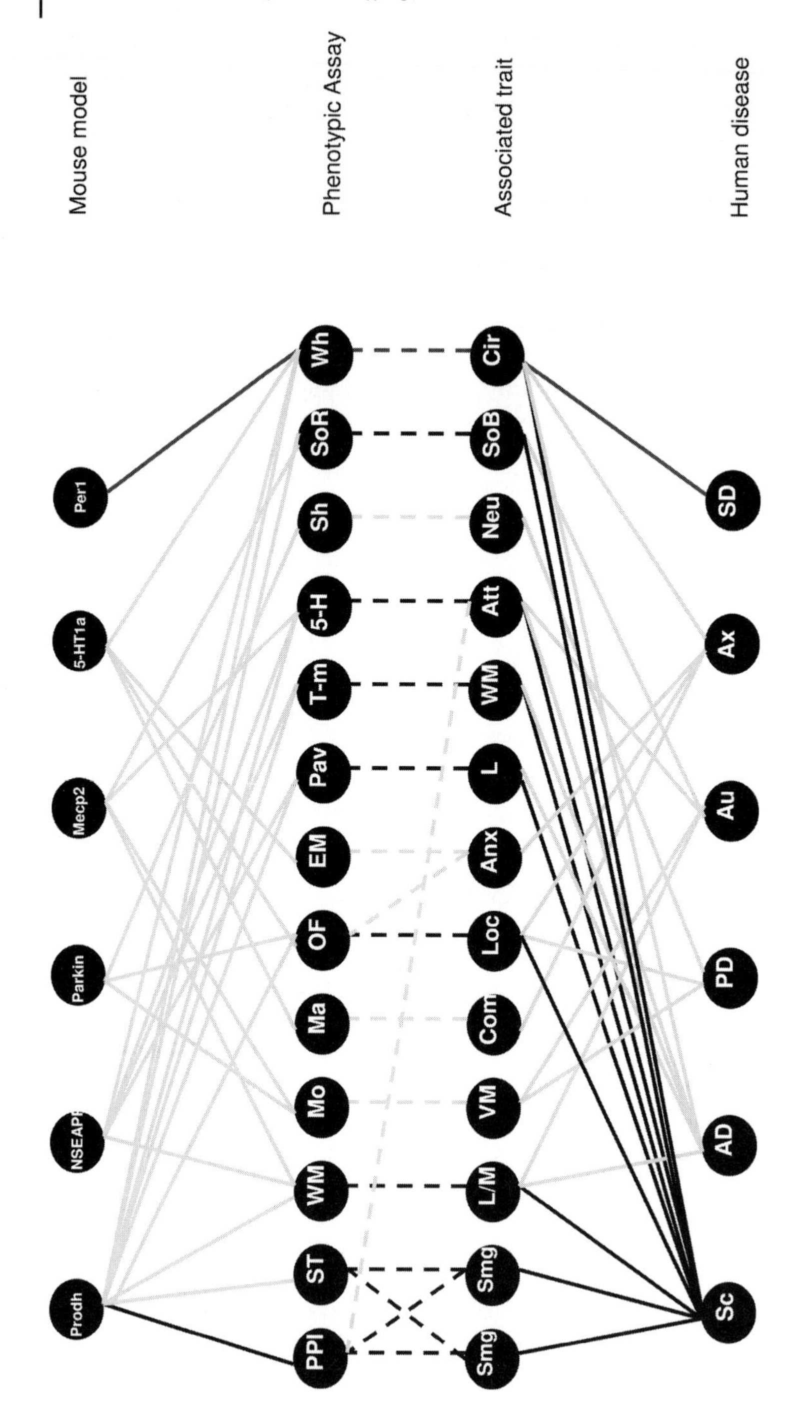
Mouse model
Phenotypic Assay
Associated trait
Human disease
Per1
5-HT1a
Mecp2
Parkin
NSEAP
Prodh
Wh
SoR
Sh
5-H
T-m
Pav
EM
OF
Ma
Mo
WM
ST
PPI
Cir
SoB
Neu
Att
WM
L
Anx
Loc
Com
VM
L/M
Smg
Smg
SD
Ax
Au
PD
AD
Sc

same. Thus deciding whether a particular mutant provides a suitable model for a human disease can be problematic. Using the feed-forward network, this can be established systematically. Connections between units have different weights that represent the association between units. The incoming signal, the input, may be described mathematically by a number (between –1 and 1), which in turn may vary across different cases (e. g. patients with the same disease can show a range of severities in each of the associated traits). The weight of a higher unit depends on the signals it receives from different units at lower levels. Thus the pattern of active and inactive units (e. g. the disease, its mechanism, the associated assay and all the weights between them) determines how robust the output is (i. e. the mouse model). This general model is derived using a neural network approach that has a number of advantages: *robustness* (is resistant to noise, e. g. the high variability of the human model), *flexibility* (may well represent different cases and diseases), *generalization* (once there are a sufficient number of cases the network is able to make accurate predictions).

6.4 Behavioral and Neurological Screening Protocols in the Mouse

Behavioral phenotyping has so far been characterized by a number of creative and complex apparatus and protocols. Due to limitations of space a review of all such protocols is precluded, therefore we will describe only a few which are currently the most widely used in mice (see Tab. 6.2). Many of the procedures described in the *primary screening* sections and some in the *secondary screening* sections have been adopted and cross validated by the Eumorphia European consortium (www.eumorphia.org). The majority of behavioral analyses are carried out on adult mice (8 weeks or older).

◁ **Fig. 6.1** Use of a "feed-forward network" to investigate mouse models of human disease. A feed-forward network is presented showing four groups (levels): human diseases (level 1), associated traits (level 2), phenotypic assays (level 3) and mouse models (level 4). Each layer consists of multiple units (nodes): the "input" level of units (human disease) is connected to the level of "hidden" units (associated trait and phenotypic assay), which is connected to a level of "output" units (the mouse model). See text, Section 6.3.6 for full description of the network. Input: Sc, schizophrenia; AD, Alzheimer disease; PD, Parkinson disease; Au, autism; Ax, anxiety-related disorders; SD, sleep disorders. Hidden nodes: (*Associated traits*) Smg, sensorimotor gating; L/M, learning and memory; VM, voluntary movement; Com, compulsive behavior; Loc, locomotor activity; Anx, anxiety; L, learning; WM, working memory; Att, attention; Neu, neurological assessment; SoB, social behavior; Cir, circadian parameters. (*Phenotypic assays*) PPI, pre-pulse inhibition; ST, Acoustic Startle Response (ASR); WM, water maze; Mo, MoRaG; Ma, marble burying test; OF, open field; EM, elevated-plus maze; Pav, Pavlovian classic conditioning; T-m, T-maze; 5-H, five-hole board test; Sh, SHIRPA; SoR, social object recognition; Wh, wheel-running activity. Ouptut: A few mouse models are presented as examples. *Prodh*, *Prodh*$^{-/-}$ mice which are deficient in the gene that encodes proline dehydrogenase, and show significantly lower PPI values in comparison with wild-type controls. *Per1* , in humans the *Per* genes are implicated in sleep disorders, (e. g. familial advanced sleep phase syndrome, FASPS). *Per1* mutant mice show a shorter circadian period in compared to wild-type animals.

Table 6.2 Behavioral and neurological phenotyping procedures.

Domain	Test name	Used for the assessment of
Neurological and neuromuscular		
Primary screens		
	SHIRPA	Muscle and lower motorneuron, spinocerebellar, sensory, neuropsychiatric and autonomic function
	Grip-strength test	Muscle strength
Secondary screens		
	Rope Grip Test	Muscle strength.
	Isometric resistance	Muscle strength.
	Electromyography	Muscle response to nervous stimulation
Motor function		
Primary screens		
	Swimming test	Vestibular and motor function
	Hind Paw Footprint Ink Test	Locomotion and gait
Secondary screens		
	Rotarod test	Motor coordination and balance
	Mouse reaching and grasping performance (MoRaG)	Reaching, grasping and additional motor behaviors
	Treadmill	Motor effort
Learning and cognitive function		
Primary screens		
	Spontaneous alternation	Exploration and working memory
Secondary screens		
	The Morris water maze	Aspects of spatial learning
	Context-dependent and -independent conditioning	Conditioning responses
Social behavior		
Primary screens		
	Social Dominance Tube Test	Aggressive tendencies
Secondary screens		
	Social transfer of food preference	Aspects of cognitive function and learning
	Social (Object) Recognition Test	Exploration and learning
Emotionality		
Primary screens		
	Open-field activity	Locomotor activity and anxiety
	Tail suspension test	Depressive-like behavior
Secondary screens		
	Light–dark box	Anxiety
	The elevated-plus maze	Anxiety
	Modified Hole Board (mHB) Test	Anxiety-related behavior and cognitive processes
	Porsolt swim test	Depressive-like behavior

Table 6.2 Behavioral and neurological phenotyping procedures. (Continued)

Domain	Test name	Used for the assessment of
Central processing of sensory information		
Primary screens		
	Acoustic startle and pre-pulse inhibition	Neural processing of sensory stimuli
	Tail flick test	Nociception
Secondary screens		
	Wheel-running activity	Circadian function
Supportive screens		
	Biochemical measurements	Biochemical assays of tissue or blood samples
	Histopathology	Brain section examination at a gross morphological level
	Immunohistochemistry	Neuronal cell loss, gliosis, intracellular protein accumulation etc.
	Re-testing of aged mice	Pathological changes related to aging

6.4.1 Screens for Neurological and Neuromuscular Function

6.4.1.1 Primary Screens

The SHIRPA Test The development of this test for the phenotypic analysis of mouse mutants is outlined in [108] and has been refined to identify phenotypic outliers in ENU mutagenesis screens [109]. Subsequently, a modified version was developed within the EUMORPHIA consortium (www.eumorphia.org). Analysis of data from this screen provides a comprehensive profile and can indicate deficits in muscle and lower motorneuron, spinocerebellar, sensory, neuropsychiatric and autonomic function. The screen, involving a battery of up to 40 simple tests, can be carried out in approximately 10 min. In addition to quantifying or identifying neurological and behavioral anomalies associated with mouse mutants, additional classes of mutations have been cataloged using this procedure. Briefly, each mouse is weighed, then placed into a viewing jar (typically for 5 min) and assessed for unprovoked behavior, after which mice are transferred to the test arena for a series of observations and manipulations. Score sheets are used to record data semi-quantitatively. At the end of the procedure, mice are returned to their home cage.

Grip-strength Measurement The grip-strength test measures the muscle strength of the forelimbs as well as the combined fore and hind limb grip strength in rodents (www.eumorphia.org). Abnormal grip strength indicates improper neuromuscular functioning or may allude to defects in neural control of muscle function. Grip strength can be measured semi-quantitatively but ideally should be measured using a commercially-available Grip Strength Meter. The apparatus is

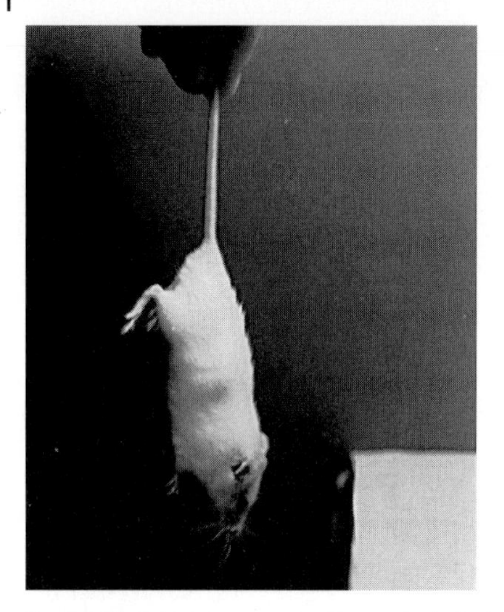
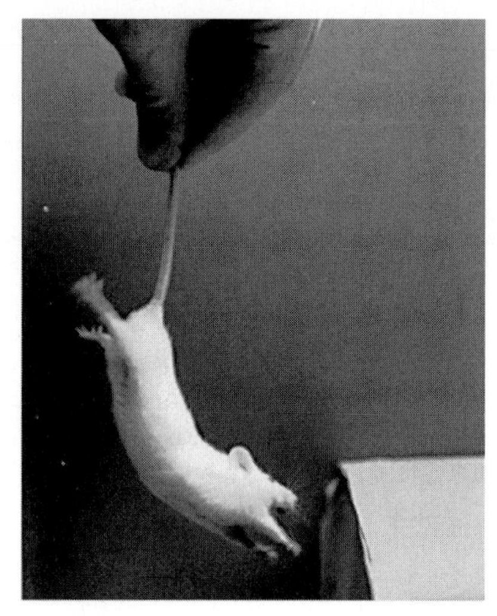

Fig. 6.2 The placing response. The typical placing response of a wild-type mouse (right) and an abnormal response as seen in the *ky* mouse (left). The *ky* mouse is characterized by a severe kyphoscoliosis and presents with a specific neuromuscular pathology.

supplied with a grid that is connected to a sensor. Animals are held by the base of the tail above the grid and gently lowered until the front paws grasp the grid. The hind limbs are kept free from contact with the grid (for forelimb-only grip testing). The animal is brought to an almost horizontal position and pulled back gently but steadily until the grip is released. The maximal force achieved by the animal is displayed on the screen and noted. Measuring grip strength on all four paws is carried out as above except that all limbs make contact with and grasp the grid initially. Typically, each animal undergoes five trials for forelimb grip strength and five trials for fore- and hindlimb grip strength measurements.

Placing Response (Upper Body Strength) Mice are suspended by the tail near an edge to trigger a spontaneous stretching response of the front legs in order to grab a frame (Fig. 6.2). The placing reflex has been used to detect mice with spinal deformities and minor scoliosis caused by weakness of postural muscles, e. g. the *ky* mouse [110].

6.4.1.2 **Secondary Screens**

Rope Grip Test This assay is conducted as described by Hall et al. [111] with some modifications described in [67]. A mouse is allowed to grip a 2-mm thick horizontal tight-rope placed 80 cm above the ground with its front legs, and the time elapsed until the mouse raises its hind legs to grip the rope is measured. If the mouse grips the rope with both hind legs within 60 s and holds the grip for at least 10 s, the test is recorded as a success. This test has been shown to distinguish several neuromuscular mutant mice, including *mdx*, from controls and represents a simple and effective test for muscle function. A simple variation of this test to evaluate muscle weakness of the limb muscles is to position the mice head-down and measure the time that elapses before they fall from the rope.

Isometric Resistance This is a measurement of the length of time a mouse can hold a gauge at a preset weight related to its own weight. This test uses a force transducer coupled to a pen recorder and a device that includes a fluid container exerting an opposite force to the mouse. The apparatus measures how long a mouse can maintain a grip on a horizontal bar. As examples, this test discriminates *wobbler* mice from normal mice [112] and the myopathic mutant Calpain3 from wild-type littermates.

Electromyography Electromyography (EMG) is a test that measures muscle response to nervous stimulation. In humans EMG is performed most often to help diagnose different diseases causing weakness. When performed in combination with a nerve conduction velocity test, EMG can determine whether a particular muscle is responding appropriately to stimulation, and whether a muscle remains inactive when not stimulated. It is well established that decreases in the amplitude and duration of spikes are associated with muscle diseases, which also show faster recruitment of other muscle fibres to compensate for weakness. Recruitment is reduced in nerve disorders. Therefore, EMG can help verify whether symptoms are due to a muscle disease or a neurological disorder. However, this analysis should be undertaken as a secondary or tertiary screen and in combination with other findings, e. g. histopathology, because the interpretation of EMG results is not a simple and requires a thorough analysis of the onset, duration, amplitude, and other characteristics of the spike patterns.

6.4.2 Screens for Motor Function

6.4.2.1 Primary Screens

Swimming Test Swimming ability in mice can be used to assess vestibular and motor function. Mice are placed in a tank containing water and their swimming ability observed (www.eumorphia.org). A score sheet is used for semi-quantitative assessment of swimming behavior. Stress is minimized during the swimming test by using tepid water and rescuing the mouse immediately if it appears unable to swim. Mice are kept warm in a clean cage with dry bedding after any swim test, and observed carefully until their coats have dried.

Hind Paw Footprint Ink Test Briefly, the soles of the hind feet are dipped in ink, and the mice are made to walk inside a long narrow track lined with paper. The resultant ink footprints are analyzed, and the average length of three strides for each leg is determined [67]. Additionally, the angle of back leg footprints when walking in an open area can be measured and this has been used to identify myopathic mutants produced from the overexpression of caveolin 3 [113].

6.4.2.2 Secondary Screens

Rotarod Assessment The rotarod is used to assess motor coordination and balance (www.eumorphia.org). A range of automated rotarod apparatuses with varying specifications are available commercially. Consequently, motor function assessments may vary across test centers. The rotarod apparatus consists of a rotating drum with a grooved surface for gripping. The speed of rotation can be set at a constant speed or can be set to accelerate at a particular rate. Under the acceleration mode, motor learning skills can be assessed. Test conditions vary although a typical paradigm is presented below. Mice are initially placed on the stationary drum for 1 min and then on the rotating drum (4 r.p.m. lowest speed) for 1 min with a 10-min break between each session. For test sessions, the rotating drum is set to accelerate from 4 to 40 r.p.m. over 300 s; there is a 30-min break between training and test sessions and a 15-min break between each test session. Up to four habituation sessions and four test sessions are recorded per animal. After four sessions, the trial is ended and the mouse returned to its home cage. If a mouse slips from the drum within 300 s, it is returned to its home cage and the process repeated for each trial.

Mouse Reaching and Grasping Performance (MoRaG) The Mouse Reaching and Grasping Performance Scale (MoRaG) consists of a qualitative and quantitative assessment of reaching, grasping and retrieval behavior in the mouse (www.eumorphia.org). Individual animals are placed in a Plexiglass chamber 10.5 cm high by 6 cm deep by 6 cm wide. Outside the front wall of each cubicle, a Plexiglass feeding platform, accessible through a 9-mm hole, is attached 5.4 cm from the floor. Small food pellets are placed on the feeding platform so that the mouse can withdraw food by using a single paw. The reaching, grasping and retrieval behavior of the animals is monitored. Scorecards are used to record semi-quantitative and quantitative behaviors. The reaction time between the mouse seeing the food pellet and reaching it (*reaching time*) is recorded for all the trials. Grasping is considered to be successful if the mouse grasps and holds the pellet. Successful retrieval is recorded if the mouse brings the food towards its mouth to consume it. Mice are deprived of food for up to 16 h prior to the assessment of their reaching and grasping performance.

Treadmill This test is used to measure the length of time that a mouse can run on a treadmill set at high speed (15–20 cm/s). The treadmill has been used to evaluate therapeutic treatment of *mdx* mice with the glucocorticoid prednisolone [114]. Regardless of the speed employed, mice no longer able to run demonstrate a characteristic fatigue profile at the rear of the treadmill facing the oncoming chamber floor. The running time is recorded once the fatigue position is reached. Blood samples may be taken from mice that demonstrate fatigue following over-exercise to assess the presence of markers of muscle injury as part of the secondary tests.

6.4.3
Screens for Learning and Cognitive Function

6.4.3.1 Primary Screens

Spontaneous Alternation This test exploits the innate tendency of mice to explore a novel environment. It can be used to assess aspects of working memory in mice but at the same time can record deficiencies in attentional traits and anxiety levels. The spontaneous alternation test is carried out using a standard T-maze or Y-maze (www.eumorphia.org). The typical T-maze consists of a start arm and two choice arms perpendicular to this. To assess T-maze alternation, mice are placed in the start arm and given a free choice to explore either of the other two arms. Once the mouse has chosen an arm, it is removed and the process repeated. Typically, the mouse enters the opposite arm on the second run. This constitutes a single alternation trial. Multiple alternation trials can be carried out over time although alternation in subsequent trials may be affected by prior experience. In one version of the test, each mouse undergoes a total of 10 alternation trials over two sessions of five trials each, usually 24 h apart. Percentage alternation over the 10 trials can be used to assess working memory. The Y-maze is made of clear Plexiglass. This will allow "piloting navigation" (allothetic map) using global landmarks in the room for orientation. The maze has three identical arms (40 × 9 × 16 cm) placed at 120 ° from each other. The center platform is a triangle with 9-cm sides. Each arm has inner walls with specific motifs. The motifs are different for each arm, allowing subjects to distinguish between them. At the beginning of a session, mice are placed facing the end of one arm and allowed to explore the apparatus freely for 5 min, with the experimenter out of the animal's sight. At the end of the session, animals are returned to their home cages. The delay in exiting the start arm and sequence and number of arm visits are recorded. Alternations are defined as successive entries into each of the three arms as overlapping triplet sets (e. g. ABC, BCA etc).

6.4.3.2 Secondary Screens

The Morris Water Maze This test [77] is one of the most widely used in the assessment of spatial learning in mice. Performance in the test may be an indicator of deficits in/enhancement of cognitive function and learning in mice although, as with all behavioral tests, performance may also depend on motor function, sensory function and emotional status of subjects. The influence of these confounding factors can be reduced by using carefully controlled experimental conditions (for example by pre-testing animals using a visible platform, measurement of thigmotaxis etc.) or by using this water maze as part of a battery of behavioral tests. In this test, mice are trained to locate a submerged platform within a tank (at least 1 m in diameter) of opaque water by using strategically placed visual cues. The performance of subjects can be measured using a visual tracking system and various parameters are measured using computer software. There are several variations in the test paradigm and only one is described here. The test is run for 10 days in total. Prior to starting the experiment, mice are placed on the platform for 1 min to allow for spatial orientation. Following this, on days 1–4, mice are placed in the water and are given a 1-min trial to locate the platform. This is repeated four times per day. Mice that find the platform within the allotted time remain there for 20 s and are removed and allowed to rest for at least 1 min before the next trial begins. Mice that do not locate the platform within the allotted time are manually placed on the platform. Over trial days, a reduction in the latency of finding the platform is expected. On day 5 the platform is removed and the mouse is placed in the tank for 1 min and the time spent in the proximity of the platform position is noted. On days 5–10, the position of the platform is changed and the process is repeated. Variations of this protocol may also be used.

Context-dependent and -independent Conditioning This test is used to assess aspects of hippocampal-dependent and -independent learning in mice (www.eumorphia.org). Mice are placed in a standard Skinner box and are exposed to a tone stimulus coupled with a mild foot-shock. Mice generally elicit a freezing response which can be reproduced in the absence of additional foot-shocks at particular times following the initial exposure. Mice are returned to their home cages and after 24 h are returned to the same box; the duration of the freezing response in subjects is measured. Mice may also be returned to the box in which the environment has been altered in some way (altered context). In this case, freezing is monitored in response to a second tone stimulus (this time not coupled to a foot-shock). A reduction in duration of the freezing response may be indicative of a learning impairment.

6.4.4 Screens for Social Behavior

6.4.4.1 Primary Screens

Social Dominance Tube Test The social dominance tube test measures aggressive tendencies, without allowing mice to injure each other. Using two mice, dominant and submissive postures are scored, along with approach/avoidance behavior during a brief pairing in a specialized chamber [115]. The social dominance tube test employs two start areas, a two-section tube and one neutral area inside the two-section tube. The apparatus is 30 cm long and 3.5 cm in diameter, built with clear Plexiglass material. Gates at the end of each section of the tube allow olfactory but not physical contact. To carry out the test, two mice of the same gender are placed at opposite ends of the tube. Both mice begin to explore in a forward direction towards the center of the tube. The gates are removed as the mice approach them, allowing the two mice to approach each other. If one mouse is dominant and the other subordinate, the dominant mouse will approach while the subordinate will back away. Each mouse is engaged in up to 10 consecutive trials with mice of either the same strain or a different strain presented in random order. A statistical hierarchy determines a subject's position on the dominance scale both within and between strains/genotypes. Mice with a null mutation in the *Dishevelled1* gene, for example, are consistently socially submissive in this test [116].

6.4.4.2 Secondary Screens

Social Transfer of Food Preference The social transfer of food preference test is used to assess aspects of cognitive function and learning in mice [117]. The screen requires the use of two sets of mice, demonstrator mice and test mice. The basic test consists of a 30-min interaction (or learning) period where naïve test mice interact with demonstrator mice who have consumed a novel "cued" flavored food. At different time intervals following the interaction, learning/memory of the tester mice is assessed over a 2-h period by measuring their preference for the "cued" food over the alternative "non-cued" food. Step 1: demonstrator mice are food deprived for up to 16 h. After this period, the mice are introduced into individual cages containing a novel (cued) food stimulus. The cued food consists of standard chow pre-soaked in a 2% cinnamon or cocoa solution. Demonstrator mice are used for the next step in the procedure if they have consumed more than 2 g of the cued food in a set time period. Step 2: demonstrator mice are introduced into cages of test mice (up to five mice per cage) to allow for periods of social interaction and transfer of olfactory cues. This step may be repeated once after 1 h. Step 3: at certain times after social interaction (up to 1 week), test mice are food deprived for up to 16 h and introduced into individual cages containing the cued food and a novel (non-cued) food. The non-cued food may also contain cinnamon or cocoa as above. After a fixed time period in the cage, test mice are assessed for food preference by determining the ratio of cued : non-cued food consumed. The nucleotide exchange factor *RasGrf1*, for example, has been implicated in this form of learning [118].

Social (Object) Recognition Test Because of the natural tendency of rodents to intensely investigate novel individuals, Thor and Holloway [119] proposed a simple laboratory test to investigate short-term, social recognition capacities. When an unfamiliar conspecific is introduced for the first time into the home cage of an adult male mouse, the resident male vigorously investigates the novel individual. If the novel animal is removed and then re-introduced to the same resident male a short time later, it will receive far less investigation during the second meeting. A variation of this social recognition paradigm consists of a social discrimination procedure between two individuals. The subject is given a choice between a previously encountered and a novel conspecific [120]. Comparing the difference in time spent investigating the familiar and the unfamiliar stimulus animal social recognition can be assessed. Another paradigm to evaluate social recognition is as follows: the subject is repeatedly exposed to a novel stimulus animal for 1 min with 10-min inter-trial intervals. Familiarity can be detected by reduced investigation on each trial. Following the fourth such exposure, a novel stimulus animal, is used to rule out the possibility that the test animal's reduced investigation is not due to fatigue or habituation [121, 122]. Based on these simple observations both social recognition and memory can be assessed. The changes in the duration of investigation during repeated pairings with the same stimulus animal represent an index of short-term memory for that individual. For object recognition, familiar/novel objects such as marbles, dice etc. are used instead of conspecifics.

6.4.5
Screens for Emotionality

6.4.5.1 Primary Screens

Open-field Activity Generally, any test that measures anxiety in mice should be carried out prior to executing other behavioral tests (www.eumorphia.org). In this way, anxiety and stress associated with novelty in experimentally naïve mice can be assessed more easily. The open field test [123] measures activity in a novel environment and can be used to assess a combination of locomotor activity, exploratory drive, neophobia, agoraphobia and other aspects of anxiety or fear in mice as well as motor function (Fig. 6.3). Testing sessions typically last up to 30 min. The apparatus consists of a Perspex arena (approximately 44 × 44 × 50 cm high) around which are infrared sensors which can detect and localize activity. Alternatively a video-tracking system can be used to monitor behavior. Individual mice are placed in the arena and their activity monitored. The test can be carried out under normal lighting conditions or under red light although it is important to document these conditions as well as noise levels within the test area. Many of these parameters can influence an animal's activity and performance. For example, open field assessment under dim lighting conditions may be used to assess activity rather than anxiety. All measurements are recorded using computer software. Using current, commercially-available software, many activity parameters can be measured including localizing activity within the arena (e. g. at the periphery of the arena or in the center), determination of speed of movement, number of bouts of activity etc. Following each test period, mice are returned to their home cages and the arena wiped with ethanol to remove olfactory cues.

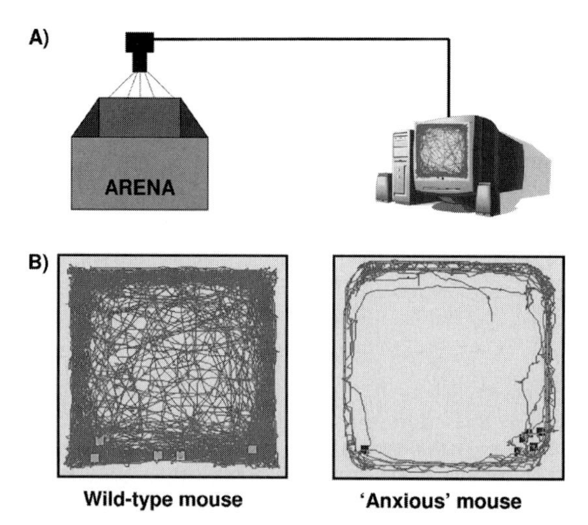

Fig. 6.3 Use of a visual-tracking system to study "anxiety" in mice. (A) A typical set-up to measure open field activity in mice. Mice are placed in a large arena and activity over a predetermined interval is monitored using an overhead camera. Computer software is used to "track" the activity of the animal. (B) Typical activities of a wild-type mouse (left) and an "anxious" mouse (right). Note the difference in center-time for the two mice.

Tail Suspension Test This test is used to assess depressive-like behavior in mice [124]. Animals are suspended by the tail in custom-designed or commercially-available apparatus. As in the Porsolt swim test, an initial period where the animal struggles is followed by prolonged periods of immobility. Again, the time spent immobile is an index of depressive-like behavior (www.eumorphia.org).

6.4.5.2 Secondary Screens

Light–Dark Box This test is used to further assess anxiety levels in mice [125]. When given a choice, mice prefer a dark to a light environment, particularly when in a novel environment. The apparatus used is the same as that employed in the open field study. In this instance, however, one half of the arena is covered with darkened opaque Perspex. A small opening at the base of this dark box allows the animal to move freely between light and dark compartments. The test can be carried out under normal lighting or red light. The test session lasts 5 min. Mice are placed inside the entrance to the dark compartment at the start of the session. Over the 5-min period, latency in entering the light compartment, the number of light compartment entries and the total time spent in the light compartment is measured. After the test session, mice are returned to their home cage.

The Elevated-plus Maze The elevated-plus maze test is also used to determine levels of anxiety in mice [126]. The apparatus consists of a narrow platform (in the shape of a "+") positioned at 1 m above floor level. One half of the apparatus (e. g. two arms of the cross) is enclosed and the other open. Again, mice prefer the safe environment of the closed arms to the "unsafe" environment of the open arms. The test simply consists of placing the mouse on the apparatus and examining its behavior over a 5-min period. Entries into and the proportion of time spent in the open and closed arms of the apparatus is noted.

Modified Hole Board (mHB) Test Animals in the Hole Board Platform are confronted with a new environment where anxiety-related behavior and cognitive processes interact (www.eumorphia.org). Thus the mHB Test allows cognitive mechanisms as well as anxiety, exploration, locomotor activity, arousal and risk assessment to be evaluated [127]. The mHB apparatus consists of an arena containing a floor of white Plexiglas with a number of holes arrayed in two dimensions. The number of holes can vary from a low (= 8) to high (= 144) spatial frequency (SF). However, the dimensions of the arena are kept constant (44 × 44 cm). The dimensions of the holes vary according to the SF. For the mHB test, all individuals are placed in the arena for an interval of time that varies from short (minimum 5 min) to long (maximum 30 min). The following behaviors can be assessed using this apparatus. *Exploration*: to assess the exploratory behavior of mice, the number of holes visited is counted over the entire session. Moreover a number of parameters, including latency of first visit, velocity, time in each zone, duration of visits etc., can be measured using an automated video tracking system. A number of these parameters are described by Ohl and collaborators [127]. The assessment of exploration (above) does not require any food restriction. For subsequent behaviors (below) mice are maintained under a controlled food restriction regimen such that they are maintained at 85–90% of their normal weight. *Cognition*: to assess cognition, several holes in the floor of the arena are baited with a 10–100% solution of condensed milk. Using this design, mice perform a visuospatial task with the operator changing the sequences of rewarded holes over the course of the test. Thus the flexibility of cognitive processes can be investigated [127].

Porsolt Swim Test Another extensively used test for antidepressant efficacy in mice is the Porsolt (forced) swim test [128]. For this test, mice are placed in a cylindrical container of tepid water from which they cannot escape. Initially mice swim vigorously as they try to escape from the container but within a short period (~ 2 min) will adopt a characteristic immobile posture. The time spent immobile is an index of depressive-like behavior and this time can be reduced by treatment with antidepressants.

6.4.6 Screens for Central Processing of Sensory Information

6.4.6.1 Primary Screens

Acoustic Startle and Pre-pulse Inhibition The acoustic startle response (ASR) is characterized by an exaggerated flinching response to unexpected auditory stimuli (Fig. 6.4; www.eumorphia.org). Abnormalities in ASR may be related to hearing

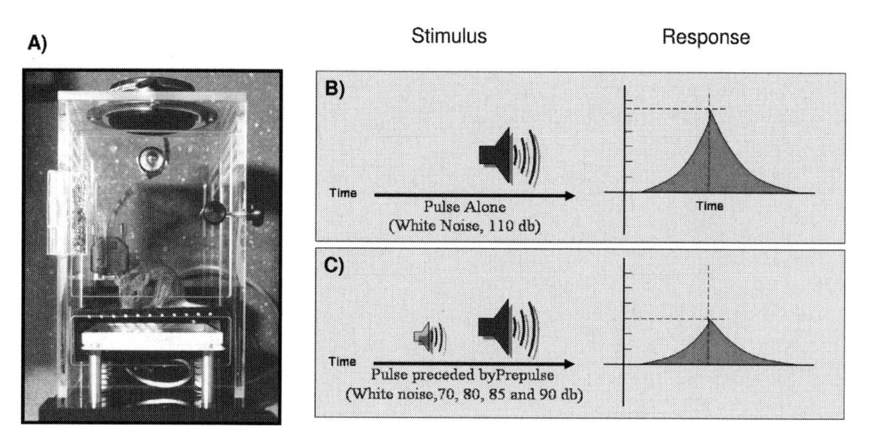

Fig. 6.4 Pre-pulse inhibition of acoustic startle response (PPI). (A) An example of a sound-attenuated chamber used to assess PPI in mice (left). A loudspeaker at the top of the chamber is programmed to emit sounds at pre-determined amplitudes and in a specific temporal order. The reaction of mice at the base of the chamber is monitored via an accelerometer and analyzed using computer software. (B) When mice are subjected to a white noise sound pulse at 110 db, they typically respond with an exaggerated startle response. A maximal response can be recorded at a discrete time following the stimulus whereas an integrated response can be recorded over a pre-set time interval. (C) When sound pulses are preceded by weaker pre-pulses (reduced amplitudes of 70, 80, 85 and 90 db) the startle response is reduced dramatically.

abnormalities or abnormalities in neural processing of sensory stimuli. The ASR can normally be attenuated when it is preceded by a weaker pre-stimulus (pre-pulse inhibition, PPI). PPI is a measure of sensorimotor gating, the subject's ability to filter and integrate sensory information appropriately. The experimental apparatus consists of an outer chamber within which is a sound-attenuated acoustic chamber with a startle platform (accelerometer) and loudspeaker. Sound pulses from the loudspeaker are regulated by a sound generator and appropriate software. The startle response is measured via the accelerometer and recorded by the computer software. Various acoustic startle chambers are available commercially and most come with associated software.

Prior to starting experimental work, appropriate conditions need to be established for each particular apparatus. Once the appropriate parameters are established, results are relatively robust across test centers. A typical startle and PPI paradigm is presented below with the entire procedure lasting approximately 30 min. The following sequences of pulses/pre-pulses are administered: (1) an initial acclimatizaton period of 5 min without a pre-pulse stimulus, an inter-trial interval of 50 ms and a pulse of white noise of 110 dB/40 ms; (2) a pre-pulse inhibition session of the following different trial types:

- A trial in which only an acoustic startle pulse is presented as white noise at 110 dB/40 ms.
- Eight different pre-pulse trials of 10 ms duration, of 70, 80, 85, or 90 dB white noise stimuli presented alone or preceding the pulse by 50 ms.
- A trial in which only background noise of 65 dB, is presented to measure baseline movement of the animal in the chamber.

Pulse combinations should be presented 10 times in pseudorandom order with an inter-trial delay varying between 20 and 30 s. The pre-pulse, along with pulse values are expressed as a percentage of the pulse response alone and values are averaged over the trial. Variations of these conditions may also be used.

Tail Flick Test The tail flick test is used to assess nociception in mice but may also give an overall index of sensory function (www.eumorphia.org). The test measures the response threshold of high-intensity heat stimuli (acute pain) when applied to the animal's tail. The latency between onset of the stimulus and a rapid flick/withdrawal of the tail from the heat source is recorded automatically using commercially-available apparatus.

6.4.6.2 Secondary Screens

Monitoring Circadian Rhythm and Entrainment Parameters Monitoring the wheel-running activity of mice maintained for the most part in constant darkness with the administration of intermittent light pulses, is the method used to assess inherent circadian function and its interaction with environmental cues [129]. Additional activity or biochemical parameters can be used to measure circadian function, a non-invasive measurement being preferred. Mice aged 8–14 weeks are housed singly in cages of conventional size but equipped with running wheels. Food and water are provided *ad libitum*. The mouse cages are kept under environmentally-controlled conditions in light-tight ventilated chambers at up to 10 cages per chamber. Mice are initially maintained for 1 week in a 12:12 light:dark cycle and subsequently switched to constant dark conditions for a maximum of 5 weeks. (In some cases, the transition from light : dark to constant darkness will be preceded by a 24-h period of constant light.) Mice are checked on a daily basis throughout this procedure using infrared viewers. During the latter period, mice receive a number of light pulses each of up to 15 min duration. Mice may be returned to a light : dark schedule after this period. As mice are housed singly, cages are changed at 2-weekly intervals so that there is only minimal disruption to the animals' behavior patterns. A variety of parameters (such as circadian period, phase shifting response, length of activity and rest phases and average wheel revolutions during light and dark periods under LD conditions) can be recorded and analyzed using commercially-available software and hardware.

6.4.7 Supportive Screens

Analysis of other screening domains can be invaluable in supporting behavioral and neurological phenotyping data. Details of the protocols used are beyond the scope of this chapter.

6.4.7.1 Biochemical Measurements

The assessment of behavioral and neurological function in mouse models can be augmented by the use of biochemical assays of tissue or blood samples. Disturbances in the balance of neurotransmitters in discrete brain regions have been identified in post-mortem brains in a number of neurological and psychiatric disorders. For example, under-activity of both the serotonergic and noradrenergic systems has been associated with the development of depression whereas over-activity in the dopamine and glutamate systems has been associated with schizophrenia. Changes in glutamate and other excitatory neurotransmitters have also been observed in neurodegenerative disorders.

Skeletal muscle injury either as a result of high-force eccentric contraction or as a consequence of an inherited muscle defect, leads to an increase in muscle proteins in the blood. The suitability of some of these markers has been analyzed in detail [130]. Useful measurements include creatine kinase (CK; EC 2.7.3.2), myoglobin and troponins.

6.4.7.2 Histopathology

In mouse models, histopathology can be an invaluable adjunct to behavioral and neurological tests. Conventional histological stains can be used initially to examine brain sections at a gross morphological level. Subsequently, immunohistochemistry can be used to assess neuronal cell loss, gliosis and the presence of intracellular or extracellular accumulation of protein.

Extensive histological characterization of skeletal and cardiac muscle from mutant phenotypes identified in primary screens, in addition to observations of whole heart morphology, is pivotal for unequivocal detection of neuromuscular phenotypes. In addition, myelination anomalies can be detected by staining leveled sections of peripheral nerve, e. g. sciatic nerve.

6.4.7.3 Re-testing of Aged Mice

It is well established that many brain, nerve and muscle defects are late onset and pathological changes manifest at a very late age. Thus it is desirable to re-test mice after aging for 12–16 months.

6.5 Implementation of Behavioral and Neurological Phenotypic Analysis

The analysis of mouse behavior is notoriously dependent on a number of inherent and external variables. When carrying out phenotypic analysis, it is important to identify and be aware of potential sources of variance. In addition, it is important that the data be analyzed appropriately when measuring complex phenotypes such as these. Different rules apply to different phenotypic analyses, some of which are discussed below.

6.5.1 Gene-driven Approach (Reverse Genetics)

The hypothesis underlying the reverse genetics approach begins with the selection of a gene and proceeds to the analysis of the phenotypic consequences of a targeted mutation in that gene. Gene targeting has been used as a tool in mouse genetics for over a decade. In addition, the use of transgenics and conditional targeting has been used to further our understanding of gene function. Phenotypic information on targeted mouse mutants is available at the Mouse Genome Informatics website coordinated by the Jackson Laboratory (http://www.informatics.jax.org). One of the disadvantages of reverse genetics is that a prior assumption is made about the function of the gene and as a consequence the phenotypic assessment carried out is biased by these preconceptions. Generally, phenotypic analysis in reverse genetics is limited to one or two tests. Undoubtedly, this approach has furthered our understanding of, for example, the molecular basis of memory [131–133], aggression [134], cocaine addiction [135], Huntington's disease [136], familial motor neuron disease [137] and Alzheimer's disease [138]. However, rather than limiting the investigation to one or two screens, it is recommended that primary screens be carried out in all domains. This approach has the potential of uncovering additional gene–function associations. When carrying out tests in targeted mutants, the most appropriate controls are littermates. It is recommended that at least six and ideally 10 mice per genotype should be studied. Mice that are heterozygous for the mutation of interest should also be studied; a total of 30 mice should be used for each mutation of interest.

6.5.2
Phenotype-driven Approach (Forward Genetics)

Forward genetics differs from reverse genetics in that its starting point is a particular phenotype (e.g. an abnormal behavior), and not a particular gene. In the late 1960s, Seymour Benzer proposed the forward genetic approach to identify mutant flies exhibiting abnormal behaviors [139]. This approach is now widely used in the mouse and has been successful in providing novel molecular insight into mammalian behaviors [103]. Phenotype-driven mutagenesis has classically employed chemical mutagenesis to efficiently introduce random mutations around the genome followed by the application of appropriate screens to recover relevant phenotypes. One of the most widely used agents for induction of mutations in forward genetics is *N*-ethyl-*N*-nitrosourea (ENU). ENU is a powerful alkylating agent that has been found to be highly mutagenic in mouse spermatagonial germ cells [140] and which induces mainly single base-pair changes in DNA. ENU has been used to generate mutations from specific regions of the genome [141], or the entire genome for dominant [63, 64] or recessive [142] mutations. The simplest form of screen is the identification of dominant ENU-induced mutations. With this approach, males are treated with ENU and subsequently mated with wild-type females. Progeny from these crosses, which can each potentially carry mutations at up to 100 different loci, are then screened for abnormal phenotype. After confirming inheritance of a phenotype, mutant genes can be identified using a conventional positional cloning approach. Screens for recessive phenotype, a far more laborious prospect involving three generations of breeding prior to phenotypic screening, are also being carried out. Details of ENU treatment and screening protocols have been discussed elsewhere [63, 143].

6.5.3
Phenotype-driven Screens: A Short Guideline

In this chapter we have discussed the benefit of using forward genetics for studying behavior. Here we define a few critical points which should be borne in mind when designing and implementing behavioral mutagenesis screens:

1. In contrast to the gene-driven approach, dominant phenotype-driven screens rely on the successful detection of single outliers among large screening populations. In recessive screens, successful detection of mutant pedigrees also relies on the systematic analysis of screened populations.
2. In general, the screen should be quick, robust and easy to conduct. A rapid screen allows for a greater number of mice to be screened, which in turn increases the chance of identifying heritable phenotypic traits.
3. Time-consuming screens may be appropriate if they measure a large number of parameters/information. Careful measurement of each parameter can potentially uncover novel mutant lines.
4. It is advantageous to ensure that screening is hierarchical, consisting of primary and secondary screens. The secondary screen, which is often more time consuming than the primary screen, leads to enhanced phenotypic assessment, confirm-

ing or dismissing the presence of an abnormal phenotype. This limits unnecessary breeding of mice.

5. Behavioral phenotypes by their very nature are intrinsically variable. A successful screen, however, should make all attempts to limit variability so as to facilitate identification of outliers. This can be achieved by implementing simple strategies such as ensuring that the same experimenter conducts the phenotypic screen, that phenotyping is conducted at the same time of day, and limiting external environmental influences.
6. Close attention should be paid to the variation between inbred mouse strains. Current approaches require the introduction of a second inbred strain to permit positional cloning of any mutation. In conducting mapping crosses, care should be taken to select an inbred strain that behaves in a similar manner to that which carries the mutation. This requires the acquisition of control (or baseline) data for various inbred strains using the same screening conditions [144].
7. If it is decided that a battery of tests (e. g. open field, PPI etc) would be the best method of assessing multiple phenotypes, careful consideration should be given to the order in which they are conducted to ensure that the behavior of a mouse is not influenced by its exposure to previous testing procedures.
8. Although screens such as these are not hypothesis-driven, the hypothesis that drives behavioral screening should be clear and well defined at the start. This not only helps to reduce the variability of the results but also facilitates interpretation of the mechanisms involved.

6.5.4
Environmental and Genetic Influences on Mutant Behavior

As with human disease traits, the expression of mouse behavioral phenotypes is notoriously susceptible to environmental and genetic influences. Extensive reports and reviews concerning the effects of the environment on behavioral performance and implications for the development of mouse behavioral models have been dealt with in detail elsewhere [145–147]. Arguments both for the standardization and for the diversification of behavioral testing environments are equally valid although standardization may allow for a more systematic assessment of mutant behavioral phenotypes. Standardization of the testing environment may also be critical, for example, in forward-genetics screens where the endeavor is to identify abnormal behavior in one individual within a population of animals. Of equal importance, however, is the effect of the genetic background on behavior in mouse mutants. Mutants of one genetic background that can express a particular behavior, such as increased aggression or a deficit in learning [148–150], may not express this behavior when the genetic background is mixed. Considering the number of behavioral studies conducted on a mixed genetic background, it is probable that researchers may be overlooking significant behavioral phenotypes in our existing catalog of mouse mutants.

6.5.5
Standardization of Screening

With the recent increase in the use of mouse models in functional genomics studies, the need for qualitative and quantitative descriptions of mutant phenotypic traits requires the use of standard phenotyping protocols. This is even more pertinent in a phenotype-driven approach, in which large numbers of mice are screened and the success of the screening depends primarily on the reduction of variability in the parameters being screened (statistical error). Currently, efforts to deal with this problem include a European research program concerned with the development of new approaches in phenotyping, mutagenesis and informatics leading to improved characterization of mouse models for the understanding of human physiology and disease. The focus of this program, as with others, is the development, standardization and dissemination of primary and secondary phenotyping protocols for all body systems in the mouse. The European Mouse Phenotyping Resource for Standardized Screens (EMPReSS) comprises a single searchable database of Standard Operating Procedures (SOPs) for the characterization of mouse mutants. Included within this program is a group of procedures devoted to the identification, standardization and optimization of behavioral, neurological and neuromuscular tests that reliably detect dysfunctions in mutant mice which are relevant to human neurological and psychiatric disorders. For example, the SHIRPA screening method includes a series of individual tests to assess the performance of mice in a wide range of tasks. These simple tests are designed to detect defects in, for example, the functions of lower and upper motor neurons, muscles, sensory neurons and behavior. As a result of the standardization of the SHIRPA protocol, direct comparisons can be made between animals over a given period of time and between groups. Recently the SHIRPA test has been cross-validated between different laboratories in Europe using a number of inbred strains. Results from the cross-validation support the idea that SHIRPA can be standardized and is a reproducible primary screen for mice. Detailed descriptions of SHIRPA SOPs and other validated SOPs are available on the EUMORPHIA website (www.eumorphia.org) or refer to Chapter 13 for further information.

6.6
Outlook

The potential for phenotyping behavioral and neurological characteristics in mice has progressed dramatically over the past decade. In fact a large proportion of targeted and randomly-generated mutants show at least some form of neurological, neuromuscular or behavioral anomaly. However, it is our belief that improvements in phenotyping and in our analysis of phenotypic data should continue to expand our understanding of both the fields of neuroscience and functional genomics. The three areas where we envisage major developments are discussed below.

6.6.1 Use of Imaging Technology

Non-invasive imaging can be used as a valuable tool for screening mutant mice or for investigating anatomical disturbances associated with, for example, cognitive dysfunctions or muscular dystrophies. Magnetic resonance imaging (MRI) and positron emission tomography (PET) are non-invasive imaging techniques that have an important place in phenotypic characterization. MRI has the potential of revealing histological lesions or morphological changes in a non-invasive manner and, as more cost-effective equipment adapted to mouse analysis becomes available, it may be feasible to incorporate this technology into first or second line screens. MRI imaging techniques have already been used to characterize abnormalities in the brains of mutant mice [151, 152], to follow dystrophy and regeneration in muscles of dystrophic mice [153, 154] and to investigate mouse mutant cardiomyopathies [155]. Furthermore, animals can be repeatedly imaged over time to assess whether progressive changes occur. PET is a versatile and informative imaging technology although it requires some development prior to its widespread use in small animal imaging. For PET, the low resolution relative to the size of a mouse (as well as the lack of cooperation from the subjects) can lead to difficulties in quantitating data from mouse PET images. At a basic level, flurodeoxyglucose can be used to study oxygen utilization and hence assess, for example, alterations in brain-region activities in mutant animals [156]. Moreover, the versatility of the technology allows the use of labeled ligands or reporter genes to study specific cell types or molecular aberrations in brains of mutant mice [157, 158].

6.6.2 Investigation of Complex Traits in Compound Mutants: Sensitized Screens

Human diseases where susceptibility is determined by the inheritance of alleles at many loci present particular difficulties for mapping and cloning. This is particularly relevant in neurological and behavioral disorders where the onset, progression and manifestations of the disease are remarkably heterogeneous. These observations are also evident in mouse studies. For example, several compound mutants may show strong specific phenotypes, whereas the respective single gene mutations do not exhibit any anomalies or only weak phenotypes at most. Similarly, in an extreme situation, a particular mutant phenotype can be severe on one genetic background and absent on another. One approach in dissecting these diseases and phenotypes is to sensitize an ENU mutagenesis screen by the use of other mutations (see for example Nelms and Goodnow [159]). Given the wide range of knockout mutations available and knowledge of the physiology of a system, this approach could be applied to many disease areas. The application of this approach to the study of complex human disease traits provides just one of many exciting opportunities for the development of ENU-based gene function studies. Using a similar strategy, the use of sensitized screens for behavioral phenotypes that combine random mutagenesis with pharmacological manipulations may yield mutations that affect a defined neurochemical pathway.

6.6.3
Use of Reporter Strains

The use of reporter strains in the characterization of mutant phenotypes adds a new degree of sophistication to phenotyping batteries. It is now possible to combine high resolution confocal imaging with the use of reporter strains carrying, for example, fluorescent motoneurons [160]. Thus, an instant image of abnormal axon branching patterns and neuromuscular junction aberrations can be observed and scored. The use of a less specific promoter driving the expression of the reporter gene [161] should also lead to the detection of morphological changes in other tissues including muscle, and provide exciting prospects in the field of mouse mutagenesis. Novel methods that allow tissue evaluation *in situ* are currently being developed following major breakthroughs in imaging technologies [162]. The latter include fluorescence endoscopy, optical coherence tomography, confocal microendoscopy, and molecular imaging. Thus, it is now feasible to carry out *in vivo* confocal observations on mouse lines carrying fluorescent markers labeling specific cell types. The increased resolution achieved by these technologies allows for instant diagnosis, provides a realistic opportunity for tissue phenotyping as a first-line screen and offers great potential for high-throughput analysis of mutagenized mice.

References

1 Quon, D., Wang, Y., Catalano, R., Scardina, J. M., Murakami, K., Cordell, B. **1991**, *Nature* 352, 239–241.
2 Moran, P. M., Higgins, L. S., Cordell, B., Moser, P. C. **1995**, *Proc. Natl Acad. Sci. USA* 92, 5341–5345.
3 Koistinaho, M., Ort, M., Cimadevilla, J. M., Vondrous, R., Cordell, B., Koistinaho, J., Bures, J., Higgins, L. S. **2001**, *Proc. Natl Acad. Sci. USA*, 98, 14675–14680. Epub 12001 Nov 14627.
4 Mudher, A., Lovestone, S. **2002**, *Trends Neurosci.* 25, 22–26.
5 Higgins, G. A., Jacobsen, H. **2003**, *Behav. Pharmacol.* 14, 419–438.
6 Polymeropoulos, M. H., Lavedan, C., Leroy, E., Ide, S. E., Dehejia, A., Dutra, A., Pike, B., Root, H., Rubenstein, J., Boyer, R. et al. **1997**, *Science* 276, 2045–2047.
7 Kruger, R., Kuhn, W., Muller, T., Woitalla, D., Graeber, M., Kosel, S., Przuntek, H., Epplen, J. T., Schols, L., Riess, O. **1998**, *Nat. Genet.* 18, 106–108.
8 Spillantini, M. G., Schmidt, M. L., Lee, V. M., Trojanowski, J. Q., Jakes, R., Goedert, M. **1997**, *Nature* 388, 839–840.
9 Farrer, M., Maraganore, D. M., Lockhart, P., Singleton, A., Lesnick, T. G., de Andrade, M., West, A., de Silva, R., Hardy, J., Hernandez, D. **2001**, *Hum. Mol. Genet.* 10, 1847–1851.
10 Chung, K. K., Dawson, V. L., Dawson, T. M. **2003**, *J. Neurol.* 250, III15–24.
11 Herbst, D. S., Miller, J. R. **1980**, *Am. J. Med. Genet.* 7, 461–469.
12 Reeves, R. H., Baxter, L. L., Richtsmeier, J. T. **2001**, *Trends Genet.* 17, 83–88.
13 Gropp, A., Kolbus, U., Giers, D. **1975**, *Cytogenet. Cell Genet.* 14, 42–62.
14 Dierssen, M., Fillat, C., Crnic, L., Arbones, M., Florez, J., Estivill, X. **2001**, *Physiol. Behav.* 73, 859–871.
15 World Health Organization **1992**, *International Classification of Diseases*, 10th edn, WHO, Geneva.
16 American Psychiatric Association **1994**, *Diagnostic and Statistical Manual of Mental Disorders*, 4th edn, *DSM-IV*), American Psychiatric Association, Washington, DC.
17 de Castro, J. M. **2004**, *Br. J. Nutr.* 92, S59–S62.

18 Johansen, J. E., Fetissov, S., Fischer, H., Arvidsson, S., Hokfelt, T., Schalling, M. **2003**, *Eur. J. Pharmacol.* 480, 171–176.
19 Siegfried, Z., Berry, E. M., Hao, S., Avraham, Y. **2003**, *Physiol. Behav.* 79, 39–45.
20 Veenstra-VanderWeele, J., Cook, E. H., Jr. **2004**, *Mol. Psychiatry* 9, 819–832.
21 Dawson, G., Webb, S., Schellenberg, G. D., Dager, S., Friedman, S., Aylward, E., Richards, T. **2002**, *Dev. Psychopathol.* 14, 581–611.
22 Moy, S. S., Nadler, J. J., Perez, A., Barbaro, R. P., Johns, J. M., Magnuson, T. R., Piven, J., Crawley, J. N. **2004**, *Genes Brain Behav.* 3, 287–302.
23 Owen, M. J., Williams, N. M., O'Donovan, M. C. **2004**, *Mol. Psychiatry* 9, 14–27.
24 Harrison, P. J., Weinberger, D. R. **2005**, *Mol. Psychiatry* 10, 40–68.
25 Angelucci, F., Mathe, A. A., Aloe, L. **2004**, *Prog. Brain Res.* 146, 151–165.
26 Alleva, E., Della Seta, D., Cirulli, F., Aloe, L. **1996**, *Prog. Neuropsychopharmacol. Biol. Psychiatry* 20, 483–489.
27 Strange, P. G. **2001**, *Pharmacol. Rev.* 53, 119–133.
28 Altar, C. A. **1999**, *Trends Pharmacol. Sci.* 20, 59–61.
29 Duman, R. S. **2004**, *Biol. Psychiatry* 56, 140–145.
30 Siuciak, J. A., Lewis, D. R., Wiegand, S. J., Lindsay, R. M. **1997**, *Pharmacol. Biochem. Behav.* 56, 131–137.
31 Gordon, J. A., Hen, R. **2004**, *Annu. Rev. Neurosci.* 27, 193–222.
32 Turri, M. G., Henderson, N. D., DeFries, J. C., Flint, J. **2001**, *Genetics* 158, 1217–1226.
33 Gershenfeld, H. K., Paul, S. M. **1997**, *Genomics* 46, 1–8.
34 Flint, J., Corley, R., DeFries, J. C., Fulker, D. W., Gray, J. A., Miller, S., Collins, A. C. **1995**, *Science* 269, 1432–1435.
35 Flint, J. **2003**, *J. Neurobiol.* 54, 46–77.
36 Yalcin, B., Willis-Owen, S. A., Fullerton, J., Meesaq, A., Deacon, R. M., Rawlins, J. N., Copley, R. R., Morris, A. P., Flint, J., Mott, R. **2004**, *Nat. Genet.* 36, 1197–1202. Epub 2004 Oct 1117.
37 Heisler, L. K., Chu, H. M., Brennan, T. J., Danao, J. A., Bajwa, P., Parsons, L. H., Tecott, L. H. **1998**, *Proc. Natl Acad. Sci. USA* 95, 15049–15054.
38 Parks, C. L., Robinson, P. S., Sibille, E., Shenk, T., Toth, M. **1998**, *Proc. Natl Acad. Sci. USA* 95, 10734–10739.
39 Ramboz, S., Oosting, R., Amara, D. A., Kung, H. F., Blier, P., Mendelsohn, M., Mann, J. J., Brunner, D., Hen, R. **1998**, *Proc. Natl Acad. Sci. USA* 95, 14476–14481.
40 Toth, M. **2003**, *Eur. J. Pharmacol.* 463, 177–184.
41 Hedera, P., Petty, E. M., Bui, M. R., Blaivas, M., Fink, J. K. **2003**, *Arch. Neurol.* 60, 1321–1325.
42 Onengut, S., Ugur, S. A., Karasoy, H., Yuceyar, N., Tolun, A. **2004**, *Neuromuscul. Disord.* 14, 4–9.
43 Campbell, K. P., Kahl, S. D. **1989**, *Nature* 338, 259–262.
44 Monaco, A. P., Neve, R. L., Colletti-Feener, C., Bertelson, C. J., Kurnit, D. M., Kunkel, L. M. **1986**, *Nature* 323, 646–650.
45 Hoffman, E. P., Brown, R. H., Jr., Kunkel, L. M. **1987**, *Cell* 51, 919–928.
46 Allamand, V., Campbell, K. P. **2000**, *Hum. Mol. Genet.* 9, 2459–2467.
47 Laval, S. H. and Bushby, K. M. **2004**, *Neuropathol. Appl. Neurobiol.* 30, 91–105.
48 Moll, J., Barzaghi, P., Lin, S., Bezakova, G., Lochmuller, H., Engvall, E., Muller, U., Ruegg, M. A. **2001**, *Nature* 413, 302–307.
49 Lynch, G. S., Cuffe, S. A., Plant, D. R., Gregorevic, P. **2001**, *Neuromuscul. Disord.* 11, 260–268.
50 Barton, E. R., Morris, L., Musaro, A., Rosenthal, N., Sweeney, H. L. **2002**, *J. Cell Biol.* 157, 137–148.
51 Spencer, M. J., Guyon, J. R., Sorimachi, H., Potts, A., Richard, I., Herasse, M., Chamberlain, J., Dalkilic, I., Kunkel, L. M., Beckmann, J. S. **2002**, *Proc. Natl Acad. Sci. USA* 99, 8874–8879.
52 Monani, U. R., Coovert, D. D. and Burghes, A. H. **2000**, *Hum. Mol. Genet.* 9, 2451–2457.
53 Hafezparast, M., Klocke, R., Ruhrberg, C., Marquardt, A., Ahmad-Annuar, A., Bowen, S., Lalli, G., Witherden, A. S., Hummerich, H., Nicholson, S. et al. **2003**, *Science* 300, 808–812.
54 Xu, H., Wu, X. R., Wewer, U. M., Engvall, E. **1994**, *Nat. Genet.* 8, 297–302.
55 Bittner, R. E., Anderson, L. V., Burkhardt, E., Bashir, R., Vafiadaki, E., Ivanova, S., Raffelsberger, T., Maerk, I., Hoger, H., Jung, M. et al. **1999**, *Nat. Genet.* 23, 141–142.

56 Bridges, L. R., Coulton, G. R., Howard, G., Moss, J., Mason, R. M. **1992**, *Muscle Nerve* 15, 172–179.
57 Hino, H., Araki, K., Uyama, E., Takeya, M., Araki, M., Yoshinobu, K., Miike, K., Kawazoe, Y., Maeda, Y., Uchino, M. et al. **2004**, *Hum. Mol. Genet.* 13, 181–190.
58 Corbett, M. A., Robinson, C. S., Dungli-son, G. F., Yang, N., Joya, J. E., Stewart, A. W., Schnell, C., Gunning, P. W., North, K. N., Hardeman, E. C. **2001**, *Hum. Mol. Genet.* 10, 317–328.
59 Hagiwara, Y., Sasaoka, T., Araishi, K., Imamura, M., Yorifuji, H., Nonaka, I., Ozawa, E., Kikuchi, T. **2000**, *Hum. Mol. Genet.* 9, 3047–3054.
60 Jayasinha, V., Nguyen, H. H., Xia, B., Kammesheidt, A., Hoyte, K., Martin, P. T. **2003**, *Neuromuscul. Disord.* 13, 365–375.
61 Bansal, D., Miyake, K., Vogel, S. S., Groh, S., Chen, C. C., Williamson, R., McNeil, P. L., Campbell, K. P. **2003**, *Nature* 423, 168–172.
62 Rogers, D. C., Peters, J., Martin, J. E., Ball, S., Nicholson, S. J., Witherden, A. S., Hafezparast, M., Latcham, J., Robinson, T. L., Quilter, C. A. et al. **2001**, *Neurosci. Lett.* 306, 89–92.
63 Nolan, P. M., Peters, J., Strivens, M., Rogers, D., Hagan, J., Spurr, N., Gray, I. C., Vizor, L., Brooker, D., Whitehill, E. et al. **2000**, *Nat. Genet.* 25, 440–443.
64 Hrabé de Angelis, M. H., Flaswinkel, H., Fuchs, H., Rathkolb, B., Soewarto, D., Marschall, S., Heffner, S., Pargent, W., Wuensch, K., Jung, M. et al. **2000**, *Nat. Genet.* 25, 444–447.
65 Avraham, K. B., Schickler, M., Sapoznikov, D., Yarom, R., Groner, Y. **1988**, *Cell* 54, 823–829.
66 Dal Canto, M. C., Gurney, M. E. **1995**, *Brain Res.* **676**, 25–40.
67 Peled-Kamar, M., Lotem, J., Wirguin, I., Weiner, L., Hermalin, A., Groner, Y. **1997**, *Proc. Natl Acad. Sci. USA* 94, 3883–3887.
68 Rafael, J. A., Nitta, Y., Peters, J., Davies, K. E. **2000**, *Mamm. Genome* 11, 725–728.
69 Heller, A. H., Eicher, E. M., Hallett, M., Sidman, R. L. **1982**, *J. Neurosci.* 2, 924–933.
70 Entrikin, R. K., Abresch, R. T., Sharman, R. B., Larson, D. B., Levine, N. A. **1987**, *Muscle Nerve* 10, 293–298.
71 Kennel, P. F., Fonteneau, P., Martin, E., Schmidt, J. M., Azzouz, M., Borg, J., Guenet, J. L., Schmalbruch, H., Warter, J. M., Poindron, P. **1996**, *Neurobiol. Dis.* 3, 137–147.
72 Haase, G., Kennel, P., Pettmann, B., Vigne, E., Akli, S., Revah, F., Schmalbruch, H., Kahn, A. **1997**, *Nat. Med.* 3, 429–436.
73 Zielasek, J., Toyka, K. V. **1999**, *Ann. NY Acad. Sci.* 883, 310–320.
74 Shettleworth, S. J. **2004**, *Nature* 430, 732–733.
75 Stamps, J. A., Krishnan, V. V. **1999**, *Quart. Rev. Biol.* 74, 291–318.
76 Hughes, R. N. **2004**, *Neurosci. Biobehav. Rev.* 28, 497–505.
77 Morris, R. G., Garrud, P., Rawlins, J. N., O'Keefe, J. **1982**, *Nature* 297, 681–683.
78 Barnes, C. A. **1979**, *J. Comp. Physiol. Psychol.* 93, 74–104.
79 Walker, J. A., Olton, D. S. **1979**, *Physiol. Behav.* 23, 11–15.
80 Croll, S. D., Suri, C., Compton, D. L., Simmons, M. V., Yancopoulos, G. D., Lindsay, R. M., Wiegand, S. J., Rudge, J. S., Scharfman, H. E. **1999**, *Neuroscience* 93, 1491–1506.
81 Crawley, J. N., Belknap, J. K., Collins, A., Crabbe, J. C., Frankel, W., Henderson, N., Hitzemann, R. J., Maxson, S. C., Miner, L. L., Silva, A. J. et al. **1997**, *Psychopharmacology (Berl.)* **132**, 107–124.
82 Hurst, J. L., Payne, C. E., Nevison, C. M., Marie, A. D., Humphries, R. E., Robertson, D. H., Cavaggioni, A., Beynon, R. J. **2001**, *Nature* 414, 631–634.
83 Humphries, R. E., Robertson, D. H., Beynon, R. J., Hurst, J. L. **1999**, *Anim. Behav.* 58, 1177–1190.
84 Nevison, C. M., Barnard, C. J., Beynon, R. J., Hurst, J. L. **2000**, *Proc. R. Soc. Lond. B Biol. Sci.* 267, 687–694.
85 Liebenauer, L. L., Slotnick, B. M. **1996**, *Physiol. Behav.* 60, 403–409.
86 Gray, S. J., Hurst, J. L. **1998**, *Anim. Behav.* 56, 1291–1299.
87 Rodgers, R. J. **2001**, *Behav. Pharmacol.* 12, 471–476.
88 Pellow, S., File, S. E. **1986**, *Pharmacol. Biochem. Behav.* 24, 525–529.
89 Crawley, J. N. **1981**, *Pharmacol. Biochem. Behav.* 15, 695–699.
90 Porsolt, R. D. **2000**, *Rev. Neurosci.* 11, 53–58.
91 Cryan, J. F., Mombereau, C. **2004**, *Mol. Psychiatry* 9, 326–357.

92 Geyer, M. A., Swerdlow, N. R., Mansbach, R. S., Braff, D. L. **1990**, *Brain Res. Bull.* 25, 485–498.

93 Braff, D. L., Grillon, C., Geyer, M. A. **1992**, *Arch. Gen. Psychiatry* 49, 206–215.

94 Gogos, J. A., Santha, M., Takacs, Z., Beck, K. D., Luine, V., Lucas, L. R., Nadler, J. V., Karayiorgou, M. **1999**, *Nat. Genet.* 21, 434–439.

95 Johansson, C., Willeit, M., Smedh, C., Ekholm, J., Paunio, T., Kieseppa, T., Lichtermann, D., Praschak-Rieder, N., Neumeister, A., Nilsson, L. G. et al. **2003**, *Neuropsychopharmacology* 28, 734–739.

96 von Zerssen, D., Barthelmes, H., Dirlich, G., Doerr, P., Emrich, H. M., von Lindern, L., Lund, R., Pirke, K. M. **1985**, *Psychiatry Res.* 16, 51–63.

97 Lenox, R. H., Gould, T. D., Manji, H. K. **2002**, *Am. J. Med. Genet.* 114, 391–406.

98 Harper, D. G., Stopa, E. G., McKee, A. C., Satlin, A., Harlan, P. C., Goldstein, R., Volicer, L. **2001**, *Arch. Gen. Psychiatry* 58, 353–360.

99 Giubilei, F., Patacchioli, F. R., Antonini, G., Sepe Monti, M., Tisei, P., Bastianello, S., Monnazzi, P., Angelucci, L. **2001**, *J. Neurosci. Res.* 66, 262–265.

100 Toh, K. L., Jones, C. R., He, Y., Eide, E. J., Hinz, W. A., Virshup, D. M., Ptacek, L. J., Fu, Y. H. **2001**, *Science* 291, 1040–1043.

101 Katzenberg, D., Young, T., Finn, L., Lin, L., King, D. P., Takahashi, J. S., Mignot, E. **1998**, *Sleep* 21, 569–576.

102 Vitaterna, M. H., King, D. P., Chang, A. M., Kornhauser, J. M., Lowrey, P. L., McDonald, J. D., Dove, W. F., Pinto, L. H., Turek, F. W., Takahashi, J. S. **1994**, *Science* 264, 719–725.

103 King, D. P., Zhao, Y., Sangoram, A. M., Wilsbacher, L. D., Tanaka, M., Antoch, M. P., Steeves, T. D., Vitaterna, M. H., Kornhauser, J. M., Lowrey, P. L. et al. **1997**, *Cell* 89, 641–653.

104 Lowrey, P. L., Shimomura, K., Antoch, M. P., Yamazaki, S., Zemenides, P. D., Ralph, M. R., Menaker, M., Takahashi, J. S. **2000**, *Science* 288, 483–492.

105 Cannon, T. D., Gasperoni, T. L., van Erp, T. G., Rosso, I. M. **2001**, *Am. J. Med. Genet.* 105, 16–19.

106 Braff, D. L., Geyer, M. A., Light, G. A., Sprock, J., Perry, W., Cadenhead, K. S., Swerdlow, N. R. **2001**, *Schizophr. Res.* 49, 171–178.

107 Castellanos, F. X., Tannock, R. **2002**, *Nat. Rev. Neurosci.* 3, 617–628.

108 Rogers, D. C., Fisher, E. M., Brown, S. D., Peters, J., Hunter, A. J., Martin, J. E. **1997**, *Mamm. Genome* 8, 711–713.

109 Nolan, P. M. **2000**, *Pharmacogenomics* 1, 243–255.

110 Blanco, G., Coulton, G. R., Biggin, A., Grainge, C., Moss, J., Barrett, M., Berquin, A., Marechal, G., Skynner, M., van Mier, P. et al. **2001**, *Hum. Mol. Genet.* 10, 9–16.

111 Hall, E. D., Wolf, D. L., Althaus, J. S., Von Voigtlander, P. F. **1987**, *Brain Res.* 435, 174–180.

112 Smith, J. P., Hicks, P. S., Ortiz, L. R., Martinez, M. J., Mandler, R. N. **1995**, *J. Neurosci. Methods* 62, 15–19.

113 Galbiati, F., Volonte, D., Chu, J. B., Li, M., Fine, S. W., Fu, M., Bermudez, J., Pedemonte, M., Weidenheim, K. M., Pestell, R. G. et al. **2000**, *Proc. Natl Acad. Sci. USA* 97, 9689–9694.

114 Hudecki, M. S., Pollina, C. M., Granchelli, J. A., Daly, M. K., Byrnes, T., Wang, J. C., Hsiao, J. C. **1993**, *Res. Commun. Chem. Pathol. Pharmacol.* 79, 45–60.

115 Messeri, P., Eleftheriou, B. E., Oliverio, A. **1975**, *Physiol. Behav.* 14, 53–58.

116 Long, J. M., LaPorte, P., Paylor, R. Wynshaw-Boris, A. **2004**, *Genes Brain Behav.* 3, 51–62.

117 Bunsey, M., Eichenbaum, H. **1995**, *Hippocampus* 5, 546–556.

118 Giese, K. P., Friedman, E., Telliez, J. B., Fedorov, N. B., Wines, M., Feig, L. A., Silva, A. J. **2001**, *Neuropharmacology* 41, 791–800.

119 Thor, D. H., Holloway, W. R. **1981**, *Anim. Learn. Behav.* 9, 561–565.

120 Engelmann, M., Wotjak, C. T., Landgraf, R. **1995**, *Physiol. Behav.* 58, 315–321.

121 Dluzen, D. E., Kreutzberg, J. D. **1993**, *Brain Res.* 609, 98–102.

122 Winslow, J. T., Camacho, F. **1995**, *Psychopharmacology (Berl.)* 121, 164–172.

123 Crusio, W. E. **2001**, *Behav. Brain Res.* 125, 127–132.

124 Dalvi, A., Lucki, I. **1999**, *Psychopharmacology (Berl.)* 147, 14–16.

125 Bourin, M., Hascoet, M. **2003**, *Eur. J. Pharmacol.* 463, 55–65.

126 Lister, R. G. 1987, *Psychopharmacology (Berl.)* 92, 180–185.

127 Ohl, F., Roedel, A., Binder, E. Holsboer, F. **2003**, *Eur. J. Neurosci.* 17, 128–136.
128 Porsolt, R. D., Le Pichon, M., Jalfre, M. **1977**, *Nature* 266, 730–732.
129 Bacon, Y., Ooi, A., Kerr, S., Shaw-Andrews, L., Winchester, L., Breeds, S., Tymoska-Lalanne, Z., Clay, J., Greenfield, A. G., Nolan, P. M. **2004**, *Genes Brain Behav.* 3, 196–205.
130 Sorichter, S., Puschendorf, B., Mair, J. **1999**, *Exerc. Immunol. Rev.* 5, 5–21.
131 Silva, A. J., Paylor, R., Wehner, J. M. Tonegawa, S. **1992**, *Science* 257, 206–211.
132 Silva, A. J. **2003**, *J. Neurobiol.* 54, 224–237.
133 Reisel, D., Bannerman, D. M., Schmitt, W. B., Deacon, R. M., Flint, J., Borchardt, T., Seeburg, P. H., Rawlins, J. N. **2002**, *Nat. Neurosci.* 5, 868–873.
134 Saudou, F., Amara, D. A., Dierich, A., LeMeur, M., Ramboz, S., Segu, L., Buhot, M. C., Hen, R. **1994**, *Science* 265, 1875–1878.
135 Kelz, M. B., Chen, J., Carlezon, W. A., Jr., Whisler, K., Gilden, L., Beckmann, A. M., Steffen, C., Zhang, Y. J., Marotti, L., Self, D. W. et al. **1999**, *Nature* 401, 272–276.
136 Mangiarini, L., Sathasivam, K., Seller, M., Cozens, B., Harper, A., Hetherington, C., Lawton, M., Trottier, Y., Lehrach, H., Davies, S. W. et al. **1996**, *Cell* 87, 493–506.
137 Gurney, M. E., Pu, H., Chiu, A. Y., Dal Canto, M. C., Polchow, C. Y., Alexander, D. D., Caliendo, J., Hentati, A., Kwon, Y. W., Deng, H. X. et al. **1994**, *Science* 264, 1772–1775.
138 Shen, J., Bronson, R. T., Chen, D. F., Xia, W., Selkoe, D. J., Tonegawa, S. **1997**, *Cell* 89, 629–639.
139 Benzer, S. **1971**, *JAMA* 218, 1015–1022.
140 Hitotsumachi, S., Carpenter, D. A., Russell, W. L. **1985**, *Proc. Natl Acad. Sci. USA* 82, 6619–6621.
141 Hagge-Greenberg, A., Snow, P., O'Brien, T. P. **2001**, *Mamm. Genome* 12, 938–941.
142 Herron, B. J., Lu, W., Rao, C., Liu, S., Peters, H., Bronson, R. T., Justice, M. J., McDonald, J. D., Beier, D. R. **2002**, *Nat. Genet.* 30, 185–189.
143 Nolan, P. M., Peters, J., Vizor, L., Strivens, M., Washbourne, R., Hough, T., Wells, C., Glenister, P., Thornton, C., Martin, J. et al. **2000**, *Mamm. Genome* 11, 500–506.
144 Tarantino, L. M., Gould, T. J., Druhan, J. P., Bucan, M. **2000**, *Mamm. Genome* 11, 555–564.
145 Wurbel, H. **2001**, *Trends Neurosci.* 24, 207–211.
146 Wahlsten, D., Metten, P., Phillips, T. J., Boehm, S. L., 2nd, Burkhart-Kasch, S., Dorow, J., Doerksen, S., Downing, C., Fogarty, J., Rodd-Henricks, K. et al. **2003**, *J. Neurobiol.* 54, 283–311.
147 Francis, D. D., Szegda, K., Campbell, G., Martin, W. D., Insel, T. R. **2003**, *Nat. Neurosci.* 6, 445–446.
148 Gerlai, R. **1996**, *Trends Neurosci.* 19, 177–181.
149 Graves, L., Dalvi, A., Lucki, I., Blendy, J. A., Abel, T. **2002**, *Hippocampus* 12, 18–26.
150 Young, K. A., Berry, M. L., Mahaffey, C. L., Saionz, J. R., Hawes, N. L., Chang, B., Zheng, Q. Y., Smith, R. S., Bronson, R. T., Nelson, R. J. et al. **2002**, *Behav. Brain Res.* 132, 145–158.
151 Greene, N. D., Lythgoe, M. F., Thomas, D. L., Nussbaum, R. L., Bernard, D. J., Mitchison, H. M. **2001**, *Eur. J. Paediatr. Neurol.* 5, 103–107.
152 Redwine, J. M., Kosofsky, B., Jacobs, R. E., Games, D., Reilly, J. F., Morrison, J. H., Young, W. G., Bloom, F. E. **2003**, *Proc. Natl Acad. Sci. USA* 100, 1381–1386.
153 McIntosh, L. M., Baker, R. E., Anderson, J. E. **1998**, *Biochem. Cell. Biol.* 76, 532–541.
154 Dunn, J. F., Zaim-Wadghiri, Y. **1999**, *Muscle Nerve* 22, 1367–1371.
155 Park, D. S., Woodman, S. E., Schubert, W., Cohen, A. W., Frank, P. G., Chandra, M., Shirani, J., Razani, B., Tang, B., Jelicks, L. A. et al. **2002**, *Am. J. Pathol.* 160, 2207–2217.
156 Valla, J., Chen, K., Berndt, J. D., Gonzalez-Lima, F., Cherry, S. R., Games, D., Reiman, E. M. **2002**, *Neuroimage* 16, 1–6.
157 Vowinckel, E., Reutens, D., Becher, B., Verge, G., Evans, A., Owens, T., Antel, J. P. **1997**, *J. Neurosci. Res.* 50, 345–353.
158 Sun, X., Annala, A. J., Yaghoubi, S. S., Barrio, J. R., Nguyen, K. N., Toyokuni, T., Satyamurthy, N., Namavari, M., Phelps, M. E., Herschman, H. R. et al. **2001**, *Gene Ther.* 8, 1572–1579.
159 Nelms, K. A., Goodnow, C. C. **2001**, *Immunity* 15, 409–418.

160 Feng, G., Mellor, R. H., Bernstein, M., Keller-Peck, C., Nguyen, Q. T., Wallace, M., Nerbonne, J. M., Lichtman, J. W., Sanes, J. R. **2000**, *Neuron* 28, 41–51.

161 Hadjantonakis, A. K., Nagy, A. **2001**, *Histochem. Cell. Biol.* 115, 49–58.

162 Wang, T. D., Van Dam, J. **2004**, *Clin. Gastroenterol. Hepatol.* 2, 744–753.

163 Galef, B. G., Jr., Whiskin, E. E. **2003**, *Learn. Behav.* 31, 160–164.

164 Gallistel, C. R. **2003**, *Behav. Processes* 62, 89–101.

165 Hofer, M. A., Shair, H. N., Masmela, J. R., Brunelli, S. A. **2001**, *Dev. Psychobiol.* 39, 231–246.

7
Cardiovascular Disorders: Insights into *In Vivo* Cardiovascular Phenotyping

Laurent Monassier and André Constantinesco

7.1
Introduction

Cardiovascular diseases constitute one of the leading causes of morbidity and mortality in industrialized countries with heart failure as the primary killer. Recently, a complex new challenge was presented to physiologists and pathophysiologists i. e. to identify new pharmacological targets from data obtained from mouse mutagenesis. At first the solution to the problem appeared to be quite simple: first identify the gene which encodes a particular protein and then observe the changes in the role of that protein when the gene is ablated or overexpressed. This information would then be passed to the pharmacochemists who would use it in the design of new drugs. This would lead to rapid preclinical development. In fact, the hurdle was significantly higher than was first believed and researchers were faced with numerous obstacles, the first being the immense gap between mouse and human physiology. Moreover, a large number of pharmacological cardiovascular studies had previously been carried out in the dog and the rat, which from a functional point of view are quite dissimilar to the mouse. The second problem encountered was the need for extensive phenotyping in the mouse. Since a wide range of transgenic animals have been generated, phenotyping centers should be able to provide precise and rapid analyses with a low rate of false negatives. Moreover, due to problems of fertility in some transgenic strains, the animals need to be kept alive throughout the phenotyping process and therefore non-invasive cardiovascular tests are required. The third major obstacle to cardiovascular drug development is the difficulty in choosing an experimental technique. The use of available techniques is limited by the unique aspects of mouse physiology, the size of the animals and by the spatial, temporal, functional and molecular resolutions of the available equipment particularly that dedicated to imaging. The present chapter will therefore focus on the main techniques used for *in vivo* cardiovascular phenotypic characterization of the mouse with particular emphasis on the cautious interpretation of data provided by *in vivo* imaging methods and on functional comparisons between mouse and human.

Standards of Mouse Model Phenotyping. Edited by Martin Hrabé de Angelis, Pierre Chambon, and Steve Brown

ISBN: 3-527-31031-2

7.2
In Vivo Imaging for Mouse Cardiovascular Phenotyping: Interests and Limits

In this section we will first give a short technical definition and indicate the main advantages and limitations of currently available *in vivo* tomographic imaging methods (Tab. 7.1). Since optical imaging is not an established method for cardiovascular phenotyping, it will be excluded from our discussions. The use of contrast agents and some crucial aspects of nuclear imaging will be discussed later in this section.

7.2.1
Echography

Ultrasound imaging is based on the path and reflection of acoustic mechanical waves through tissues. In the mouse good resolution can be obtained by using high frequency ultrasound (15–50 MHz). At the present time this low-cost method produces images of selected 2D slices in real time with high spatial (100 μ/pixel, slice thickness of 1 mm at 15 MHz) and good temporal (nine to 20 images/RR depending on the acquisition conditions) resolutions and together with the rapid acquisition time, these features make the method suitable for rapid cardiac investigations in anaesthetized and conscious animals [1] (Fig. 7.1). Nevertheless, echography is limited by its 2D-based reconstruction modality. Therefore, some researchers have developed 3D echography by conjugating 2D image acquisition with the translation of the mouse holder or transducer. Without having to create a geometrical model, this method will provide accurate ventricular and wall volume measurements. Using the same slice thickness as that for 2D echography, 3D echography has been made possible by the use of isotropic image voxels. Nevertheless, this improvement has led to a loss of up to 5 mm in resolution [2].

7.2.2
Magnetic Resonance Imaging (MRI)

The principle of MRI is based on the effect of a magnetic field on the ^{1}H nuclei of the water in tissue. The spatial resolution of cardiac imaging in the mouse can be improved with the use of high and very high magnetic fields (up to 11.7 T). Although the technique is time consuming (45 min to 1 h per examination) it provides very high spatial resolution (50–100 μ/pixel, slice thickness 1 mm) and good time resolution (15–25 images/beat), depending on the imaging sequences used. Because of the variable NMR properties of different tissues, MRI offers the unique opportunity to produce images with various contrasts. For example, (1) the T1 and T2 relaxation properties in tissue lead to a contrast between the images of anatomical structures and edema respectively and (2) the contrast in angiographic images is due to the phase difference in the flow velocity of protons in the blood. Because standard MR acquisition processes have facilitated the detection of heart rate and motion, all phases of the cardiac cycle can now be analyzed in detail. Nevertheless, accurate physiological monitoring of these phases can only be achieved if the ani-

Table 7.1 Properties and limitations of the main tomographic *in vivo* small animal imaging modalities.

Imaging method	Mean examination duration [mn]	2D/ 3D	Isotropic	In plane spatial resolution [mm]	Temporal res. (time bins/RR)	ECG gating	Resp. gating	LV volumes	Wall motion	Wall perfusion and metabolism	Relative cost
Echo	5	2D/3D	N/Y	01–0.5	9–20	Y	N	F	S	N	++(+)
MRI	45	2DC (2D)*	N	0.05–0.1	15–25	Y	Y	Y	S, T ?	NA	+++++
SPECT	45	3D	Y	0.9–1	10	Y	NA	Y	T	Y	+(+)
PET	45	2DC, (3D)	N	1.2–2	NA	NA	NA	NA	NA	Y	+++
CT	40	2DC, 3D	Y (3D)	0.045–0.1	NA	NA	Y	NA	NA	NA	++(+)

( ): work in progress, unpublished data; ( )*: 2D in the x, y, z directions; 2DC: 2D contiguous slices. F: formula based on diameter measurements; Y: yes; N: no; S: shortening; T: thickening; NA: no data available.

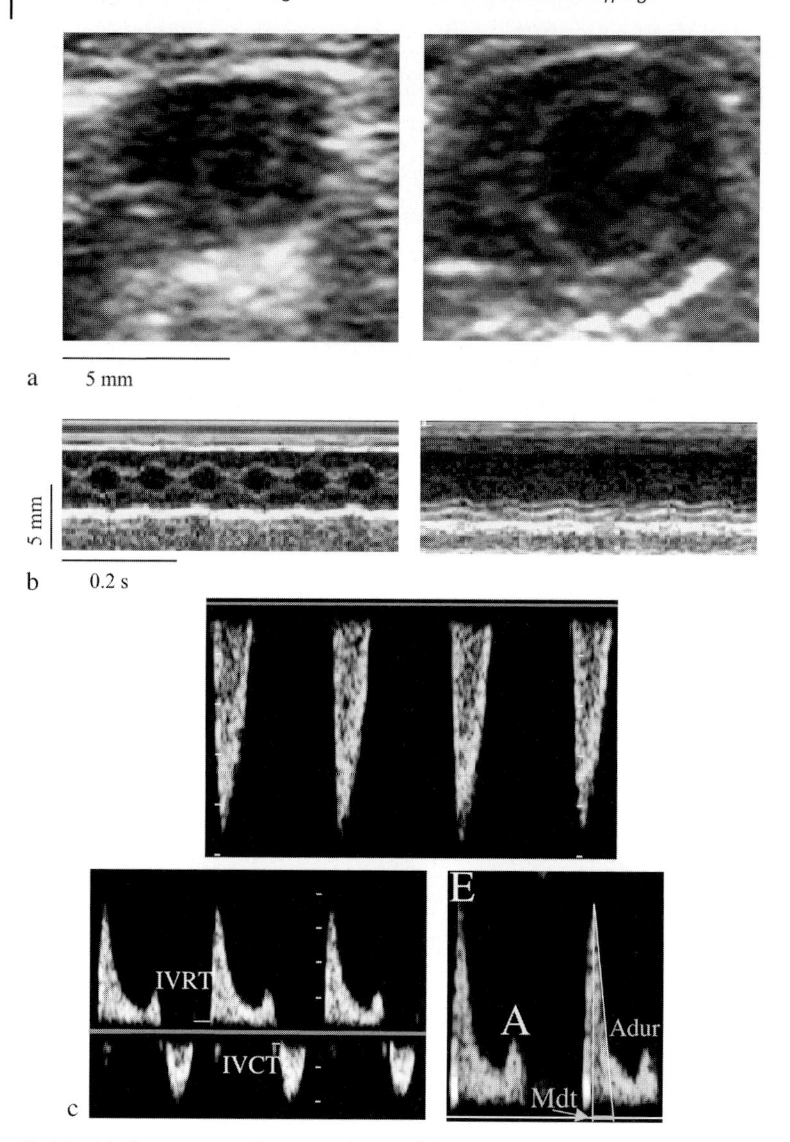

Fig. 7.1 (a) Short-axis two-dimensional view of the mouse heart obtained at the level of papillary muscles in a 7-week-old conscious mouse. Left, normal mouse; right, mouse exhibiting a ventricular dilatation due to heart failure (from [1]). (b) Corresponding M-mode imaging to (a). (c) Doppler analysis of the mouse heart. Upper panel, aortic flow; left lower panel, simultaneous recording of mitral and aortic flows (IVRT, isovolumetric relaxation time; IVCT, isovolumetric contraction time); right lower panel, mitral inflow pattern (E, E wave; A, A wave; Mdt, mean deceleration time of the E wave; Adur, duration of the A wave).

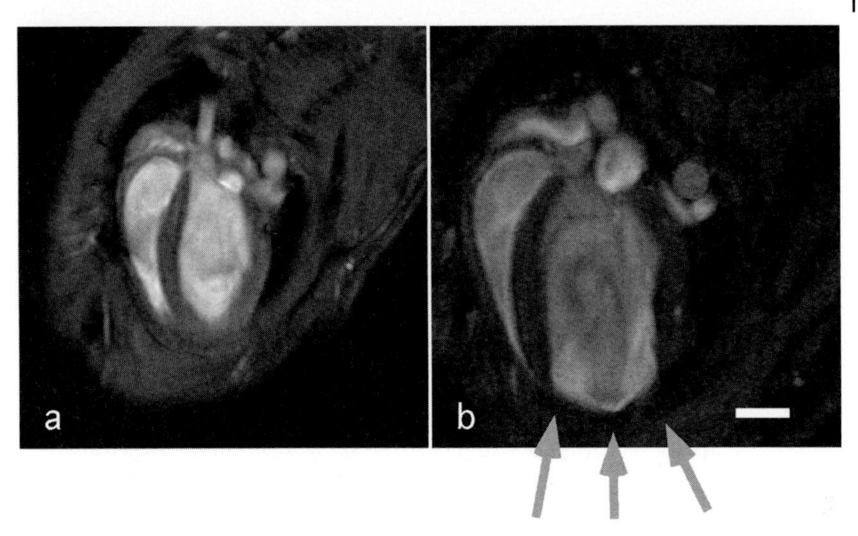

Fig. 7.2 Mid-ventricular end-diastolic MRI (11.7 T) images in the four-chamber long-axis orientation. (a) Normal mouse heart. (b) Post-myocardial infarction failure in mouse heart. Arrows indicate the infarcted area which is characterized by akinesis and thinning of the wall. Pixel size, 0.05 × 0.05 × 1 mm; scale bar, 2 mm. (Courtesy of Dr J. Schneider, Dr S. Grieve and Professor S. Neubauer, Department of Cardiovascular Medicine, John Radcliffe Hospital, Oxford, UK).

mal is placed in the correct position at the center of the magnet [3]. Over recent years, in addition to the visualization of coronary arteries and heart valves, measurements of left ventricular mass, end-diastolic, systolic and stroke volumes, ejection fraction, wall thicknesses and cardiac output [4–8] have been carried out [9] and validated in both normal and diseased mice (Fig. 7.2). The reliability and reproducibility of these measurements have also been assessed. For optimization purposes, a chart of relaxation times can be computed. Importantly, the transthoracic imaging of the right ventricle is limited with echocardiography but its anatomy and function can be investigated using MRI [10]. Although essential for investigating the structure and function of the right ventricle, more importantly, MRI can be used to detect increases in right ventricular pressure due to congenital pulmonary artery stenosis and plumonary hypertension (primary or secondary to a change in the left ventricle). Interestingly, as in humans, the response of the mouse myocardium to a pharmacological stressor can now be analyzed by MRI [11]. This is an important achievement in phenotyping studies because cardiac investigations using MRI require anesthesia, which by itself, because of the reduction in cardiac contractility and the loading conditions, could affect the results of the phenotyping analysis. Moreover, some ischemic phenotypes can only be revealed by the induction of cardiovascular stress. In this context, a reduction in overall function can be detected but in many cases the alteration does not affect the entire organ but only parts of the muscular wall. Therefore, it is necessary to identify and analyze anomalies in the various segments. Abnormalities of ventricular motion can now be studied by tagging, displacement encoding or velocity mapping sequences in the mouse [12–14]. Such MRI studies should be complemented by spin labeling sequence studies to provide important information regarding the perfusion of myocardial tissue [15]. Contrast agents such as the widely used gadolinium, iron oxide nanoparticles or manganese, which combine the properties of being an NMR contrast agent and a calcium analog can be used as novel contrast agents for tissue perfusion, thus providing a method for the assessment of metabolism that is complementary to standard MRI morphologic measurements [16–18]. Coupled to the analysis of the coronary blood flow and segmental motion, it has recently been demonstrated that the study of energetic metabolism is now possible using ^{31}P NMR spectroscopy [19, 20]. Taken together, these data indicate that the use of MRI provides a unique opportunity to collect data in relation to the contractility (at rest and during pharmacological stress test), perfusion and metabolic activity of a particular spatially well-defined area of the myocardium.

MRI is now well established as the current method for studying cardiac function in the mouse. Nevertheless, establishing magnet fields and optimizing the experimental environment requires the expertise of a skilled technician in order to produce reliable results [21–24]. Investigators should take into account that MRI is probably the most versatile of all the imaging modalities but also the most complex as it is a technique that requires skill and a great deal of experience in its implementation. Moreover it is a high-cost technique with low throughput. Because of these factors the use of MRI tends to be limited to specialized investigations following the initial diagnosis of a phenotype.

7.2.3 Single Photon Emission Computed Tomography (SPECT)

SPECT is a functional and metabolic imaging method based on the detection by scintillators of gamma rays emitted by radio-labeled tracers. The half-lives of the tracers are of several hours duration which makes it possible to study the physiological equilibrium of tissues. The image magnification needed for cardiac applications in mice is achieved with very small aperture pinhole collimators made of lead, tungsten or gold, this results in micro-SPECT and offers the advantage of sub-millimeter/millimeter spatial resolution with 3D isotropic voxels [25]. This method is time consuming (45 min to 1 h) and can only be carried out in anesthetized animals. ECG-gated micro-SPECT gives a time resolution of about 10–16 time bins/RR which is somewhat less than other imaging modalities and could therefore affect the accuracy of measurements of the diastolic parameter. However, micro-SPECT offers the advantage of measuring mean values of cardiac parameters (tissue perfusion, ejection fraction, end-diastolic volume, end-systolic volume, cardiac output wall thickening) over hundreds of cardiac cycles under steady-state physiological conditions (Fig. 7.3). Micro-SPECT is well suited to pharmacological rest and stress studies as it is able to reveal not only reversible segmental ischemia but also to detect viable myocardial tissue because of the intracellular myocyte specificity of the single-photon tracers used.

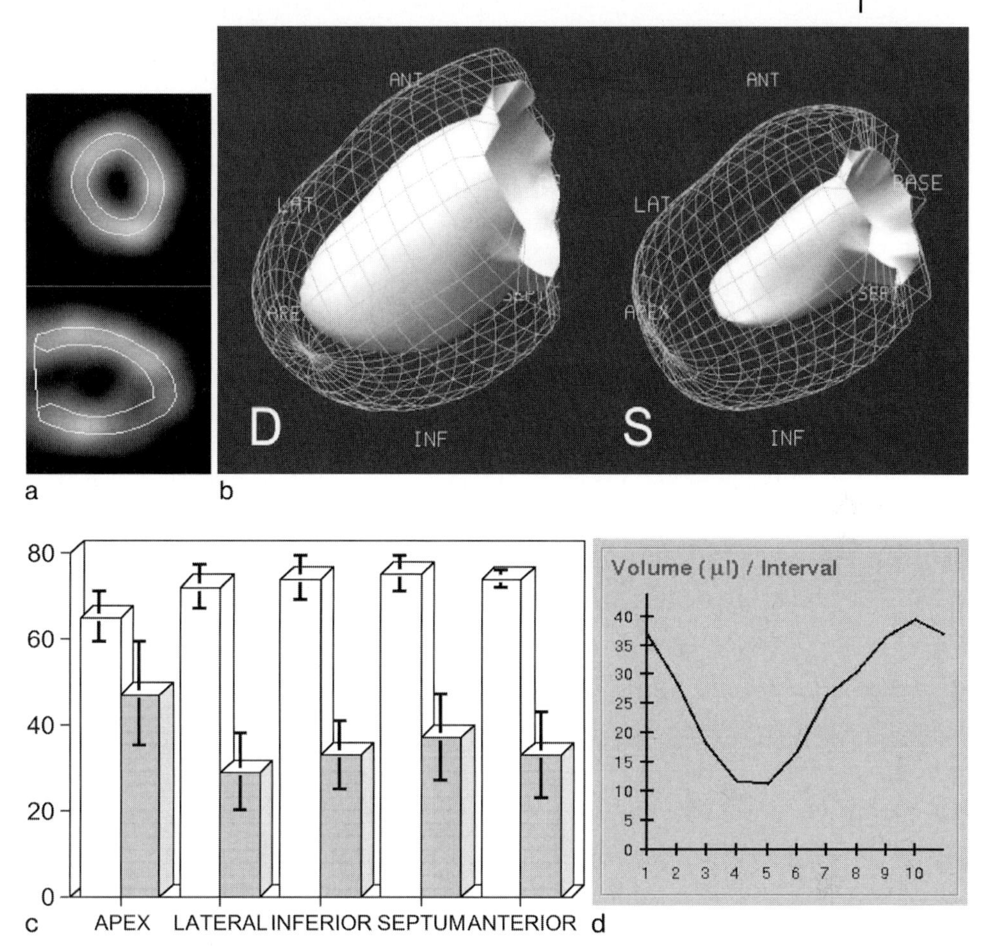

Fig. 7.3 Perfusion and function of left ventricular wall in a normal adult mouse obtained in our laboratory by micro-gated SPECT after i. v. injection of ^{99m}Tc-Tetrofosmin (Amersham, GE Healthcare, USA). (a) Examples of mouse perfusion at the levels of the mid-left ventricular short-axis and horizontal axis and computed endocardial and epicardial contours. (b) Computed end-diastolic (D) and end-systolic (S) left ventricular volumes corresponding to (a). (c) Mean values expressed as a percentage and standard deviations of segmental perfusion (white) and wall thickening (grey) in a series of normal mice (adult females CD1, $n = 8$). (d) Typical left ventricular ejection curve in a normal mouse.

7.2.4
Positron Emission Tomography (PET) Imaging

PET imaging involves dual gamma (511 keV) emission at 180° following annihilation of a positron in the tissue. Micro-PET systems have been developed and used for cardiac imaging in small animals [26]. Despite the high level of sensitivity resulting from the high photon density count compared to pinhole SPECT, there are fundamental limitations to the spatial resolution in PET due to random movement of positrons in tissues before annihilation with an electron and detection of dual 511 keV high-energy photons. This random movement is also dependent on the isotope being used and therefore the spatial resolution of micro-PET is actually between 1.2 and 2 mm in the center of the field of view (FOV) and less within 5 cm of the center of the FOV. Nevertheless, the ultimate 1-mm resolution is expected to be available within the next few years due to progress being made in image reconstruction algorithms. ECG-gated micro-PET has not yet been undertaken in mice. However PET imaging has unrivalled advantages linked to the fact that many positron emitters correspond to isotopes of atoms which are fundamental to basic biological processes such as ^{11}C, ^{15}O and ^{13}N. Moreover, these tracers can be used to study a wide range of metabolic pathways such as the glucose pathway using ^{18}F-2-deoxy-2-fluoro-glucose (FDG). PET also offers the advantage of rapid serial (or dynamic) acquisition of images at a frame rate in the range of 2–10 s per image which is appropriate for the generation of time activity curves, and compatible with radiotracer kinetic studies and the evaluation of the effects of pharmacological interventions [26].

7.2.5
X-Ray Computed Tomography (CT)

X-ray CT (known as micro-CT for concerned applications) offers a unique high contrast between bone and soft tissues, as well as high resolution nAQ5 nof 100 µm side length utilizing voxels [27]. A higher resolution can be achieved with longer acquisition times but only with longer exposure to ionizing radiations. In every study, this key issue should be addressed because too high a level of radiation may jeopardize the safety of the study. With regard to soft tissues such as the blood, the contrast is quite poor. Therefore, to allow for the long acquisition time, specific iodinated contrast compounds such as pool agents (Fenestra VC, Alerion Biomedical Inc, San Diego, CA, USA, www.alerionbio.com) have been developed (Fig. 7.4 a). They allow for excellent contrast between the blood pool and myocardial tissue and facilitate the visualization of blood vessels. Access to 3D acquisition and precise reconstruction of cardiac structures and the vascular tree is now accessible to micro-CT for which 3D-based reconstruction is the rule. For this purpose, a recently developed new prototype seems quite promising [28]. Its use for studying tumor vascularization has now to be extended to other cardiovascular applications. Similar to echography, micro-CT appears to be a simpler technique than MRI or nuclear imaging. This modality probably allows the higher throughput and its use in large-scale phenotypic cardiovascular studies seems realistic. In the very near future, new dedicated devices will probably facilitate the complete morphological evaluation of heart and blood vessels in the mouse.

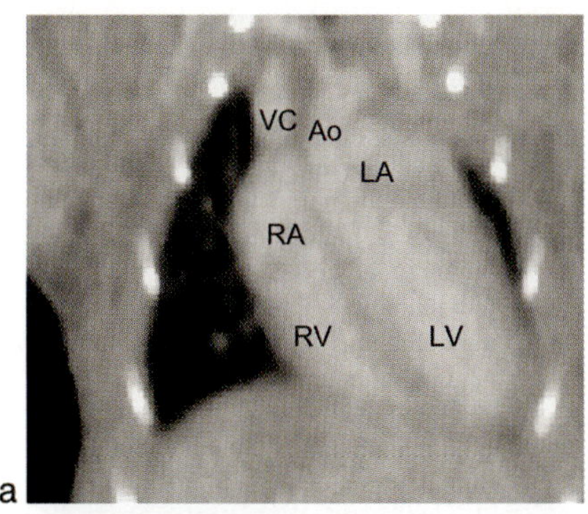

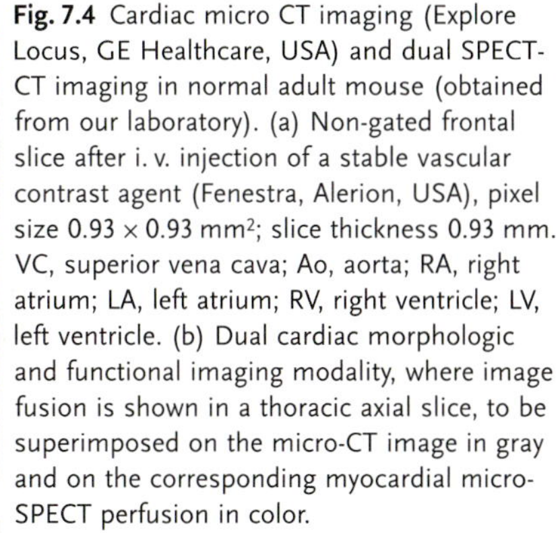

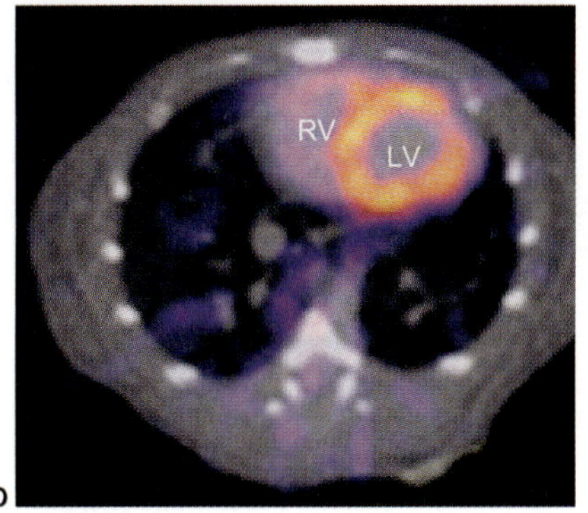

Fig. 7.4 Cardiac micro CT imaging (Explore Locus, GE Healthcare, USA) and dual SPECT-CT imaging in normal adult mouse (obtained from our laboratory). (a) Non-gated frontal slice after i. v. injection of a stable vascular contrast agent (Fenestra, Alerion, USA), pixel size 0.93×0.93 mm^2; slice thickness 0.93 mm. VC, superior vena cava; Ao, aorta; RA, right atrium; LA, left atrium; RV, right ventricle; LV, left ventricle. (b) Dual cardiac morphologic and functional imaging modality, where image fusion is shown in a thoracic axial slice, to be superimposed on the micro-CT image in gray and on the corresponding myocardial micro-SPECT perfusion in color.

7.2.6
Limitations in Studies Using Contrast Agents: Particular Aspects of Nuclear-based Imaging

Nowadays, nearly all imaging technologies have been adapted from human to mouse species and commercially-available systems are now being developed. Of course the selection of a particular modality will depend on the application but there is an important issue to consider. It is possible to divide all the techniques into two groups: (1) the methods in which contrast as visualized through adequate physical stimulation (X-ray CT-scan, MRI, echography), is an intrinsic property of the tissue and organs; such property can be enhanced by contrast agents and (2) those for which no contrast can be obtained and as a consequence no image can be produced without the administration of a tracer which interacts closely with specific molecules, cells, tissues or organs (SPECT, PET). For the second group, the quality of the image will be highly dependent on the strength of the signal emanating from the tissue and, therefore, on the concentration of the tracer or contrast agent that has been administered. For methods which fall into the first group, although the spontaneous contrast is a real advantage, in cases where it is necessary to use a contrast agent, their use will be limited by the high doses which would be required. For MRI and CT-scans, obtaining a good signal with these agents is a significant technical challenge because the minimum detectable concentrations of tracer or contrast agent are respectively 40 μM for Gd or Fe and 2 mM for iodine, corresponding to approximately 3.10^7 and 10^9 atoms/cell respectively. On the other hand for PET and SPECT, these values are several orders of magnitude lower than those for radio-labeled tracers, 10 pM for ^{18}F, ^{64}Cu, ^{99m}Tc for instance corresponds to approximately 100 atoms/cell. In this respect nuclear imaging techniques represent the most sensitive tomographic imaging methods available to date. Pinhole SPECT and micro-PET enables the *in vivo* imaging of a wide variety of radio-labeled tracers in mice. Application of these techniques to the study of the heart is often complementary and the method used in practice will depend on numerous factors including use of either 2D contiguous slices or 3D isotropic imaging, spatial and temporal or molecular resolutions, count sensitivity, gating capabilities, duration of the observation, availability of required radio-molecules adapted to the particular investigation, data quantification and finally, financial aspects (Tab. 7.1). Another important factor is the availability of a cyclotron and specific hot-lab facilities in the vicinity of laboratories which want to employ PET methods. These facilities are necessary because of the very short half-life (few minutes to tenths of minutes) of the majority of PET tracers. In contrast however, SPECT imaging is not limited by these requirements. A vast array of relatively low-cost isotopes commonly used in many nuclear medicine departments is available for labeling cells, ligands, nucleic acids, peptides, and antibodies in addition to a large variety of radio-pharmaceuticals which can be directly used in cardiovascular imaging for the study of tissue perfusion and viability (^{99m}Tc-Sestamibi, ^{99}Tc-Tetrofosmin, ^{201}Tl), sympathetic function (^{123}I-MIBG) or fatty acid metabolism (^{123}I-BMIPP). Using SPECT, an *in vivo* spatial resolution of 1 mm can now be achieved routinely as shown in Fig. 7.2. This resolution is not limited by physical properties of single-photon emitters but by the magnification system and the low sensitivity of count density due to the small aperture of the pinhole and low sensitivities of the scintillators. In order to improve spatial resolution without a loss in sensitivity, multi-pinhole and coded aperture strategies which eliminate the need for rotating detectors are currently being developed. We can now expect an ultimate resolution of 150 μm with these new generation systems. Moreover, the relatively low spatial resolution of current small animal nuclear imaging systems can be efficiently counterbalanced by image fusion with micro-CT (Fig. 7.4 b), thus resulting in potentially very high CT morphologic resolution together with the high sensitivity of nuclear imaging. Taking into consideration that nuclear imaging can provide functional data in regard to all aspects of the cardiac cycle (systolic and diastolic volumes, motion, thickening) as well as information on tissue metabolism (perfusion, glucose metabolism for instance), dual imaging is of particular interest for the simultaneous analysis of the different metabolic processes (fatty acids, glucose, perfusion) and functions (motion and thickening) of the myocardial wall.

7.3 Exploring the Heart in Living Animals

7.3.1 Anatomy

Echocardiography (Echo) is a common method used to monitor myocardial remodeling during long-term studies. It has been validated in numerous mammalian species such as the mouse, rat, cat, dog, and of course, human. With this technique, we recently identified and characterized dilated and hypertrophic cardiomyopathies in mice in which the serotoninergic 5-HT_{2B} receptors were either knocked out or over-expressed [29, 30]. High frequency transducers (12–50 MHz) are now commonly used for this purpose, and give good resolution for the small mouse heart even in the conscious state (Fig. 7.1). The left ventricular mass (LVM) is usually estimated according to the following equation: $LVM = 1.055 \times [(S_d + PW_d + LVEDD)^3 - (LVEDD)^3]$ where S_d is the septal thickness in diastole, PW_d the posterior wall thickness in diastole and LVEDD the left ventricular end-diastolic diameter. In is important to note that this method has been validated by direct measurements in hearts where the shape of the ventricle was not modified [31]. In the case of geometric remodeling (end-stages of heart failure or myocardial infarction (MI)), this method is limited by its two-dimensional-based reconstruction. A three-dimensional-based reconstruction was applied to the MI model allowing a volumetric evaluation of the left ventricular cavity and infarcted wall [32]. This method gives mass values that are close to those generated by direct histological evaluation, but is difficult to apply in routine examinations because consecutive serial parasternal short-axis sections are required. This is difficult to achieve with currently available devices but the development of probes fixed on rail systems for tomographic reconstructions or high frequency three-dimensional matritial probes will address this problem. However, it seems that even after MI, a two-dimensional reconstruction will give similar results to direct histological measurement in the mouse. This surprising observation is probably a result of the model of infarction used in this species where the usual permanent occlusion of the sole left coronary artery induces a very reproducible necrosis among animals [33]. Finally, transthoracic Echo gives poor resolution for the right ventricle. Transesophageal echocardiography with an intravascular ultrasound catheter has been shown to be a simple, accurate, and reproducible method for studying the size and function of the right ventricular in mice [34]. This is very important in the detection of primary and secondary pulmonary hypertension, an important goal of many pharmacological and mutagenesis cardiovascular studies.

The measurement of ventricular mass and volumes has also been evaluated using ECG and respiratory-gating high-field magnetic resonance imaging. This method successfully estimates the ventricular mass in concentric hypertrophy [35] and is particularly useful in the case of geometric remodeling. Use of this technique was reported in infarcted mice that were over-expressing the AT_2 receptor for angiotensin in the myocardium [36].

Nuclear imaging offers the opportunity to evaluate cardiac mass through myocardial perfusion. Micro single photon emission computed tomography (SPECT) imaging can be performed with widely available and long half-life single photon radiotracers for perfusion such as ^{201}Tl, and ^{99m}Tc-Sestamibi or ^{99m}Tc-Tetrofosmin (72 hours for ^{201}Tl and 6 hours for ^{99m}Tc) and can also be used to study sympathetic cardiac function using ^{123}I-meta-iodo-benzyl-guanidine (MIBG) or fatty acid metabolism using ^{123}I-15 p-iodo-phenyl-3R,S-methyl-pentadecanoic acid (BMIPP).

7.3.1.1 **Systolic Function and Hemodynamics**

The evaluation of systolic function can be successfully achieved with left ventricular catheterization, Echo, MRI and gated SPECT.

7.3.1.2 **Global Systolic Function**

Ventricular catheterization studies are used in hemodynamic investigations. The catheter can be placed into the left ventricle after catheterization of the carotid artery or direct transthoracic puncture of the ventricle can also be performed. This technique provides important data regarding the left ventricular end-diastolic pressure; the first derivative of left ventricular pressure taken at its maximal value (dP/dt_{max} or $+dP/dt$) is a parameter of contractility, the maximum pressure achieved and dP/dt_{min} (or $-dP/dt$) and τ are the parameters of relaxation. Moreover, some systems can now measure pressure/volume loops at rest and during pharmacological interventions. Nevertheless, the results given by this method can only be obtained under anesthesia after the surgical insertion of the probe, and interpretation of the results should take this limitation into account. Imaging determination of systolic function is based on changes in ventricular size and volume as cardiac output is only modified in the end stages of heart failure. Using Echo, the left ventricular volumes can be obtained by the modified Simpson method but it is necessary to use a high frequency frame acquisition methodology due the high heart rate of the mouse. With Echo, when no significant abnormalities in regional wall motion are observed, fractional shortening (FS) is usually measured using M-mode imaging (high frequency) on a short-axis view after optimizing endocardial definitions (Fig. 7.1). This method accurately detects systolic dysfunction but, in some cases, the FS can appear normal despite true systolic dysfunction [37]. In fact, this index is highly dependent on loading conditions and the activity of the autonomic nervous system [38]. Moreover, mouse hemodynamics are highly sensitive to anesthesia [39]. This effect can be so pronounced that mice with heart failure may not tolerate anesthesia. Therefore, we recently explored systolic function in conscious animals only [1]. However, some phenotypes may be more easily revealed in anesthetized animals, as mice with mild systolic defects usually show a greater decrease in cardiac function under anesthesia than normal animals. Particular caution is recommended when comparing catheterization and Echo values using different anesthetics. In a recent study of mice over-expressing adenylyl cyclase 8 in the myocardium, Echo parameters were not modified (heart rate, FS, ventricular diameters) by Avertin* anesthesia in contrast to a marked increase in heart rate, LV systolic pressure and LV dP/dt_{max} when mice were anesthetized with ketamine/xylazine [40]. This example illustrates that the data produced by invasive and non-invasive investigations in the mouse should be interpreted with caution. The evaluation of cardiac contractility by echocardiography can also be affected by the method used for the calculation of FS. FS on a short-axis view only explores the contraction of circular midlayer fibers. In the mouse, an alteration of axial contractile fibers, especially in the subendocardium, appears early in the progression of systolic dysfunction. It is therefore useful to calculate the velocity of circumferential shortening (Vcf) which is the ratio of FS to the ejection time, an index which integrates into global contractility and can decrease before the FS changes [41] or to analyze the peak of the S wave recorded by pulsed-tissue Doppler on the mitral annulus in an apical view.

The assessment of systolic function can be improved by using such methods as color tissue Doppler, acoustic quantification and contrast echocardiography. Using ECG gating, a temporal resolution of 8-time bins per cardiac cycle has been demonstrated justifying the employment of micro-SPECT perfusion images to perform global systolic analysis of the LV. However we have shown that 10-time bins per cardiac cycle is achievable with the long half-life tissue perfusion emitters, the concentrations of which can be considered stationary at physiological equilibrium. This then allows the measurements of EF, ESV, EDV and CO (Fig. 7.3). MRI has also demonstrated its ability to accurately determine the same parameters [8]. In systole, the twisting motion of the ventricle has recently also been measured by tagged cine-MRI showing that the magnitude and systolic time-course of ventricular torsion were equivalent in mouse and humans [12].

7.3.1.3 **Regional Function**

Analysis of the motion of the left ventricular regional wall is based on grading the contractility of the individual wall segments. With Echo, this can be achieved by visualizing wall subdivision in two to three views and scoring the contractility of each individual segment. Automatic analysis can be used in the mouse with automatic border detection based on acoustic quantification and tissue Doppler or by perfusion-gated micro SPECT. Cine-tagged MRI now offers a very effective opportunity to study segmental contractility at rest and during pharmacological stress in the mouse post-myocardial infarction [42].

7.3.1.4 **Diastolic Function**

Normal diastolic function is characterized by adequate filling of the ventricles without abnormal increases in pressure. Diastolic filling is achieved and limited by *myocardial relaxation, elastic ventricular recoil* and *chamber compliance.* In the mouse, diastole can be studied by the analysis of Doppler flow velocities (Fig. 7.1 c) and diastolic tissue motion. In order to reduce the heart rate, the diastolic function is only assessed in anesthetized mice because transmitral E and A waves fuse at rapid heart rates. It is noteworthy that the heart rate must never been reduced to less than 300 beats/min because below this value, diameters in the left ventricle dramatically increase and this is accompanied by the impairment of early ventricular relaxation. The challenge is to determine whether the adult mouse diastolic filling pattern in this scenario is comparable to the human pattern. Increasing the number of registered time bins per cardiac cycle in ECG-gated micro-SPECT up to 16 phases should also lead to measurements of filling parameters in the near future.

7.3.1.5 Impaired Myocardial Relaxation Pattern

An impaired myocardial relaxation pattern is the initial abnormality of diastole observed in the early stages of many cardiac diseases. It has recently been described in an elegant study targeting the initial relaxation of the ventricle. Mice over-expressing a mutant form of phospholamban (PLB/N27A), resulting in a "superinhibition" of the sarcoplasmic reticulum Ca^{++}-ATPase, exhibited a true primary anomaly in relaxation with a reduced peak E wave, E/A ratio, color M-mode Doppler flow propagation velocity (Vp) and an increase in isovolumic relaxation time (IVRT) and mean deceleration time of the E wave (Mdt) [41]. A similar transmitral inflow pattern was identified in the early stages of cardiac hypertrophy induced by aortic banding and confirmed with pulsed tissue Doppler, which identified a reduction of peak early maximal velocity [43]. Diabetic and obese mice also exhibit impaired relaxation but with a pattern in which the reduction of the E/A ratio is mainly due to an isolated increase in the peak of the A wave. In these cases, the E wave is not affected in peak height or duration [44, 45]. Moreover, neither the Mdt or the IVRT was affected, probably as a consequence of slight increase in left atrial pressure.

7.3.1.6 Restrictive Filling Pattern

A restrictive filling pattern is associated with a reduction in ventricular compliance and a marked increase in left atrial pressure. Mice expressing a troponin T mutation linked to familial hypertrophic cardiomyopathy (I79N) have an increased myofilamental calcium sensitivity and therefore an augmentation of baseline systolic function without change of diastolic pattern [46]. Under isoproterenol stress echocardiography, as a probable consequence of an acute myocardial ischemia, a restrictive mitral profile appeared, associated with an increase in E/A due to a marked reduction of the peak A wave, the E wave was not modified. The Mdt of the E wave was also markedly reduced.

This restrictive pattern is a common feature of advanced cardiac hypertrophy and failure. Mice overexpressing the calcium-dependent phosphatase calcineurin exhibit an important cardiac hypertrophy with fibrosis and progressive systolic and diastolic failure. At the time of systolic failure (ventricular dilatation with FS and Vcf collapse), an increase in the E/A ratio due to a decrease in the peak A wave only was observed [47].

7.3.1.7 Myocardial Perfusion, Metabolism and Gene Expression Imaging

In BALB/c mice, myocardial perfusion has recently been assessed with SPECT using ^{99m}Tc-Sestamibi, tetrofosmin and ^{201}Talium as tracers. This method led to the relative quantification of the perfusion of the left ventricular segmental tissue which correlated well with the size of the myocardial infarct following permanent ligation of the left anterior descending coronary artery [48] (Fig. 7.2).

Myocardial perfusion can also be studied with MRI which elucidates the fine detail of the regional microcirculation [17]. The study of cardiac metabolism can now be approached using positron emission tomography (PET) and image-guided ^{31}P magnetic resonance spectroscopy. Recently non-gated ^{13}N-ammonia micro-PET images of mouse heart following intraperitoneal administration of clonidine demonstrated directional changes of myocardial perfusion associated with a reduction in cardiac effort [49]. Preliminary studies have also demonstrated the usefulness of micro-PET in imaging the distribution of ^{18}F-deoxyglucose and ^{11}C-palmitate in genetically-modified mice hearts [50]. In knockout mice for the glucose transporter GLUT-4, the time-course of insulin-stimulated myocardial glucose uptake has been studied by PET, emphasizing the rapid adaptive response of myocardial glucose metabolism [51]. PET is also a reference method used for analyzing the efficiency of tissue gene transfer. In this respect, cardiac PET reporter gene imaging can be used to monitor the expression of reporter genes and probes as exemplified by recent work carried out in rat myocardium with *Herpes simplex* type-1 virus thymidine kinase (HSV1-sr39 tk) as the reporter gene and 9-(4-^{18}F-fluoro-3-hydroxymethylbutyl)-guanine (^{18}F-FHBG) as the reporter probe [52]. However the use of non-gated cardiac mouse micro-PET still remains limited due to the spatial resolution of the images (between 1.2 and 2 mm), the very short half-life of the majority of positrons emitters (2, 10 and 20 min respectively for ^{15}O, ^{13}N and ^{11}C) and the necessity of having easy access to a cyclotron.

The phosphocreatine-to-ATP ratio can be measured by MR spectroscopy which now provides new opportunities for the non-invasive study of *in vivo* myocardial bioenergetics [20].

7.3.1.8 Cardiac Conduction and Arrhythmias

The electrocardiogram (ECG) of the mouse was described many years ago and is characterized by very rapid repolarization, the T wave being fused with the QRS complex in most cases. Therefore, any prolongation of the QT interval is easy to identify [53]. Nevertheless, in the case of such a prolongation, particular caution needs to be exercised with regard to its interpretation and transposition to the human ECG because of the specificity of the channels involved in the repolarization process in the mouse. Numerous anomalies such as blocks, arrhythmias and sudden cardiac death have been identified in transgenic mice [54, 55]. An ECG can easily be obtained from animals anesthetized with tribromoethanol (Fig. 7.5 a) as this compound does not reduce the heart rate to any great extent. When necessary, particularly in the case of cardiac arrhythmias or to analyze variability in the heart rate, continuous and sustained monitoring can be achieved with telemetry (Fig. 7.5 a). All of this “standard” analyses can now be completed with “deep” electrophysiological techniques using intracardiac catheters [56, 57]. Moreover, it is now possible to induce arrhythmias at the supraventricular and ventricular levels by transesophageal and ventricular stimulations [58, 59]. These techniques are quite specialized and can be employed for extended phenotyping.

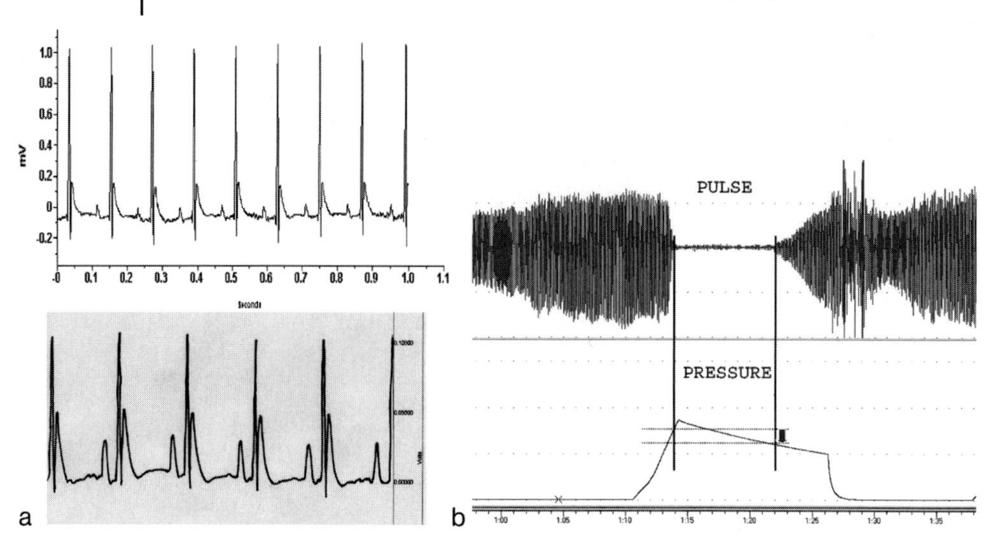
mV
1.0
0.8
0.6
0.4
0.2
0
-0.2
0.1
0.2
0.3
0.4
0.5
0.6
0.7
0.8
0.9
1.0
1.1
Seconds
PULSE
PRESSURE
a
b

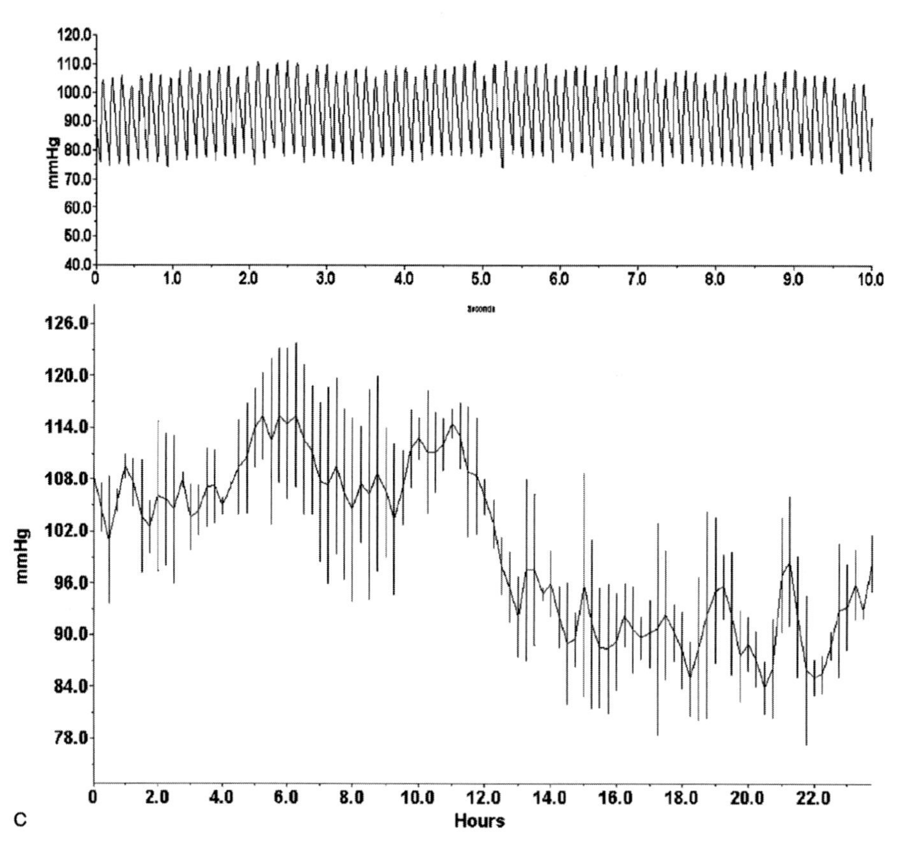
120.0
110.0
100.0
90.0
80.0
70.0
60.0
50.0
40.0
mmHg
0
1.0
2.0
3.0
4.0
5.0
6.0
7.0
8.0
9.0
10.0
Seconds
126.0
120.0
114.0
108.0
102.0
96.0
90.0
84.0
78.0
mmHg
0
2.0
4.0
6.0
8.0
10.0
12.0
14.0
16.0
18.0
20.0
22.0
Hours
c

7.3.1.9 Exploring the Great Arteries

Recording Blood Pressure In cardiovascular phenotyping, recording arterial blood pressure is crucial. Its measurement can be obtained in conscious animals by the completely non-invasive tail-cuff method or by continuous telemetric monitoring, following the insertion of a chronic catheter usually into a common carotid artery. The first method requires a training period and measurements are taken in restrained mice that have previously been exposed to warming conditions (Fig. 7.5 b). This method is time consuming (five to seven consecutive days of experiments) but very reproducible if performed under stable conditions and gives accurate values for systolic blood pressure and heart rate. However, measurements can only be obtained at given times and the method is limited by the fact that the signal is usually lost when the resting blood pressure is quite low (systolic < 70 mmHg) or markedly reduced by a particular treatment. Telemetry is probably the best technique for continuous monitoring of diastolic and systolic blood pressure and heart rate together with their short- and long-term cyclical variations (Fig. 7.5 c). Nevertheless, the following facts should be taken into account in the interpretation of the results: (1) surgery is necessary, (1) carotid artery ligation with the consequent suppression of a baroreflex loop can modify autonomic nervous system activity and (3) the mouse will be isolated on a registering platform with the consequent suppression of all physiological social interactions. Nevertheless, telemetry is the only method used on conscious mice that produces accurate measurements of cardiovascular responses to stressors such as pharmacological interventions, exercise on a treadmill or swimming.

On the other hand, invasive methods of measuring arterial and venous pressures are only undertaken in anaesthetized mice by means of micrometer-tipped catheters inserted into the great vessels such as the carotid or femoral arteries or jugular vein. Saline-filled catheters can be employed for measuring blood pressure only but cannot be used for intraventricular pressure recordings because of their low frequency response. Invasive methodology is the gold standard for the evaluation of cardiac hemodynamics and during the same experiment it is possible to record blood pressure parameters in the aorta and then to push the catheter into the left ventricle to investigate whether it is functioning normally. Nevertheless, two aspects of this method should be taken into consideration. First, all these evaluations are carried out on anaesthetized animals and all anesthetics modify at least

Fig. 7.5 (a) Electrocardiographic recordings in a conscious mouse by telemetry (upper panel) and in a tribromoethanol-anaesthetized mouse (lower panel). (b) Tail cuff method in the conscious mouse. The upper panel is the pulse recording and the lower panel the pressure. The blue arrow indicates the difference in systolic blood pressure values when measured during deflation of the cuff as compared to inflation. (c) Upper panel, instantaneous recording of blood pressure by telemetry with a probe placed in the common carotid artery of a C57Bl6 mouse. Lower panel, mean blood pressure values obtained in four mice showing day and night variations.

one important cardiovascular parameter, usually by downregulation (bradycardia, reduction of the cardiac output, fall in blood pressure, etc.). Second, this invasive method can only be used as a terminal investigation followed by euthanasia of the animal and then usually by histological examination for morphologic data.

Imaging the Great Vessels The great vessels can now be easily visualized by high frequency ultrasound imaging (Fig. 7.6). This technique can detect atherosclerosis, and blood flow can be measured by placing pulsed tissue Doppler into the lumen. High frequency ultrasound imaging is probably the simplest method for obtaining repetitive measurements of blood flow in arteries and also for standardizing models of vascular stenosis (aortic or pulmonary) by measuring the pressure gradient after the surgery. Conventional angiography was used for the assessment of the great vessels the hindlimbs and angiogenesis in the mouse [60]. Yamashita et al. reported a marked improvement in resolution using synchrotron radiation microangiography which facilitated their impressive detection of vascular lesions in the carotid, brachiocephalic, brachial and coronary arteries [61, 62] using a new X-ray-dedicated digital microangiographic camera which gave a resolution of 7 μm/pixel. However, the cost and limited access to synchrotron facilities seems to rule out this technique as a routine method of analysis of atherosclerotic plaques in the mouse. With regard to X-ray techniques, micro-CT following the intravenous administration of a reasonably stable vascular contrast agent, currently seems to offer the best resolution of approximately 0.1 mm. MRI has been successfully employed to study the development of atherosclerotic lesions in mice fed a high fat diet [23] and a 3D reconstruction of the coronary arteries with 100 μm resolution has recently been published thus establishing the use of repetitive evaluations of the coronary vasculature under various experimental conditions [9].

7.3.1.10 **Exploring Microvessels**

At present establishing the size of mouse microvessels is dependent on the resolution of current imaging modalities. Laser Doppler perfusion imaging has been successfully employed to evaluate of angiogenesis in hindlimb muscles [60]. This method can be used to evaluate perfusion in the peripheral microvascular in transgenic animals and also during pharmacological and genetic therapies.

7.4 A Scheme for Identifying the Main Cardiovascular Disorders in Genetically-modified Mice

The question of high-throughput phenotyping is a key issue in most functional genomic programs. Such programs endeavor to achieve rapid and comprehensive phenotypic detection of the main cardiovascular disorders with a low incidence of false negatives. In Europe, we have designed a flow scheme which is divided into three different levels (www.eumorphia.org) for cardiovascular screening in mice:

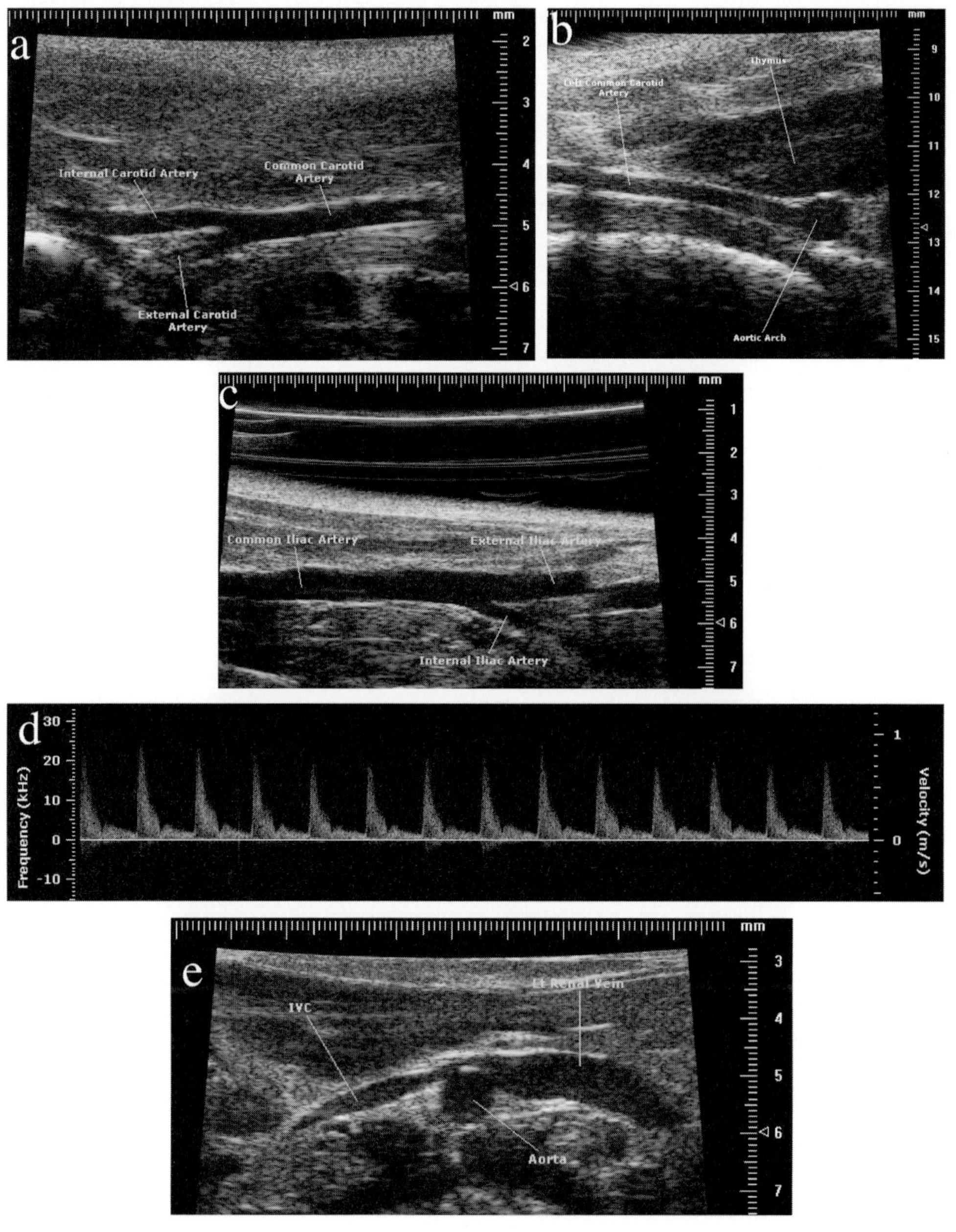

Fig. 7.6 (a) Longitudinal section of the common, internal and external carotid arteries. (b) Longitudinal section of the left common carotid artery originating from the aortic arch (sectioned transversally). (c) Common iliac artery and its two branches. (d) Pulsed-Doppler recording of the common carotid artery flow. (e) Longitudinal section of the left renal vein reaching the inferior vena cava (IVC) and transverse section of the abdominal aorta.

Table 7.2 Some phenotyping parameters obtained by our group following the flow scheme of first-line cardiovascular phenotyping established by the European Eumorphia program (non-invasive strategy).

	C57Bl6		129PAS		Balb/c		C3HbFeJ	
	M(n=6)	F(n=6)	M(n=6)	F(n=6)	M(n=6)	F(n=6)	M(n=6)	F(n=5)
SAP	92±1	91±2	91±2	86±1	102±1	98±2	96±2	92±1
HR	582±8	529±28	578±12	584±6	478±7	507±21	542±6	538±11
PR	25±1	31±2	26±1	24±1	26±1	26±1	28±1	29±1
RR	155±18	129±5	145±6	153±10	131±7	134±6	139±8	133±6
QT	53±2	58±5	66±1	66±1	55±2	59±1	51±2	51±1
LVM/BW	4.2±0.2	4.3±0.2	3.7±0.2	4.5±0.3	3.6±0.1	3.0±0.6	3.7±0.1	3.8±0.1
EF	64±2	69±1	71±3	70±1	72±3	72±3	60±3	64±3
E/A	1.9±0.1	1.8±0.1	2.2±0.2	2.2±0.1	2.1±0.3	2.0±0.3	1.4±0.1	1.5±0.1
Mdt	43±2	42±2	39±1	37±2	43±2	39±1	43±1	39±1
IVRT	22±1	22±1	22±1	28±2	24±1	24±2	32±4	23±3
CO	26±2	24±2	28±3	27±2	29±2	26±2	27±1	32±5

M: male; F: female; SAP: systolic arterial pressure; HR: heart rate; PR, RR and QT: some interval durations on the electrocardiogram, LVM/BW: left ventricular mass to body weight ratio, EF: ejection fraction, E/A: E on A wave ratio, Mdt: mean deceleration time of the E wave, IVRT: isovolumetric relaxation time, CO: cardiac output.

- First-line phenotyping: two strategies
  - Non invasive: electrocardiogram in 12-week-old (WO) tribromoethanol-anesthetized mice; blood pressure and heart rate by the tail-cuff method (12 WO) and echo-Doppler analysis of the heart in anesthetized mice (13 WO).
  - Invasive: electrocardiogram in 12-week-old (WO) tribromoethanol-anesthetized mice; left ventricular and arterial catheterization followed by histology in 13 WO animals.
- First-line extended: continuous monitoring of blood pressure, heart rate and electrocardiogram by telemetry.
- Second-line phenotyping: magnetic resonance imaging, single photon emission tomography, micro-CT, invasive electrophysiology, vascular echotracking, etc.

In this flow scheme, we consider that first- and first-extended phenotyping are part of the high-throughput phenotyping strategy and that the so-called first-extended screen mainly aims to complete a description or to determine very complex phenotypes.

References

1 Seznec, H., et al. **2004**, Idebenone delays the onset of cardiac functional alteration without correction of Fe-S enzymes deficit in a mouse model for Friedreich ataxia, *Hum. Mol. Genet.* 13 (10), 1017–1024.

2 Dawson, D., et al. **2004**, Quantitative 3-dimensional echocardiography for accurate and rapid cardiac phenotype characterization in mice, *Circulation* 110 (12), 1632–1637.

3 Cassidy, P. J., et al. **2004**, Assessment of motion gating strategies for mouse magnetic resonance at high magnetic fields, *J. Magn. Reson. Imaging* 19 (2), 229–237.

4 Slawson, S. E., et al. **1998**, Cardiac MRI of the normal and hypertrophied mouse heart, *Magn. Reson. Med.* 39 (6), 980–987.

5 Ruff, J., et al. **1998**, Magnetic resonance microimaging for noninvasive quantification of myocardial function and mass in the mouse, *Magn. Reson. Med.* 40 (1), 43–48.

6 Franco, F., et al. **1999**, Magnetic resonance imaging and invasive evaluation of development of heart failure in transgenic mice with myocardial expression of tumor necrosis factor-alpha, *Circulation* 99 (3), 448–454.

7 Ross, A. J., et al. **2002**, Serial MRI evaluation of cardiac structure and function in mice after reperfused myocardial infarction, *Magn. Reson. Med.* 47 (6), 1158–1168.

8 Schneider, J. E., et al. **2003**, Fast, high-resolution in vivo cine magnetic resonance imaging in normal and failing mouse hearts on a vertical 11.7 T system, *J. Magn. Reson. Imaging* 18 (6), 691–701.

9 Ruff, J., et al. **2000**, Magnetic resonance imaging of coronary arteries and heart valves in a living mouse: techniques and preliminary results. *J. Magn. Reson.* 146 (2), 290–296.

10 Wiesmann, F., et al. **2002**, Analysis of right ventricular function in healthy mice and a murine model of heart failure by *in vivo* MRI, *Am. J. Physiol. Heart Circ. Physiol.* 283 (3), H1065–H1071.

11 Wiesmann, F., et al. **2001**, Dobutamine-stress magnetic resonance microimaging in mice: acute changes of cardiac geometry and function in normal and failing murine hearts, *Circ. Res.* 88 (6), 563–569.

12 Henson, R. E., et al. **2000**, Left ventricular torsion is equal in mice and humans, *Am. J. Physiol. Heart Circ. Physiol.* 278 (4), H1117–H1123.

13 Zhou, R., et al. **2003**, Assessment of global and regional myocardial function in the mouse using cine and tagged MRI, *Magn. Reson. Med.* 49 (4), 760–764.

14 Streif, J. U., et al. **2003**, *In vivo* time-resolved quantitative motion mapping of the murine myocardium with phase contrast MRI, *Magn. Reson. Med.* 49 (2), 315–321.

15 Bauer, W. R., et al. **2001**, Fast high-resolution magnetic resonance imaging demonstrates fractility of myocardial perfusion in microscopic dimensions, *Circ. Res.* 88 (3), 340–346.

16 Yang, Z., et al. **2004**, Simultaneous evaluation of infarct size and cardiac function in intact mice by contrast-enhanced cardiac magnetic resonance imaging reveals contractile dysfunction in noninfarcted regions early after myocardial infarction, *Circulation* 109 (9), 1161–1167.

17 Wu, E. X., et al. **2004**, Mapping cyclic change of regional myocardial blood volume using steady-state susceptibility effect of iron oxide nanoparticles, *J. Magn. Reson. Imaging* 19 (1), 50–58.

18 Hu, T. C., et al. **2001**, Manganese-enhanced MRI of mouse heart during changes in inotropy, *Magn. Reson. Med.* 46 (5), 884–890.

19 Chacko, V. P., et al. **2000**, MRI/MRS assessment of *in vivo* murine cardiac metabolism, morphology, and function at physiological heart rates, *Am. J. Physiol. Heart Circ. Physiol.* 279 (5), H2218–H2224.

20 Omerovic, E., et al. **2000**, *In vivo* metabolic imaging of cardiac bioenergetics in transgenic mice, *Biochem. Biophys. Res. Commun.* 271 (1), 222–228.

21 De Souza, A. P., et al. **2004**, Effects of early and late verapamil administration on the development of cardiomyopathy in experimental chronic *Trypanosoma*

cruzi (Brazil strain) infection, *Parasitol. Res.* 92 (6), 496–501.
22 Stenbit, A. E., et al. **2000**, Preservation of glucose metabolism in hypertrophic GLUT4-null hearts, *Am. J. Physiol. Heart Circ. Physiol.* 279 (1), H313–H318.
23 Hockings, P. D., et al. **2002**, Repeated three-dimensional magnetic resonance imaging of atherosclerosis development in innominate arteries of low-density lipoprotein receptor-knockout mice, *Circulation* 106 (13), 1716–1721.
24 Bove, C. M., et al. **2004**, Nitric oxide mediates benefits of angiotensin ii type 2 receptor overexpression during post-infarct remodeling, *Hypertension*.
25 Wu, M. C., B. H. Hasegawa, M. W. Dae **2002**, Performance evaluation of a pinhole SPECT system for myocardial perfusion imaging of mice, *Med. Phys.* 29 (12), 2830–2839.
26 Schelbert, H. R., M. Inubushi, R. S. Ross **2003**, PET imaging in small animals, *J. Nucl. Cardiol.* 10 (5), 513–520.
27 Ford, N. L., M. M. Thornton, D. W. Holdsworth **2003**, Fundamental image quality limits for microcomputed tomography in small animals, *Med. Phys.* 30 (11), 2869–2877.
28 Kiessling, F., et al. **2004**, Volumetric computed tomography (VCT): a new technology for noninvasive, high-resolution monitoring of tumor angiogenesis, *Nat. Med.* 10 (10), 1133–1138.
29 Nebigil, C. G., et al. **2001**, Ablation of serotonin 5-HT(2B) receptors in mice leads to abnormal cardiac structure and function, *Circulation* 103 (24), 2973–2979.
30 Nebigil, C. G., et al. **2003**, Overexpression of the serotonin 5-HT2B receptor in heart leads to abnormal mitochondrial function and cardiac hypertrophy, *Circulation* 107 (25), 3223–3229.
31 Collins, K. A., et al. **2001**, Accuracy of echocardiographic estimates of left ventricular mass in mice, *Am. J. Physiol. Heart Circ. Physiol.* 280 (5), H1954–H1962.
32 Scherrer-Crosbie, M., et al. **1999**, Three-dimensional echocardiographic assessment of left ventricular wall motion abnormalities in mouse myocardial infarction, *J. Am. Soc. Echocardiogr.* 12 (10), 834–840.
33 Wu, J. C., et al. **2003**, Influence of sex on ventricular remodeling after myocardial infarction in mice, *J. Am. Soc. Echocardiogr.* 16 (11), 1158–1162.
34 Scherrer-Crosbie, M., et al. **1998**, Determination of right ventricular structure and function in normoxic and hypoxic mice: a transesophageal echocardiographic study, *Circulation* 98 (10), 1015–1021.
35 Wiesmann, F., et al. **2000**, Developmental changes of cardiac function and mass assessed with MRI in neonatal, juvenile, and adult mice, *Am. J. Physiol. Heart Circ. Physiol.* 278 (2), H652–H657.
36 Yang, Z., et al. **2002**, Angiotensin II type 2 receptor overexpression preserves left ventricular function after myocardial infarction, *Circulation* 106 (1), 106–111.
37 Nemoto, S., et al. **2002**, Effects of changes in left ventricular contractility on indexes of contractility in mice, *Am. J. Physiol. Heart Circ. Physiol.* 283 (6), H2504–H2510.
38 Tan, T. P., et al. **2003**, Assessment of cardiac function by echocardiography in conscious and anesthetized mice: importance of the autonomic nervous system and disease state, *J. Cardiovasc. Pharmacol.* 42 (2), 182–190.
39 Rottman, J. N., et al. **2003**, Temporal changes in ventricular function assessed echocardiographically in conscious and anesthetized mice, *J. Am. Soc. Echocardiogr.* 16 (11), 150–157.
40 Lipskaia, L., et al. **2000**, Enhanced cardiac function in transgenic mice expressing a Ca(2+)-stimulated adenylyl cyclase, *Circ. Res.* 86 (7), 795–801.
41 Schmidt, A. G., et al. **2002**, Evaluation of left ventricular diastolic function from spectral and color M-mode Doppler in genetically altered mice, *J. Am. Soc. Echocardiogr.* 15 (10 Pt 1), 1065–1073.
42 Epstein, F. H., et al. **2002**, MR tagging early after myocardial infarction in mice demonstrates contractile dysfunction in adjacent and remote regions, *Magn. Reson. Med.* 48 (2), 399–403.
43 Schaefer, A., et al. **2003**, Evaluation of left ventricular diastolic function by pulsed Doppler tissue imaging in mice, *J. Am. Soc. Echocardiogr.* 16 (11), 1144–1149.
44 Semeniuk, L. M., A. J. Kryski, D. L. Severson **2002**, Echocardiographic assessment

of cardiac function in diabetic db/db and transgenic db/db-hGLUT4 mice, *Am. J. Physiol. Heart Circ. Physiol.* 283 (3), H976–H982.

45 Christoffersen, C., et al. **2003**, Cardiac lipid accumulation associated with diastolic dysfunction in obese mice, *Endocrinology* 144 (8), 3483–3490.

46 Knollmann, B. C., et al. **2001**, Inotropic stimulation induces cardiac dysfunction in transgenic mice expressing a troponin T (I79N) mutation linked to familial hypertrophic cardiomyopathy, *J. Biol. Chem.* 276 (13), 10039–10048.

47 Semeniuk, L. M., et al. **2003**, Time-dependent systolic and diastolic function in mice overexpressing calcineurin, *Am. J. Physiol. Heart Circ. Physiol.* 284 (2), H425–H430.

48 Wu, M. C., et al. **2003**, Pinhole single-photon emission computed tomography for myocardial perfusion imaging of mice. *J. Am. Coll. Cardiol.* 42 (3), 576–582.

49 Inubushi, M., et al. **2004**, Nitrogen-13 ammonia cardiac positron emission tomography in mice: effects of clonidine-induced changes in cardiac work on myocardial perfusion, *Eur. J. Nucl. Med. Mol. Imaging* 31 (1), 110–116.

50 Schelbert, H. R., et al. **2003**, PET myocardial perfusion and glucose metabolism imaging: Part 2-Guidelines for interpretation and reporting, *J. Nucl. Cardiol.* 10 (5), 557–571.

51 Simoes, M. V., et al. **2004**, Delayed response of insulin-stimulated fluorine-18 deoxyglucose uptake in glucose transporter-4-null mice hearts, *J. Am. Coll. Cardiol.* 43 (9), 1690–1697.

52 Inubushi, M., et al. **2003**, Positron-emission tomography reporter gene expression imaging in rat myocardium, *Circulation* 107 (2), 326–332.

53 Folco, E., et al. **1997**, A cellular model for long QT syndrome. Trapping of heteromultimeric complexes consisting of truncated Kv1.1 potassium channel polypeptides and native Kv1.4 and Kv1.5 channels in the endoplasmic reticulum, *J. Biol. Chem.* 272 (42), 26505–26510.

54 Donoghue, M., et al. **2003**, Heart block, ventricular tachycardia, and sudden death in ACE2 transgenic mice with downregulated connexins, *J. Mol. Cell. Cardiol.* 35 (9), 1043–1053.

55 Berul, C. I., et al. **2000**, Progressive atrioventricular conduction block in a mouse myotonic dystrophy model, *J. Interv. Card. Electrophysiol.* 4 (2), 351–358.

56 Berul, C. I. **2003**, Electrophysiological phenotyping in genetically engineered mice, *Physiol. Genomics* 13 (3), 207–216.

57 Saba, S., P. J. Wang, N. A. Estes, 3rd **2000**, Invasive cardiac electrophysiology in the mouse: techniques and applications, *Trends Cardiovasc. Med.* 10 (3), 122–132.

58 Schrickel, J. W., et al. **2002**, Induction of atrial fibrillation in mice by rapid transesophageal atrial pacing, *Basic Res. Cardiol.* 97 (6), 452–460.

59 Wakimoto, H., et al. **2001**, Induction of atrial tachycardia and fibrillation in the mouse heart, *Cardiovasc. Res.* 50 (3), 463–473.

60 Silvestre, J. S., et al. **2002**, Antiangiogenic effect of angiotensin II type 2 receptor in ischemia-induced angiogenesis in mice hindlimb, *Circ. Res.* 90 (10), 1072–1079.

61 Yamashita, T., et al. **2002**, *In vivo* angiographic detection of vascular lesions in apolipoprotein E-knockout mice using a synchrotron radiation microangiography system, *Circ. J.* 66 (11), 1057–1059.

62 Yamashita, T., et al. **2002**, Images in cardiovascular medicine. Mouse coronary angiograph using synchrotron radiation microangiography. *Circulation* 105 (2), E3–E4.

8
Phenotyping of Host–Pathogen Interactions in Mice

Andreas Lengeling, Werner Müller, and Rudi Balling

8.1
Introduction

Infectious diseases are still a major cause of morbidity and mortality worldwide. One of the major challenges in contemporary infection research is the development of suitable animal models to further advance our understanding of host–pathogen interactions. There is an urgent need to better understand (1) the strategies used by pathogens to evade the host immune response; and (2) the mechanisms that the host immune system employs to fight off infectious agents. This is a prerequisite for improving the current prevention and therapeutic strategies that are used in the battle against infectious diseases.

Susceptibility or resistance of the host to infection is determined by a complex interplay of environmental, host, and pathogen factors. In the past, our approaches to understanding infectious diseases focused mainly on the study of either individual environmental or pathogen related factors while the role of the host was limited to studies on pathogenesis and adaptive immune responses. A few years ago this changed and now more attention is directed towards analyzing the contribution of host genetics to the process of infection. In addition, our knowledge of the mechanisms of natural or innate immunity has been largely extended mainly due to genetic studies in model organisms [1–4]. The mouse as a genetically tractable mammal has played a pivotal role in defining new pathways and gene functions that are important for host defense. A crucial requirement for the genetic dissection of the host immune response to pathogens in mice is the development of robust and standardized phenotyping assays [5]. Without standardization and comprehensive phenotyping it is not possible to identify relationships between genes and phenotypes and thus to discover new gene functions.

After giving a brief summary of the history of approaches that have been used in mice to analyze host defenses we will shortly review how these studies helped to improve our understanding of the mechanisms of infection susceptibility in man. We will give a few examples of how mouse genetics has facilitated the identification of critical host proteins that are involved in immune defense. It is impossible to adequately review all of the work that has been carried out using different classes of pathogens in mice. Therefore, we will focus on mouse models of bacterial infec-

Standards of Mouse Model Phenotyping. Edited by Martin Hrabé de Angelis, Pierre Chambon, and Steve Brown

ISBN: 3-527-31031-2

tion, which are also under investigation in our laboratories and at our research center. For further reading on mouse models of viral and parasitic infection, we can recommend some of the excellent reviews that have been published recently [6–8].

8.2 Looking Back and Forward: History and State-of-the-Art of Mouse Infection Phenotyping and Studies of Genetic Infection Susceptibility

The influence of genetic factors on resistance and susceptibility to bacterial infection in the mouse were first analyzed systematically in the early work of Leslie Webster at the Rockefeller Institute for Medical Research in New York [9, 10]. In fact, back in the early 1930 s, he was the first to establish what is today called "baseline-data" for different inbred strains of mice. He started to develop a mouse model of human typhoid fever by infecting mice orally with *Bacillus enteritidis* [9], now known as *Salmonella enterica* Serovar Typhimurium (or sometimes as the shortened but incorrect form, *Salmonella typhimurium*). He standardized many critical parameters of his experiments and thereby pioneered the reproducible investigation of host responses in animal models under controlled conditions (e. g. temperature, diet, age, weight and sex of mice [10]). Webster defined new standards of mice handling and housing by paying careful attention to the monitoring and maintenance of hygiene conditions in his mouse colonies [9]. He analyzed the influence of different routes of *Salmonella* inoculation (e. g. oral, intravenous, and subcutaneous application of the pathogen) on the outcome of infection and thoroughly investigated parameters such as dose of infection and kinetics of survival over time [10]. His studies enabled him to set up a mouse typhoid infection model by selective inbreeding of high-mortality and low-mortality lines of mice. He started to explore the heritability of infection susceptibility and demonstrated that "genetic factors" segregated in the backcross progenies of his newly established salmonella-resistant and -sensitive mouse strains. He concluded for the first time that "inherited components of resistance affect the response of the host to infection" [9]. This conclusion was possible because Webster tried to minimize the effects of interfering environmental factors. In particular, he worked with mice that were free from pre-existing infections with *Salmonella* or other pathogens, conditions to which not many of his colleagues paid attention to at that time. Today, a "specific pathogen-free (SPF)" standard in mouse breeding facilities is still an important requirement for any immunological study. Below we will discuss the impact that hygiene conditions can have on the outcome of infection challenge experiments and the standards necessary for the analysis of immune responses in mice under infection challenge conditions (see Sections 8.4 and 8.5.4).

Another well-established model is murine listeriosis. It has been studied for the past four decades to examine basic aspects of innate and acquired cellular immunity. Listeriosis is caused by *Listeria monocytogenes* which has emerged as a remarkably tractable pathogen with which to investigate basic aspects of intracellular pathogenesis. Infection challenge experiments with *L. monocytogenes* have been proven to be one of the most successful experimental models in history for defining mechanisms that underlie immunity and host defense to infectious diseases. Fundamental concepts in immunology, such as macrophage activation [11, 12], the role of $CD4^+$- and $CD8^+$-T cells [13], major histocompatibility (MHC) restriction [14], adoptive transfer of T cell-mediated immunity [15], and the function of cytokines (for a detailed review see [16]), were derived from or further explored in this model.

L. monocytogenes is a facultative intracellular, Gram-positive bacterium that causes sepsis and meningitis in immunocompromised patients and devastating maternal/fetal infection in pregnant women. In 1962, George Mackaness described the first experimental model of listeriosis in mice [11]. Mackaness was interested in the immunological basis of non-humoral and acquired resistance to infection and investigated the pathogenesis of listeriosis in mice. He established assays for the examination of macrophage responses to *L. monocytogenes* in mice by isolating the cells from the peritoneal cavity and infecting them with *Listeria in vitro*. His readout assays for the bactericidal activity of the macrophages were quite simple and effective; he counted lytic plaques in cellular monolayers and enumerated ingested bacteria in macrophages by microscopy after staining with May–Grunwald–Giemsa [11]. More importantly, he developed a delayed-type hypersensitivity assay for *Listeria* by injecting a sterile filtrate of *L. monocytogenes* into the hind footpads of previously sub-lethal infected mice and afterwards examined the swelling of the footpads over time. This technique enabled him to monitor the response of mice to re-infection with *L. monocytogenes* [11] and it was one of the crucial methods that allowed him to define the general basis of cellular immunity to pathogens. By combining the delayed-type hypersensitivity assay with the transfer of lymphoid spleen cells from *Listeria*-infected mice into naïve mice, he was able to demonstrate that the transferred cells conferred protection against subsequent *L. monocytogenes* infection [15]. This was the starting point for the investigation of mechanisms of cellular immunity.

The genetic control of immune responses to infection with *L. monocytogenes* in different inbred strains of mice soon became a subject of intensive studies [17–19]. C57BL/6J, C57BL/10Sn, and B10.A mice were reported to be resistant to *L. monocytogenes* infection, whereas A/J, BALB/c, CBA/J, C3H/HeJ, DBA/1J and DBA/2J mice have been shown to be sensitive to infection with this pathogen [17, 18]. In Emil Skamene's laboratory genetic linkage studies using the AXB and BXA recombinant inbred (RI) lines of mice that were established from A/J and C57BL/6J parental strains of mice led to the mapping of two loci involved in the control of bacterial growth in the group of the susceptible RI strains [20]. Co-segregation with alleles at the *hemolytic complement* (*Hc*) locus on mouse chromosome 2 suggested early on that a deficiency in complement factor C5 was responsible for the susceptibility of the A/J parental strain. This was further supported by the observation that A/J mice were protected from *Listeria* infection when they were reconstituted with C5-rich serum from the $C5^+$ B10.D2/nSN strain [20]. More recently, the genetic susceptibility to *L. monocytogenes* infection in C57BL/6ByJ and BALB/cByJ mice was further refined by the mapping of two loci controlling systemic infection to mouse chromosomes 5 and 13 [21].

Today, the use of RI strains has again become very popular in the study of the complex genetics of pathogen defense. A powerful new genetic tool will be developed within the next years that will probably allow a new roadmap of interactions between genes and polygenic networks to be defined. The complex trait consortium will establish the 1K-Collaborative-Cross (1KCC) of RI strains. It is planned to generate about 1000 new RI strains of mice from eight different parental inbred strains that can be used for phenotyping and high-resolution trait mapping [22]. One of the main attractions of the 1KCC system is the possibility of combining traditional genetic methods with novel systems-biology approaches.

The identification of the *Lsh/Ity/Bcg* locus is an impressive early example of the positional cloning of a single host gene in the mouse that is responsible for the susceptibility of different inbred strains to taxonomically unrelated intracellular pathogens. This locus mediates resistance of mice to *Leishmania donavani* (*Lsh* locus), *Salmonella enterica* Serovar Typhimurium (*Ity* locus), and to a number of different Mycobacterial species such as *Mycobacterium bovis* BCG (*Bcg* locus). Genetic susceptibility to these pathogens is caused by a mutation in the *Slc11a1* gene (previously known as natural resistance-associated macrophage protein 1, Nramp1) [23], which has been suggested to be a pH-dependent divalent cation efflux pump at the phagosomal membrane of macrophages (for a review see [24]). In phagocytes the Slc11a1 protein has been implicated in acidification and maturation processes of the phagosome, which are important for intracellular, bactericidal host defense.

Screening the responses of mice to bacterial products instead of using living pathogens can also reveal host factors involved in pathogen resistance. Variation in inflammatory responses in inbred strains of mice after challenge with purified lipopolysaccharide (LPS) had already been found 40 years ago (for review see [25]). LPS is an abundant glycolipid present in the outer membrane of Gram-negative bacteria, which can provoke generalized, inflammatory responses in the infected host. Hyporesponsiveness to LPS can render mice highly susceptible to Gram-negative infections with pathogens such as *Salmonella* and *Klebsiella* because macrophage activation is impaired. Defects in LPS signaling in C3H/HeJ and C57BL/10ScN mice were found to be under control of the *Lps* locus on mouse chromosome 4. This locus was identified as the *Toll-like receptor 4* gene and the positional cloning of *Lps* first demonstrated that mutations in this class of pathogen-recognition receptors can profoundly affect susceptibility to infectious agents [26, 27]. The Toll-like receptor (TLR) gene family now comprises 11 members in mice (*Tlr1-Tlr11*, [28]) and is one of the best-studied immune sensors for invading pathogens. The signaling pathways triggered after recognition of *pathogen-associated molecular patterns* (PAMPs), which are the evolutionary conserved products of microbial metabolism initiate innate immunity and help to strengthen adaptive immune responses. Several important Tlr-adaptor molecules and downstream Tlr-pathway regulator proteins were identified through *N*-ethyl-*N*-nitrosourea (ENU)-mutagenesis screens in mice [29] or via gene targeting approaches and the subsequent phenotyping of PAMP responses in *Tlr*-deficient mice (for reviews see [30, 31]). In addition, other PAMP recognition receptors such as the Cd36 molecule were linked with hypersusceptibility to Gram-positive pathogens (e. g. *Staphylococcus aureus*) using ENU mutagenesis approaches in the mouse [32]. Here, the laboratory of Bruce Beutler at the Scripps Institute in San Diego has played a pivotal role in establishing specialized ENU mutagenesis screens in mice which allow the systematic genetic investigation of the mouse immune and host defense system [29, 32, 33].

Within the last few years many valuable new techniques have been developed for phenotyping immune responses in infected mice. In particular, many new non-invasive techniques have been designed to image host defense responses in mice. These are very interesting developments because they have the potential for infection phenotyping of mice in high-throughput primary screens. Currently, the characterization of infection susceptibility is very laborious, time consuming, and usually requires many animals to monitor the kinetics of pathogen dissemination and to examine the organ pathology of the mice after infection. With new diagnostic tools in imaging this might change in the future. Such efforts need support because non-invasive imaging techniques will be crucial tools for experimental pharmacological research and antimicrobial drug development. Examples of modern imaging technologies are two-photon microscopy [34], magnetic resonance imaging (MRI) [35], and *in vivo* bioluminescence imaging of pathogen dissemination in the entire body of an infected mouse over time with sensitive charge-coupled device (CCD) cameras [36, 37]. Another interesting new imaging technique is the *in vivo* reporter enzyme assay which uses radiolabeled substrates that allow the *in situ* detection of pathogens with single-photon emission computed tomography (SPECT) [38].

A very important method for monitoring the induction and maintenance of T-cell responses to infectious agents is the tetramer staining technique which facilitates investigation of the function of T cells in recall infection experiments [39, 40]. Most T cells recognize peptides derived from pathogens that are bound to MHC molecules on the surface of target cells or antigen-presenting cells (APCs). This recognition is specific for both the MHC allele and the pathogen-derived peptide. Flow cytometry can be used to detect soluble peptide–MHC complexes attached to fluorochromes thus making it possible to identify antigen-specific T cells. However, due to the poor binding of monomeric peptide–MHC complexes to the T cell receptor, the use of multimers which are typically tetramers, is required. For staining, the purified MHC molecules are biotinylated and then added to fluorescently-labeled streptavidin complexes in solution. The tetramer staining technique was essential to study for example, the transition from primary effector T cell to memory T cell responses after *L. monocytogenes* infection [41] and for the characterization of $CD8^+$ T cell responses to this pathogen [42].

8.3 The Impact of Mouse Genetics on the Understanding of Human Infectious Diseases

Primary immunodeficiency diseases in humans consist of a group of more than 100 inherited clinical manifestations that can predispose individuals to different infectious diseases, allergy, autoimmunity and cancer [43]. Many of the identified genes that have been associated with abnormal or deficient immune responses in patients had been identified before using the mouse as a model system (see also Chapter 10 of this volume). Mutations in genes that lead to primary immunodeficiencies are often associated with recurrent infections in patients that are caused by very diverse microorganisms. A good example of such a generalized immunodeficiency is chronic granulomatous disease (CGD), which is characterized by frequent infections with pathogens such as *Staphylococcus aureus*, *Aspergillus fumigatus*, *Salmonella* (non-typhoid serovars), *Serratia marcescens* and *Burkholderia cepacia* (for a review see [44]). CGD can be caused by inherited mutations in genes that encode the gp91-phox, p47-phox, p22-phox or p67-phox subunits of the NADPH-dependent phagocyte oxidase (for a review see OMIM database http://www.ncbi.nlm.nih.gov/entrez/query.fcgi?db=OMIM and [44]). Murine models of CGD that lack functional *Cybb* (gp91^{phox}; [45]) or *Ncf1* (gp47^{phox}; [46]) alleles, recapitulate the immune defects that are observed in man and can be considered as geno- and phenocopies of the human disease.

Mutations and disease-associated polymorphisms in TLR-pathway genes or in the TLR-encoding genes that are themselves responsible for the predisposition of patients to infectious diseases, have also been identified. This was achieved through the extensive work that had previously been carried out in mice to unravel the underlying mechanisms of host defense. A dominant *TLR5* stop codon mutation that abolishes the recognition of bacterial flagellin could be linked to susceptibility to Legionnaires' disease. This was confirmed after a large outbreak of the disease at a flower show in the Netherlands in 1999 by the subsequent screening of affected individuals for sequence variants in the *TLR5* gene [47, 48]. *TLR5* was a candidate gene because the molecular and functional characterization of this receptor had previously been established using knockout mice [49]. Other examples of mutated human genes that are involved in TLR signaling are deficiencies in *IRAK4* which cause hyporesponsiveness to LPS and a predisposition to pyogenic bacterial infections [50, 51] and variations in the *TLR4* sequence that may be linked to the development of meningococcal sepsis in patients [52] and to resistance to Legionnaires' disease [53].

Mendelian susceptibility to mycobacterial diseases (MSMD) is a rare syndrome of severe infections caused by low-virulence mycobacteria and *Salmonella* in patients carrying mutations in five genes of the IL-12-IFN-γ-STAT1 signaling axis. Interestingly, these patients seem to be resistant to other pathogens, while knockout mice with deficiencies in the homologous genes are widely susceptible to many different microorganisms (for a detailed review see [54]).

Since the early studies of Emil Skamene on complement factor C5 deficiency in mice and their susceptibility to *Listeria* infections, it would seem that the complement-activation product C5 a in particular, may be associated with the development of sepsis in humans (see the recent detailed review by Peter Ward [55]).

The mouse has been instrumental in accelerating the advances in immunological research which have been made in recent decades [56, 57]. However, it must be remembered that discoveries made using mice do not necessarily lead to corresponding insights in humans. Different selection pressures (many caused by parasites and pathogens) over the last 65 million years of evolution have left their imprints on the human and rodent genomes. These are also responsible for the differences between mouse and human immunology [58]. Nevertheless, mice will remain the prime experimental model of choice for immunological research and defined differences in host defense genes between both organisms might be tackled in the future by directed transgenic approaches (e. g. humanization of mice through "knock in" procedures or replacement with human genes).

8.4
Phenotyping at the GBF-Mouse Infection Challenge Platform (ICP)

At our research institution we are working primarily with three bacterial pathogens to characterize the innate and adaptive immune responses of mice. The infection experiments are performed under controlled "SPF" conditions (see also Section 8.5.4). Mice are housed in individually ventilated cages (IVCs) and are handled only in protected areas (under laminar flow hoods). All material, which is brought into the animal infection unit, is either autoclaved or sterilized with H_2O_2. An extensive sentinel program is used to screen the unit every 3 months for the presence of unwanted microorganisms. New mouse strains from external sources are imported into the facility via embryo transfer. In addition, the "altered Schaedler flora (ASF)" is used to colonize the gastrointestinal tracts of transferred mice with a defined microflora [59]. To accomplish this, germ-free foster mice that have previously been colonized with the ASF are used for embryo transfers.

We use *Streptococcus pyogenes* as an extracellular, Gram-positive pathogen to investigate cellular mechanisms and molecules that are linked to the induction of bacteremia and sepsis [60]. Within the last few years, this infection model has proven very useful in the elucidation of immune mechanisms underlying disease susceptibility to streptococcal-induced sepsis in the mouse (for more information see [61]).

To investigate mucosal immune responses to pathogenic Gram-negative bacteria in the gut we established a low-virulence infection model with *Yersinia enterocolitica*. Here, the strain *Y. enterocolitica* E40 serotype O:9 is employed to specifically analyze responses to local infection in the intestine and Peyer's patches. Mice are infected with *Y. enterocolitica* by the natural route (oral) and immune responses are monitored at 3, 9, and 21 days after infection. The model is very informative but also quite labor intensive because it involves histology as one of the major out-read-systems for infection susceptibility and resistance (Frischmann, U. and Müller, W. in preparation). Its main advantage over other mouse infection models is that the presence of infectious lesions and the influx of various immune effector cells after infection can be monitored effectively at very high resolution (e. g. the cellular level). To test new mutant lines of mice for general defects in host defense we use the intracellular pathogen *Listeria monocytogenes*, which has been mentioned above. The screening protocols used for this pathogen will be listed below. Additional infection protocols for the other ICP-pathogens, *S. pyogenes* and *Y. enterocolitica* can be accessed at the EUMORPHIA website (http://www.eumorphia.org). Within the last 3 years we have established standardized operation procedures (SOPs) for infection phenotyping of mice in the framework of EUMORPHIA that can be used for reference. In addition to the SOPs for mouse infection challenge with *S. pyogenes*, *L. monocytogenes*, and *Y. enterocolitica*, other associated documents and primary-extended screens for macrophages (e. g. PAMP responses) can be found on this website.

8.4.1 Sreening Protocols

8.4.1.1 Infection with Listeria monocytogenes

The mouse model of *L. monocytogenes* infection has been used extensively to investigate immune responses to bacterial infection [16]. The response of the mouse to *L. monocytogenes* infection can be divided into an early non-specific inflammatory response, which is initiated after recognition of the pathogen by the innate immune system, and a more delayed specific immune response which is mediated by the adaptive immune system (Fig. 8.1). *L. monocytogenes* can be administered through different routes of infection. Inoculation of *L. monocytogenes* into mice can be carried out orally, intraperitoneally (i. p.), subcutaneously (s.q.) or intravenously (i. v.). However, although infecting mice orally is the natural route of infection for *L. monocytogenes,* high doses of the bacterium are required for the successful establishment of a systemic infection. This is thought to be due to the species-specific interaction of the *Listeria* protein internalin A on the surface of the bacterium with its host cell receptor E-cadherin [62]. The interaction of both proteins mediates the adhesion and invasion of *Listeria* into the epithelial cells of the intestine. Humans and mice have a different amino-acid residue at a critical site in the E-cadherin protein [63]. It has been suggested that this difference in E-cadherin is responsible for the host-specific tropism of *Listeria* in the gut. In our laboratory we use the intravenous route of infection. This inoculation method is highly reproducible and results in a rapid systemic infection. Different strains of *L. monocytogenes* can vary in virulence. For our infection protocols we use the *L. monocytogenes* strain EGD, which is commonly used for phenotyping and has an intermediate to high virulence in mice when compared to other available strains of *Listeria* [64]. The advantage of this strain is that a European consortium has sequenced its genome [65] and many isogenic *L. monocytogenes* mutants are available in this genetic background. This allows the use of bacterial mutants, which are attenuated in virulence due to the deficiency of critical virulence factors. Therefore, these *Listeria* mutants can be used to phenotype the host response of very susceptible mouse mutant strains that would otherwise immediately succumb to the infection (for instance, *interferon-γ* gene knockout mice). In general infection doses ranging from 10^2 (sublethal) to 10^5 colony forming units (cfu) are used, depending on the LD_{50} of the mouse strain under investigation. As already mentioned, the most commonly used inbred mouse strains differ significantly in their susceptibility to *Listeria* infection [17]. Therefore, it is advisable to test the LD_{50} of a particular mouse mutant strain before starting extensive experiments, especially when the mouse strain under investigation is maintained on a mixed genetic background. Together with the laboratory of Dirk Busch at the German Mouse Clinic (see also Chapter 10 in this volume) we have recently established extensive baseline data sets for the C57BL/6J, C3H/HeN, BALB/c and CBA/J inbred mouse strains that can be used for reference [66]. In the course of this project we made the interesting observation that infection of mice with *L. monocytogenes* is associated with a sex-dependent susceptibility pattern. Independent of the genetic background, female mice are in general more susceptible to the infection than male mice [66] and this should also be taken into account when undertaking infection challenge experiments with this pathogen. Following i. v. infection of mice, *L. monocytogenes* is first confronted by macrophages and neutrophils in the spleen and the liver. Within the next 6–12 h of infection 60 to 90 % of the bacteria are killed and the organism can no longer be detected in the peripheral blood of immunocompetent mice. The early pro-inflammatory response in the liver is initiated by the interaction of the pathogen or its products with pattern recognition receptors (e. g. Tlr2 and Tlr5) on the surface of Kuppfer cells [67], which are the macrophages of the liver. The interaction of the bacterial PAMPs with TLRs induces signal transduction pathways that lead to the activation of transcription factors (e. g. NF-κB), which in turn promotes the production and release of pro-inflammatory cytokines involved in innate host defense mechanisms (e. g. IL-1, IL-6 and TNF-α). These events promote the accumulation of neutrophils at the local infection sites. These immune effector cells are the principal microbicidal population in the liver during the first 24 to 48 h of *Listeria* infection [68]. They eradicate most of the bacteria and produce additional cytokines and chemokines that attract and further stimulate additional immune effector cells, such as monocytes, granulocytes, natural killer (NK) cells and T cells. The cytokine IFN-γ which is mainly produced by NK cells and T cells of the T-helper cell type-1 type (Th1), in turn activates macrophages that phagocytose and kill the pathogen inside the infected tissue. These responses substantially reduce the infectious burden of the animal and are essential for the survival of the infected mice. If these initial host reactions fail, the mice will eventually die within the first 2–3 days after infection. In these cases, histopathological analysis will reveal necrotic, multifocal granuloma in the liver and spleen and necrotic lesions caused by multiplying *Listeria* in the bone marrow (Fig. 8.2). During the normal infection process in an immunocompetent host, a fraction of the *L. monocytogenes* taken up by the liver escapes the antimicrobial activity of neutrophils and macrophages and will spread to hepatocytes, in which the bacteria will further replicate intracellularly. In immunocompetent mice, T cells will finally mediate the clearance of the pathogen and will eventually provide long-term immunity. This specific immune response to the pathogen occurs 5 to 7 days after infection and is mainly mediated by cytolytic $CD8^+$ T cells (for a review see [69]). If these T cell-mediated adaptive immune responses fail the mice will die from the infection 7 to 10 days after inoculation. To test T cell immunity specifically to *Listeria monocytogenes* we recommend the recall infection experiments described in Chapter 10 of this volume.

When live *Listeria* are used for infection challenge experiments in mice, it should be remembered that the bacteria are pathogenic organisms that can cause severe, sometimes lethal infections in immunocompromised patients or in individuals receiving immunosuppressive drugs. Therefore, all work carried out using *L. monocytogenes* must be performed under Biosafety Level 2 (BL-2) conditions. Investigators, who decide to work with this pathogen, should consult their local Biological Safety Department for guidance and approval prior to planning any experiments. If standard precautions are observed and followed, work with this microorganism can be carried out quite safely. The fact that detailed microbial and immunological characterization has been carried out on *L. monocytogenes*, makes it one of the safest and most popular pathogens to be used in experimental immunology worldwide.

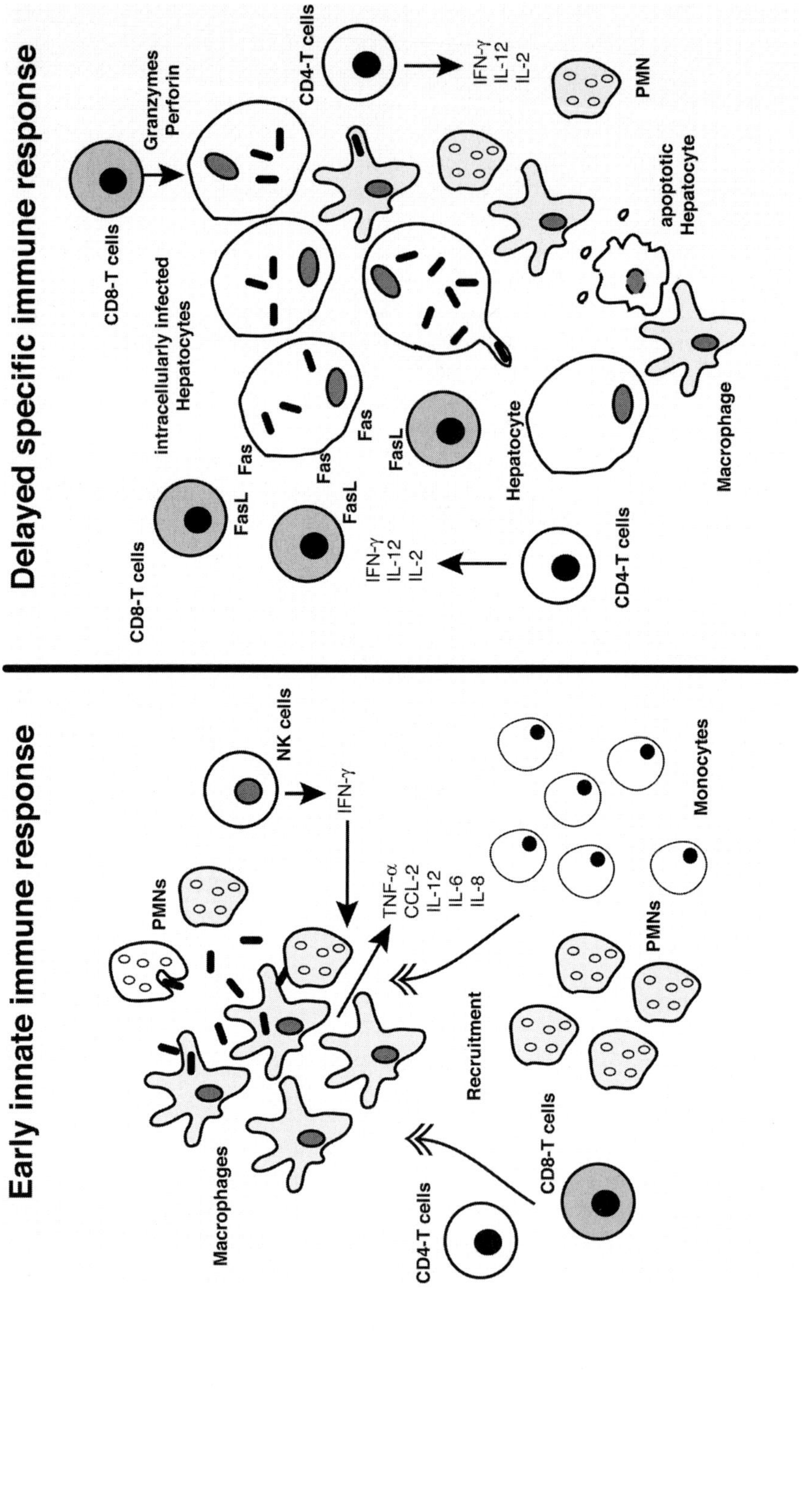

Early innate immune response
Macrophages
PMNs
NK cells
IFN-γ
TNF-α
CCL-2
IL-12
IL-6
IL-8
Monocytes
PMNs
Recruitment
CD4-T cells
CD8-T cells
Delayed specific immune response
CD8-T cells
Granzymes
Perforin
CD4-T cells
IFN-γ
IL-12
IL-2
PMN
intracellularly infected
Hepatocytes
apoptotic
Hepatocyte
CD8-T cells
FasL
Fas
Fas
FasL
Fas
FasL
Hepatocyte
Macrophage
IFN-γ
IL-12
IL-2
CD4-T cells

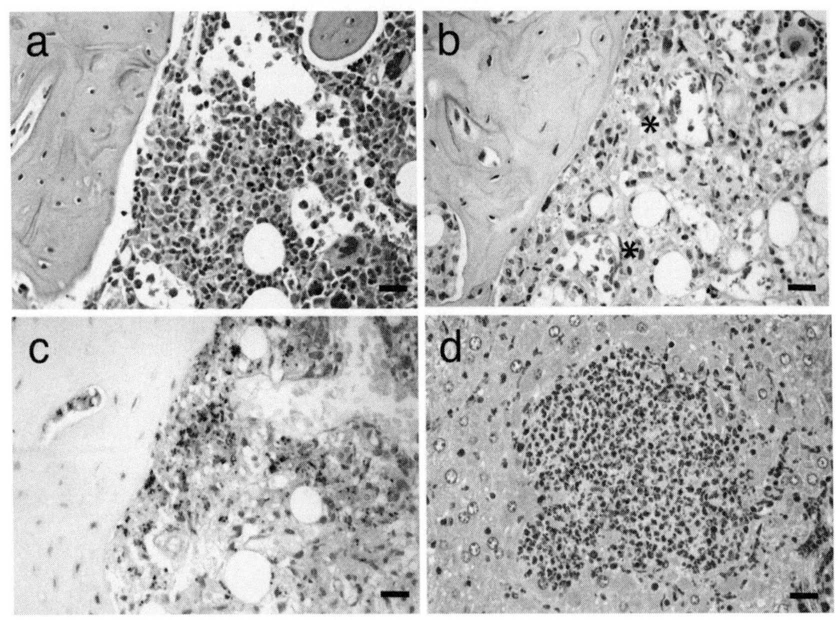

Fig. 8.2 Histopathological analysis of *Listeria monocytogenes* infections in mice. (a) Transverse section of the bone marrow from a CBA/J male 3 days after infections (*L. monocytogenes* EGD, 1×10^4 cfu , i. v.). (b) Transverse bone marrow section of a CBA/J female 3 days after infection with the same infection dose as in (a); note in (b) in comparison to (a) the acute and focal necrosis in the epiphysis of the bone marrow (indicated by stars). Female mice are more susceptible to *L. monocytogenes* infection than male mice. Sections in (a) and (b) were stained with hematoxylin and eosin. (c) Bone marrow section from the female shown in (b) stained with an anti-*Listeria* antibody. Bacteria are multiplying in the bone marrow at the sites of cellular necrosis (brown color). (d) Transverse liver section of a CBA/J female 1 day after i. v. infection with 1×10^4 cfu *L. monocytogenes* EGD. Granulomatous hepatitis with central necrosis and infiltration of neutrophilic granulocytes are visible. Scale bars = 12.5 μm.

◁ **Fig. 8.1** Early innate and delayed specific host immune responses to *Listeria monocytogenes* in the liver. Important effector cells involved in the immune response and released cytokines and chemokines are shown (PMNs, polymorphonuclear neutrophils; NK, natural killer cells). Cytolytic CD8-T cells which are involved in the specific immune response to *L. monocytogenes* kill infected hepatocytes through induction of apoptosis via the FAS ligand (FasL)/FAS receptor pathway or the perforin and granzyme pathway.

8.5
Practical Guidelines

8.5.1
Growing Log-phase Cultures of *Listeria monocytogenes* EGD for Mouse Infection

Reagents Phosphate buffered saline (PBS) pH 7.0, Brain Heart Infusion Medium (BHI from Becton and Dickinson, MD, USA; dissolve 37 g BHI powder in 1 l H_2O, autoclave and store at 4 °C), BHI plates (add 15 g agar to 1000 ml BHI medium, autoclave, pour plates and store at 4 °C for a maximum of 4 weeks), 0.4% Trypan blue (dilute 400 mg Trypan blue (Sigma) in 100 ml H_2O, sterile filter (0.22 µm) and store at room temperature.
Equipment Laboratory centrifuge (with temperature control), incubator set at 37 °C, shaking incubator (37 °C, 110 r.p.m.), photometer set at a wavelength of 600 nm, microscope (upright, no inverse) with 10 × ocular and 20 × (brightfield) and 40 × (phase contrast) objectives and phase contrast condenser, Thoma chamber (normal chamber depth, 0.1 mm), laminar-flow bench (BL2 level).

Procedure

1. Day 1: plate *Listeria* from a fresh glycerol stock on a pre-warmed (37 °C) BHI plate. Let bacteria grow for 24 h at 37 °C.
2. Day 2: inoculate a single colony from the BHI plate into 6 ml of BHI medium in a 14-ml snap cap tube. Let bacteria grow overnight at 37 °C in a shaking incubator (110 r.p.m.).
3. Day 3: measure OD_{600} of the overnight culture, OD should be around 1.0. Dilute overnight culture 1 : 10 in BHI medium (3 ml bacteria in 27 ml BHI) in an Erlenmeyer flask. Measure OD_{600} of the dilution. Let the bacteria grow at 37 °C in a shaking incubator (110 r.p.m.).
4. After 2 h of growth start to measure OD_{600} in 5–10-min intervals until an OD_{600} of 0.5 is reached. Centrifuge bacteria for 5 min at 1600 *g*, 4 °C. Pour off the supernatant and add 10 ml of ice-cold PBS. Re-suspend the pellet carefully using a pipette. Centrifuge for 5 min at 1600 *g*, 4 °C. Pour off the supernatant and re-suspend the pellet in 25 ml of ice-cold PBS. Dilute 100 µl of bacteria in 900 µl of 0.4% Trypan blue. Count the cells under a microscope using a Thoma chamber. Dilute *Listeria* to a concentration of 1×10^6 cfu/ml in PBS.
5. Prepare 10^{-1}, 10^{-2}, 10^{-3}, 10^{-4} dilutions of the *Listeria* using PBS. Plate 100 µl of the 10^{-2}, 10^{-3} and 10^{-4} dilution on BHI plates, each on three separate plates. Incubate overnight at 37 °C. Count the colonies on the following day and calculate the concentration of *Listeria* inoculum (cfu) used in the experiment.

8.5.2
Infection of Mice with *Listeria monocytogenes* EGD

Reagents Phosphate buffered saline (PBS), pH 7.0.

Equipment Infrared light (e. g. Philips Infrared PAR-38, 150 W), mouse strainer (e. g. Plas-Labs, #551-BSSR), syringe with 29 G needle (e. g. B.Braun Omnican®-F), Laminar-flow bench (BL2 level).

Procedure

1. Use the *L. monocytogenes* inoculum prepared from the protocol above. Dilute bacteria to 1.5×10^5 cfu/ml in PBS. An injection of 100 µl is equivalent to an infection dose of 1.5×10^4 cfu per mouse which is the LD_{50} for female C57BL/6J mice. If the mouse strain of interest is more susceptible or resistant in an initial experiment, repeat the infection procedure using a lower or higher dose respectively.
2. In a laminar flow hood open the animal cage under infrared light for at least 2 min. The distance between the top of the cage and the bulb should be approximately 20 cm.
3. Pre-warm mice individually by placing each mouse on top of the cage under the bulb for 15–30 s. Place the mouse in a mouse restraining device.
4. Inoculate 100 µl diluted *L. monocytogenes* into the lateral tail vein (veins are visible on both sides of a pre-warmed mouse tail). Start injecting near the tip of the tail.
5. Survival rates can then be determined with a daily health check of the infected animals for a period of 14 days (for more detailed instructions see "SOP Health monitoring of mice in infection experiments", Workpackage 6 accessible at http://www.eumorphia.org, or upon request). Animals that survive for 14 days are ranked as "resistant".

8.5.3 Quantification of Bacterial Growth in Spleen and Liver after *L. monocytogenes* Infection

Reagents Phosphate buffered saline (PBS) pH 7.0, Brain Heart Infusion Medium (BHI from Becton and Dickinson, MD, USA; dissolve 37 g BHI powder in 1 l H_2O, autoclave and store at 4 °C), BHI plates (add 15 g agar to 1000 ml BHI medium, autoclave, pour plates and store at 4 °C for a maximum of 4 weeks).

Equipment Laboratory centrifuge (with temperature control), incubator set at 37 °C, automated tissue homogenizer, Laminar-flow bench (BL2-level).

Procedure

1. Sacrifice mice by cervical dislocation on day 2 or 3 after infection (depending on the survival curve of the previous experiment; use time points prior to the deaths of the first mice).
2. Aseptically remove liver and spleen. Weigh all organs and put in a 14-ml Snap cap tube containing 5 ml pre-cooled (4 °C) PBS. Store all organs on ice until all mice have been sacrificed and dissected.
3. Homogenize tissues for 30 s using an automatic homogenizer. Clean out homogenizer between two samples by squirreling with PBS (2×5 s), 70% ethanol (1×5 s) and PBS (1×5 s).

4. Prepare dilutions of tissue homogenates (10^{-1}, 10^{-2}, 10^{-3}, 10^{-4}, and 10^{-5} dilutions) using PBS and snap cap tubes. Plate dilutions onto BHI plates in triplicate. Store dilutions at 4 °C for further dilutions (if necessary).
5. Next day, count colonies and calculate total cfu/organ and cfu/mg organ. If the 10^{-5}-dilution plates are still overgrown, prepare further dilutions (10^{-6}, 10^{-7}, and 10^{-8}) and plate again in triplicate.

8.5.4 Troubleshooting

Critical parameters for infection challenge experiments and immunological phenotyping are as follows:

- "Specific pathogen free" (SPF) housing conditions for mice. Background infections can severely alter the immune status of mice and can therefore have a deep impact on the outcome of infection experiments [70]. In our mouse facility we follow the recommendations of the Federation of European Laboratory Animal Science Association (FELASA) to maintain standardized hygiene conditions [71].
- Standardization of experiments and inclusion of internal controls. For infection experiments age- and sex-matched animals should be used. To control for inoculum quality and technical effectiveness of infection we use mice from known susceptible inbred strains as internal controls. These are infected together with the experimental group of mice and the results in this control group should match the previously recorded baseline data for this particular strain.
- It is advisable to use standardized microbiology laboratory practice to exclude contamination of bacterial stocks, inoculum, and experimental samples. Protective clothing should be worn when handling and manipulating bacteria, samples and animals and all procedures should be carried out in a laminar-flow hood.

8.6 Outlook

The identification and characterization of infectious disease loci in mice cannot be successfully accomplished without *in vivo* infection experiments combined with sophisticated and detailed phenotypic analysis. To further investigate the molecular and cellular mechanisms of host–pathogen interactions in the future there is a need to extend the currently available genetic toolbox for the mouse. We need to establish new transgenic mouse lines that allow a spatial and temporal depletion of critical immune effector cells during the host defense response. For example, conditional expression of cholera toxin could be used to specifically deplete the immune system of cells of the myeloid or lymphoid cell lineage. This would help to characterize the contributions of specialized or activated cells (such as different types of macrophages) to the host defense response.

We also need to further miniaturize the readout assays and systems that are used to monitor the host response in a non-invasive manner. New advances in proteomics technology should be combined with immunology. The new emerging field of "immunoproteomics" is very promising in this respect [72, 73]. New techniques should help to identify novel surrogate markers of immune responses that can subsequently be used to establish new high-throughput diagnostic tools (e. g. antibody arrays for the detection of antigen-specific immunoglobulins or new acute-phase serum proteins). New imaging technologies will definitely have a deep impact on the phenotyping of infection responses. They will elucidate the complexity and dynamics of infection processes *in vivo* without disturbing or interfering with the integrity of the different host immune compartments that are involved in defense mechanisms against pathogens. Some examples of emerging new imaging technologies have already been mentioned in this chapter and in Chapter 10.

The challenge of the future is to identify and understand the diverse intersection points where pathogen virulence factors interfere with host defense and metabolism. Systems biology approaches combined with new quantitative readout assays for "immunophenotypes" might be the key to understanding regulatory and signaling networks that are most critical for defense against pathogens and drug development.

Acknowledgement

We are grateful to Dr. Bastian Pasche for his contributions to the development of the *Listeria* infection protocols and to Drs Evi Wollscheid-Lengeling and Jens Böse for discussions and comments on the manuscript. We thank Professor Dr. Achim Gruber and Silke Mateika for analyzing and providing images of *L. monocytogenes*-infected mouse tissues. Our work is supported by the EU-project EUMORPHIA (Grant QLG2-CT-2002–00930) and the German National Genome Network (NGFN Grant 01GR0439).

References

1 Hoffmann, J. A. **2003**, The immune response of Drosophila, *Nature* 426, 33–38.
2 Beutler, B., Hoebe, K., Georgel, P., Tabeta, K., Du, X. **2004**, Genetic analysis of innate immunity: TIR adapter proteins in innate and adaptive immune responses, *Microbes Infect.* 6, 1374–1381.
3 Beutler, B., Du, X., Hoebe, K. **2003**, From phenomenon to phenotype and from phenotype to gene: forward genetics and the problem of sepsis, *J. Infect. Dis.* 187 (Suppl. 2), S321–S326.
4 Janeway, C. A., Jr., Medzhitov, R. **2002**, Innate immune recognition, *Annu. Rev. Immunol.* 20, 197–216.
5 Buer, J., Balling, R. **2003**, Mice, microbes and models of infection, *Nat. Rev. Genet.* 4, 195–205.
6 Beutler, B., Crozat, K., Koziol, J. A., Georgel, P. **2005**, Genetic dissection of innate immunity to infection: the mouse cytomegalovirus model, *Curr. Opin. Immunol.* 17, 36–43.
7 Sacks, D., and Noben-Trauth, N. **2002**, The immunology of susceptibility and resistance to *Leishmania major* in mice, *Nat. Rev. Immunol.* 2, 845–858.
8 Sacks, D., Sher, A. **2002**, Evasion of innate immunity by parasitic protozoa, *Nat. Immunol.* 3, 1041–1047.

9 Webster, L. **1933**, Inherited and acquired factors in resistance to infection: I. Development of resistant and susceptible lines of mice through selective breeding, *J. Exp. Med.* 57, 793–817.

10 Webster, L. **1933**, Inherited and acquired factors in resistance to infection: II. A comparison of mice inherently resistant or susceptible to *Bacillus enteritidis* infection with respect to fertility, weight, and susceptibility to various routes and types of infection, *J. Exp. Med.* 57, 819–843.

11 Mackaness, G. B. **1962**, Cellular resistance to infection, *J. Exp. Med.* 116, 381–406.

12 Bancroft, G. J., Schreiber, R. D., Bosma, G. C., Bosma, M. J., Unanue, E. R. **1987**, A T cell-independent mechanism of macrophage activation by interferon-gamma, *J. Immunol.* 139, 1104–1107.

13 Mielke, M. E., Niedobitek, G., Stein, H., Hahn, H. **1989**, Acquired resistance to *Listeria monocytogenes* is mediated by Lyt-2+ T cells independently of the influx of monocytes into granulomatous lesions, *J. Exp. Med.* 170, 589–594.

14 Ladel, C. H., Flesch, I. E., Arnoldi, J., and Kaufmann, S. H. **1994**, Studies with MHC-deficient knock-out mice reveal impact of both MHC I- and MHC II-dependent T cell responses on *Listeria monocytogenes* infection, *J. Immunol.* 153, 3116–3122.

15 Mackaness, G. B. **1969**, The influence of immunologically committed lymphoid cells on macrophage activity *in vivo*, *J. Exp. Med.* 129, 973–992.

16 Pamer, E. G. **2004**, Immune responses to *Listeria monocytogenes, Nat. Rev. Immunol.* 4, 812–823.

17 Cheers, C., McKenzie, I. F. **1978**, Resistance and susceptibility of mice to bacterial infection: genetics of listeriosis, *Infect. Immun.* 19, 755–762.

18 Skamene, E., Kongshavn, P. A., Sachs, D. H. **1979**, Resistance to *Listeria monocytogenes* in mice: genetic control by genes that are not linked to the H-2 complex, *J. Infect. Dis.* 139, 228–231.

19 Skamene, E., Kongshavn, P. A. **1979**, Phenotypic expression of genetically controlled host resistance to *Listeria monocytogenes, Infect. Immun.* 25, 345–351.

20 Gervais, F., Stevenson, M., Skamene, E. **1984**, Genetic control of resistance to *Listeria monocytogenes*: regulation of leukocyte inflammatory responses by the Hc locus, *J. Immunol.* 132, 2078–2083.

21 Boyartchuk, V. L., Broman, K. W., Mosher, R. E., D'Orazio, S. E., Starnbach, M. N., Dietrich, W. F. **2001**, Multigenic control of *Listeria monocytogenes* susceptibility in mice, *Nat Genet* 27, 259–260.

22 Churchill, G. A., Airey, D. C., Allayee, H., Angel, J. M., Attie, A. D., Beatty, J., Beavis, W. D., Belknap, J. K., Bennett, B., Berrettini, W., Bleich, A., Bogue, M., Broman, K. W., Buck, K. J., Buckler, E., Burmeister, M., Chesler, E. J., Cheverud, J. M., Clapcote, S., Cook, M. N., Cox, R. D., Crabbe, J. C., Crusio, W. E., Darvasi, A., Deschepper, C. F., Doerge, R. W., Farber, C. R., Forejt, J., Gaile, D., Garlow, S. J., Geiger, H., Gershenfeld, H., Gordon, T., Gu, J., Gu, W., de Haan, G., Hayes, N. L., Heller, C., Himmelbauer, H., Hitzemann, R., Hunter, K., Hsu, H. C., Iraqi, F. A., Ivandic, B., Jacob, H. J., Jansen, R. C., Jepsen, K. J., Johnson, D. K., Johnson, T. E., Kempermann, G., Kendziorski, C., Kotb, M., Kooy, R. F., Llamas, B., Lammert, F., Lassalle, J. M., Lowenstein, P. R., Lu, L., Lusis, A., Manly, K. F., Marcucio, R., Matthews, D., Medrano, J. F., Miller, D. R., Mittleman, G., Mock, B. A., Mogil, J. S., Montagutelli, X., Morahan, G., Morris, D. G., Mott, R., Nadeau, J. H., Nagase, H., Nowakowski, R. S., O'Hara, B. F., Osadchuk, A. V., Page, G. P., Paigen, B., Paigen, K., Palmer, A. A., Pan, H. J., Peltonen-Palotie, L., Peirce, J., Pomp, D., Pravenec, M., Prows, D. R., Qi, Z., Reeves, R. H., Roder, J., Rosen, G. D., Schadt, E. E., Schalkwyk, L. C., Seltzer, Z., Shimomura, K., Shou, S., Sillanpaa, M. J., Siracusa, L. D., Snoeck, H. W., Spearow, J. L., Svenson, K. **2004**, The Collaborative Cross, a community resource for the genetic analysis of complex traits, *Nat. Genet.* 36, 1133–1137.

23 Vidal, S. M., Malo, D., Vogan, K., Skamene, E., Gros, P. **1993**, Natural resistance to infection with intracellular parasites: isolation of a candidate for Bcg, *Cell* 73, 469–485.

24 Lam-Yuk-Tseung, S., Gros, P. **2003**, Genetic control of susceptibility to bacterial infections in mouse models, *Cell. Microbiol.* 5, 299–313.

25 Beutler, B., Rietschel, E. T. **2003**, Innate immune sensing and its roots: the story of endotoxin, *Nat. Rev. Immunol.* 3, 169–176.

26 Poltorak, A., He, X., Smirnova, I., Liu, M. Y., Van Huffel, C., Du, X., Birdwell, D., Alejos, E., Silva, M., Galanos, C., Freudenberg, M., Ricciardi-Castagnoli, P., Layton, B., Beutler, B. **1998**, Defective LPS signaling in C3H/HeJ and C57BL/10ScCr mice: mutations in Tlr4 gene, *Science* 282, 2085–2088.

27 Qureshi, S. T., Lariviere, L., Leveque, G., Clermont, S., Moore, K. J., Gros, P., Malo, D. **1999**, Endotoxin-tolerant mice have mutations in Toll-like receptor 4 (Tlr4), *J. Exp. Med.* 189, 615–625.

28 Roach, J. C., Glusman, G., Rowen, L., Kaur, A., Purcell, M. K., Smith, K. D., Hood, L. E., Aderem, A. **2005**, The evolution of vertebrate Toll-like receptors, *Proc. Natl Acad. Sci. USA* 102, 9577–9582.

29 Hoebe, K., Du, X., Georgel, P., Janssen, E., Tabeta, K., Kim, S. O., Goode, J., Lin, P., Mann, N., Mudd, S., Crozat, K., Sovath, S., Han, J., Beutler, B. **2003**, Identification of Lps2 as a key transducer of MyD88-independent TIR signalling, *Nature* 424, 743–748.

30 Akira, S., Takeda, K. Toll-like receptor signalling, *Nat. Rev. Immunol.* **2004**, 4, 499–511.

31 Liew, F. Y., Xu, D., Brint, E. K., O'Neill, L. A. **2005**, Negative regulation of toll-like receptor-mediated immune responses, *Nat. Rev. Immunol.* 5, 446–458.

32 Hoebe, K., Georgel, P., Rutschmann, S., Du, X., Mudd, S., Crozat, K., Sovath, S., Shamel, L., Hartung, T., Zahringer, U., Beutler, B. **2005**, CD36 is a sensor of diacylglycerides, *Nature* 433, 523–527.

33 Georgel, P., Crozat, K., Lauth, X., Makrantonaki, E., Seltmann, H., Sovath, S., Hoebe, K., Du, X., Rutschmann, S., Jiang, Z., Bigby, T., Nizet, V., Zouboulis, C. C., Beutler, B. **2005**, A toll-like receptor 2-responsive lipid effector pathway protects mammals against skin infections with gram-positive bacteria, *Infect. Immun.* 73, 4512–4521.

34 Cahalan, M. D., Parker, I., Wei, S. H., Miller, M. J. **2002**, Two-photon tissue imaging: seeing the immune system in a fresh light, *Nat. Rev. Immunol.* 2, 872–880.

35 Marzola, P., Nicolato, E., Di Modugno, E., Cristofori, P., Lanzoni, A., Ladel, C. H., Sbarbati, A. **1999**, Comparison between MRI, microbiology and histology in evaluation of antibiotics in a murine model of thigh infection, *Magma* 9, 21–28.

36 Hardy, J., Francis, K. P., DeBoer, M., Chu, P., Gibbs, K., Contag, C. H. **2004**, Extracellular replication of *Listeria monocytogenes* in the murine gall bladder, *Science* 303, 851–853.

37 Zhao, M., Yang, M., Baranov, E., Wang, X., Penman, S., Moossa, A. R., Hoffman, R. M. **2001**, Spatial-temporal imaging of bacterial infection and antibiotic response in intact animals, *Proc. Natl Acad. Sci. USA* 98, 9814–9818.

38 Bettegowda, C., Foss, C. A., Cheong, I., Wang, Y., Diaz, L., Agrawal, N., Fox, J., Dick, J., Dang, L. H., Zhou, S., Kinzler, K. W., Vogelstein, B., Pomper, M. G. **2005**, Imaging bacterial infections with radiolabeled 1-(2'-deoxy-2'-fluoro-beta-D-arabinofuranosyl)-5-iodouracil, *Proc. Natl Acad. Sci. USA* 102, 1145–1150.

39 Altman, J. D., Moss, P. A., Goulder, P. J., Barouch, D. H., McHeyzer-Williams, M. G., Bell, J. I., McMichael, A. J., Davis, M. M. **1996**, Phenotypic analysis of antigen-specific T lymphocytes, *Science* 274, 94–96.

40 Klenerman, P., Cerundolo, V., Dunbar, P. R. **2002**, Tracking T cells with tetramers: new tales from new tools, *Nat. Rev. Immunol.* 2, 263–272.

41 Busch, D. H., Pilip, I., Pamer, E. G. **1998**, Evolution of a complex T cell receptor repertoire during primary and recall bacterial infection, *J. Exp. Med.* 188, 61–70.

42 Busch, D. H., Pilip, I. M., Vijh, S., Pamer, E. G. **1998**, Coordinate regulation of complex T cell populations responding to bacterial infection, *Immunity* 8, 353–362.

43 Fischer, A. **2004**, Human primary immunodeficiency diseases: a perspective, *Nat. Immunol.* 5, 23–30.

44 Fang, F. C. **2004**, Antimicrobial reactive oxygen and nitrogen species: concepts and controversies, *Nat. Rev. Microbiol.* 2, 820–832.

45 Pollock, J. D., Williams, D. A., Gifford, M. A., Li, L. L., Du, X., Fisherman, J.,

Orkin, S. H., Doerschuk, C. M., Dinauer, M. C. **1995**, Mouse model of X-linked chronic granulomatous disease, an inherited defect in phagocyte superoxide production, *Nat. Genet.* 9, 202–209.

46 Jackson, S. H., Gallin, J. I., Holland, S. M. **1995**, The p47 phox mouse knock-out model of chronic granulomatous disease, *J. Exp. Med.* 182, 751–758.

47 Den Boer, J. W., Yzerman, E. P., Schellekens, J., Lettinga, K. D., Boshuizen, H. C., Van Steenbergen, J. E., Bosman, A., Van den Hof, S., Van Vliet, H. A., Peeters, M. F., Van Ketel, R. J., Speelman, P., Kool, J. L., Conyn-Van Spaendonck, M. A. **2002**, A large outbreak of Legionnaires' disease at a flower show, the Netherlands, 1999, *Emerg. Infect. Dis.* 8, 37–43.

48 Hawn, T. R., Verbon, A., Lettinga, K. D., Zhao, L. P., Li, S. S., Laws, R. J., Skerrett, S. J., Beutler, B., Schroeder, L., Nachman, A., Ozinsky, A., Smith, K. D., Aderem, A. **2003**, A common dominant TLR5 stop codon polymorphism abolishes flagellin signaling and is associated with susceptibility to Legionnaires' disease, *J. Exp. Med.* 198, 1563–1572.

49 Hayashi, F., Smith, K. D., Ozinsky, A., Hawn, T. R., Yi, E. C., Goodlett, D. R., Eng, J. K., Akira, S., Underhill, D. M., and Aderem, A. **2001**, The innate immune response to bacterial flagellin is mediated by Toll-like receptor 5, *Nature* 410, 1099–1103.

50 Medvedev, A. E., Lentschat, A., Kuhns, D. B., Blanco, J. C., Salkowski, C., Zhang, S., Arditi, M., Gallin, J. I., Vogel, S. N. **2003**, Distinct mutations in IRAK-4 confer hyporesponsiveness to lipopolysaccharide and interleukin-1 in a patient with recurrent bacterial infections, *J. Exp. Med.* 198, 521–531.

51 Picard, C., Puel, A., Bonnet, M., Ku, C. L., Bustamante, J., Yang, K., Soudais, C., Dupuis, S., Feinberg, J., Fieschi, C., Elbim, C., Hitchcock, R., Lammas, D., Davies, G., Al-Ghonaium, A., Al-Rayes, H., Al-Jumaah, S., Al-Hajjar, S., Al-Mohsen, I. Z., Frayha, H. H., Rucker, R., Hawn, T. R., Aderem, A., Tufenkeji, H., Haraguchi, S., Day, N. K., Good, R. A., Gougerot-Pocidalo, M. A., Ozinsky, A., Casanova, J. L. **2003**, Pyogenic bacterial infections in humans with IRAK-4 deficiency, *Science* 299, 2076–2079.

52 Smirnova, I., Mann, N., Dols, A., Derkx, H. H., Hibberd, M. L., Levin, M., Beutler, B. **2003**, Assay of locus-specific genetic load implicates rare Toll-like receptor 4 mutations in meningococcal susceptibility, *Proc. Natl Acad. Sci. USA* 100, 6075–6080.

53 Hawn, T. R., Verbon, A., Janer, M., Zhao, L. P., Beutler, B., Aderem, A. **2005**, Toll-like receptor 4 polymorphisms are associated with resistance to Legionnaires' disease, *Proc. Natl Acad. Sci. USA* 102, 2487–2489.

54 Casanova, J. L., Abel, L. **2004**, The human model: a genetic dissection of immunity to infection in natural conditions, *Nat. Rev. Immunol.* 4, 55–66.

55 Ward, P. A. **2004**, The dark side of C5 a in sepsis, *Nat. Rev. Immunol.* 4, 133–142.

56 Mak, T. W., Penninger, J. M., Ohashi, P. S. **2001**, Knockout mice: a paradigm shift in modern immunology, *Nat Rev Immunol* 1, 11–19.

57 Lengeling, A., Pfeffer, K., Balling, R. **2001**, The battle of two genomes: genetics of bacterial host/pathogen interactions in mice, *Mamm. Genome* 12, 261–271.

58 Mestas, J., Hughes, C. C. **2004**, Of mice and not men: differences between mouse and human immunology, *J. Immunol.* 172, 2731–2738.

59 Dewhirst, F. E., Chien, C. C., Paster, B. J., Ericson, R. L., Orcutt, R. P., Schauer, D. B., Fox, J. G. **1999**, Phylogeny of the defined murine microbiota: altered Schaedler flora, *Appl. Environ. Microbiol.* 65, 3287–3292.

60 Medina, E., Goldmann, O., Rohde, M., Lengeling, A., Chhatwal, G. S. **2001**, Genetic control of susceptibility to group A streptococcal infection in mice, *J. Infect. Dis.* 184, 846–852.

61 Medina, E., Lengeling, A. **2005**, Genetic regulation of host responses to Group A Streptococcus in mice, *Brief. Funct. Genomic Proteomic* 4, 248–257.

62 Lecuit, M., Vandormael-Pournin, S., Lefort, J., Huerre, M., Gounon, P., Dupuy, C., Babinet, C., Cossart, P. **2001**, A transgenic model for listeriosis: role of internalin in crossing the intestinal barrier, *Science* 292, 1722–1725.

63 Schubert, W. D., Urbanke, C., Ziehm, T., Beier, V., Machner, M. P., Domann, E.,

Wehland, J., Chakraborty, T., Heinz, D. W. **2002**, Structure of internalin, a major invasion protein of *Listeria monocytogenes*, in complex with its human receptor E-cadherin, *Cell* 111, 825–836.

64 Busch, D. H., Vijh, S., Pamer, E. G. **2000**, Animal model for infection with *Listeria monocytogenes*. In *Current Protocols in Immunology*, John Wiley & Sons, pp. 19.19.11–19.19.19.

65 Glaser, P., Frangeul, L., Buchrieser, C., Rusniok, C., Amend, A., Baquero, F., Berche, P., Bloecker, H., Brandt, P., Chakraborty, T., Charbit, A., Chetouani, F., Couve, E., de Daruvar, A., Dehoux, P., Domann, E., Dominguez-Bernal, G., Duchaud, E., Durant, L., Dussurget, O., Entian, K. D., Fsihi, H., Garcia-del Portillo, F., Garrido, P., Gautier, L., Goebel, W., Gomez-Lopez, N., Hain, T., Hauf, J., Jackson, D., Jones, L. M., Kaerst, U., Kreft, J., Kuhn, M., Kunst, F., Kurapkat, G., Madueno, E., Maitournam, A., Vicente, J. M., Ng, E., Nedjari, H., Nordsiek, G., Novella, S., de Pablos, B., Perez-Diaz, J. C., Purcell, R., Remmel, B., Rose, M., Schlueter, T., Simoes, N., Tierrez, A., Vazquez-Boland, J. A., Voss, H., Wehland, J., Cossart, P. **2001**, Comparative genomics of *Listeria* species, *Science* 294, 849–852.

66 Pasche, B., Kalaydjiev, S., Franz, T. J., Kremmer, E., Gailus-Durner, V., Fuchs, H., Hrabé de Angelis, M., Lengeling, A., Busch, D. H. **2005**. Sex-dependent susceptibility to *Listeria monocytogenes* infection is mediated by differential interleukin-10 production, *Infect. Immun* 73, 5952–5960.

67 Seki, E., Tsutsui, H., Tsuji, N. M., Hayashi, N., Adachi, K., Nakano, H., Futatsugi-Yumikura, S., Takeuchi, O., Hoshino, K., Akira, S., Fujimoto, J., Nakanishi, K. **2002**, Critical roles of myeloid differentiation factor 88-dependent proinflammatory cytokine release in early phase clearance of *Listeria monocytogenes* in mice, *J. Immunol.* 169, 3863–3868.

68 Gregory, S. H., Sagnimeni, A. J., Wing, E. J. **1996**, Bacteria in the bloodstream are trapped in the liver and killed by immigrating neutrophils, *J. Immunol.* 157, 2514–2520.

69 Gregory, S. H., Liu, C. C. **2000**, CD8+ T-cell-mediated response to *Listeria monocytogenes* taken up in the liver and replicating within hepatocytes, *Immunol. Rev.* 174, 112–122.

70 Nicklas, W., Homberger, F. R., Illgen-Wilcke, B., Jacobi, K., Kraft, V., Kunstyr, I., Maehler, M., Meyer, H., Pohlmeyer-Esch, G. **1999**, Implications of infectious agents on results of animal experiments, *Lab. Anim.* 33 (Suppl. 1), 39–87.

71 Nicklas, W., Baneux, P., Boot, R., Decelle, T., Deeny, A. A., Fumanelli, M., Illgen-Wilcke, B. **2002**, Recommendations for the health monitoring of rodent and rabbit colonies in breeding and experimental units. *Lab. Anim.* 36, 20–42.

72 Chen, Z., Peng, B., Wang, S., Peng, X. **2004**, Rapid screening of highly efficient vaccine candidates by immunoproteomics, *Proteomics* 4, 3203–3213.

73 Krah, A., Jungblut, P. R. **2004**. Immunoproteomics, *Methods Mol. Med.* 94, 19–32.

9
Animal Models of Nociception

Ildikó Rácz and Andreas Zimmer

9.1
Introduction

Pain is an important signal alerting us to the possibility or occurrence of tissue injury. Chronic pain, however, is a pathological condition that does not serve any useful purpose, but results in a rather substantial loss of life quality. Indeed, chronic pain is one of the most common and costly health problems. Animal studies are essential to acquire new and clinically relevant knowledge about the mechanisms of pain and to develop novel pain therapies.

In recent years, mouse mutants, most of which have been generated by gene targeting, have become an increasingly powerful tool in the analysis of nociceptive signaling (Lariviere et al. 2001). Mouse models have been established that show alterations at virtually all levels of pain perception, including those with developmental defects in nociceptive neurons, mutants in which primary nociceptive afferents do not respond appropriately to painful stimuli (Agarwal et al. 2004), mice with defects in the central processing of pain (Kieffer and Gaveriaux-Ruff 2002), mice with disruptions of the descending modulation of pain signaling (Watanabe et al. 2005), and those in which dynamic adaptations of the nociceptive system under pathological conditions is altered (Harvey et al. 2004). These mouse strains have enabled researchers to study these different aspects of pain signaling at the molecular level and to evaluate the mechanisms of action of analgesic drugs.

As mice lack the ability to communicate their experience of pain, animal experimentation depends on the analysis and interpretation of physiological responses and motor behaviors, ranging from simple withdrawal reflexes to more complex avoidance or escape behaviors. The ideal animal paradigm should distinguish between innocuous and noxious stimuli, consider a range of responses in correlation with the stimulus intensity, and be susceptible to pharmacological manipulation.

Standards of Mouse Model Phenotyping. Edited by Martin Hrabé de Angelis, Pierre Chambon, and Steve Brown

ISBN: 3-527-31031-2

9.2 Ethical Considerations

Ethical considerations are an important aspect of all animal experiments. However, because in the study of pain it is sometimes necessary to inflict pain on conscious animals, ethical aspects have special relevance in this field of research.

All experiments intended to stimulate acute or chronic pain require careful planning in order to avoid all unnecessary pain and in order to minimize the pain that may be necessary to achieve the scientific goal. Researchers studying pain should be particularly aware of the fact that animals are not an object for exploitation, but living individuals.

The International Association for the Study of Pain has published the following *Ethical Guidelines for Investigations of Experimental Pain in Conscious Animals*:

1. It is essential that the intended experiments on pain in conscious animals be reviewed beforehand by scientists and lay-persons. The potential benefit of such experiments to our understanding of pain mechanisms and pain therapy needs to be shown. The investigator should be aware of the ethical need for a continuing justification of his investigations.
2. If possible, the investigator should try the pain stimulus on himself; this principle applies for most non-invasive stimuli causing acute pain.
3. To make possible the evaluation of the levels of pain, the investigator should give a careful assessment of the animal's deviation from normal behavior. To this end, physiological and behavioral parameters should be measured. The outcome of this assessment should be included in the manuscript.
4. In studies of acute or chronic pain in animals measures should be taken to provide a reasonable assurance that the animal is exposed to the minimal pain necessary for the purposes of the experiment.
5. An animal presumably experiencing chronic pain should be treated for relief of pain, or should be allowed to self-administer analgesic agents or procedures, as long as this will not interfere with the aim of the investigation.
6. Studies of pain in animals paralyzed with a neuromuscular blocking agent should not be performed without a general anesthetic or an appropriate surgical procedure that eliminates sensory awareness.
7. The duration of the experiment must be as short as possible and the number of animals involved kept to a minimum.

In our description of experimental procedures, we will alert the reader to measures that we have found useful for minimizing the severity and duration of pain in animals, and also for minimizing the number of animals involved. However it should be noted that in some instances it may not be possible, for scientific reasons, to follow our advice and that sometimes more effective measures may be available. This chapter endeavors to introduce new researchers to the basic methodologies of pain research. It is not intended as a laboratory manual.

Therefore, when planning experiments that inflict pain on animals, researchers should carefully evaluate the most recent literature and consult with experts to ensure that the experimental design meets the most rigorous ethical and scientific standards in the field.

9.3
General Considerations

Behavioral experiments such as those described in this chapter are influenced by a number of environmental factors that are sometimes difficult to adjust or to avoid. For example, stress produces analgesia and is one of the most important confounding factors in pain research (as well as in most other behavioral studies). Stress should therefore be minimized as much as possible before and during the behavioral experiment. This requires careful planning.

For most experiments, animals will be removed from the room in which they are housed to a procedure room. Such relocation is always stressful for a mouse. The animals should therefore have sufficient time to acclimatize to the new environment. The length of the acclimatization period depends on the specific location and may exceed several hours. The mice should be carefully observed during acclimatization. Initially, mice will show signs of unrest and increased activity. They should calm down, perhaps sleep or rest, and resume their normal home-cage activities, such as grooming and feeding. Mice should also be acclimatized to the apparatus or cage in which they are to be tested, if possible. In some instances, it may be advisable to familiarize the animals with the experimental situation, by performing the entire procedure without the nociceptive stimulation. The equipment should be carefully cleaned after each test using an appropriate cleaning solution. The rigor with which these procedures are carried out is of course, dependent on the stress sensitivity of the mouse strain under investigation.

Within the procedure room, all sources of stress, such as loud noises or bright light should be kept to an absolute minimum. It may be better to provide constant background noise such as music, rather than subjecting the animals to a quiet environment that is frequently interrupted by sharp noises from external sources which cannot be controlled. Other animal species or their odors, especially rats, should also not be present in the procedure room.

It is very important to test all animals at the same time of the day. Because mice are nocturnal animals, behavioral experiments are carried out during the dark (active) phase. We therefore house our mice for behavioral studies in a room with an inverted dark–light cycle (21.00 hours lights on, 09.00 hours lights off).

Even with the most careful planning, it may be difficult or impossible to provide constant environmental conditions over an extended period of time. Even within a normal working week there are day-to-day changes: the mice are probably left relatively undisturbed over the weekend, they may be in a soiled cage, or they may have recently been transferred into a new cage, etc. It is therefore very important to analyze animals from different groups that are to be compared, including treatment groups, on the same day. It is preferable to alternately test a small number of animals from different groups, rather than to test one complete group before starting the next.

In our experience, a group size of 10 mice is sufficient to obtain statistically significant results for most experiments. However, if the effect is small, or the responses variable, it may be necessary to include more animals.

9.4 Assays for Acute Pain Thresholds

9.4.1 Thermal Stimuli

The acute application of a thermal stimulus to the skin is one of the most commonly used procedures to assess the threshold of nociceptive perception. Thermal stimuli above 45 °C activate high-threshold sensory fibers that innervate the skin. These axons respond to temperatures that are in the range of those that produce escape behaviors, with the frequency of discharge proportional to the intensity of the stimulus to which the skin is exposed (Treede et al. 1992). These afferents activate dorsal-horn neurons, which are located in the superficial spinal lamina and project into the contralateral ventrolateral tracts to supraspinal sites where they serve to activate neurons in the medulla, mesencephalon, and thalamus (Willis and Westlund 1997).

Heat stimuli evoke an escape response with a latency that is inversely correlated to the intensity of the stimulus (Dirig and Yaksh 1995; Tsuruoka et al. 1988;). A shorter response latency in a test group of animals compared to a control group, is an indicator of increased pain sensitivity (hyperalgesia), while a longer response latency would indicate a reduced pain threshold (hypoalgesia; Agarwal et al. 2004). *It is important to limit the time that the animal is exposed to heat in order to avoid tissue damage.* The length of the cut-off period depends on the nature and intensity of the stimulus.

In the following sections we will describe three tests that are commonly used to assess thermal pain responses: the tail flick test (a thermal stimulus applied to the tail; D'Amour and Smith 1941), the Hargreaves test (application of a radiant stimulus to the paw; Hargreaves et al. 1988), and the hot plate test (animals placed on a heated metal plate; Eddy and Leimbach 1953). If carried out correctly, these tests should not produce any tissue damage or long-lasting pain. It is therefore not necessary to treat the animals with analgesic drugs after the experiment.

9.4.1.1 Tail-flick Test

When the tail of a mouse is exposed to a noxious heat, a spinally-mediated withdrawal reflex (a vigorous flexion of the tail) is elicited at the nociceptive threshold (D'Amour and Smith 1941). Although the behavioral response consists of a spinal reflex, it is affected by descending pain-modulating inputs.

Tail-flick devices are available from several equipment suppliers (e. g. Harvard Apparatus, Omnitech Electronics, Columbus). These devices produce defined radiant heat stimuli of different intensities and automatically detect tail movements. The heat source and the motion detector are usually placed in a V-shaped groove, into which the distal tip of the animal's tail is placed. When the heat source is turned on, a timer is activated. Movement of the tail stops the timer and turns off the heat source. *The cut-off time should correspond to three to five times the baseline latency.* Because the risk of tissue injury is relatively small in animals that respond within the normal range and because there is relatively little adaptation of the animal's response, it is possible to repeat the test three to four times.

In our experience, the intensity of the heat source is likely to change after prolonged use, or when a light bulb is changed. Therefore at the beginning of the test, or at least when the instrument has not been used for some time, it is necessary to re-calibrate the setting of the instrument using animals with well-characterized response times. This re-calibration should be done by *first using low intensities, followed by a slow increase* until the desired baseline latency (often 4–8 s) is obtained.

The most common variation of the radiant heat tail-flick test is the tail immersion test (Grotto and Sulman 1967). In this test a mouse is gently placed on a platform immediately adjacent to a warm (45–52 °C) water bath. The distal half of the mouse-tail is dipped into warm water and the time until a vigorous withdrawal reflex is triggered is measured manually. The main advantage of the tail immersion test is the fact that no special equipment is required, as most laboratories have water baths that are sufficiently accurate (±1 °C) for this experiment. It may be necessary to involve two people in this test, one to handle the animal, the other to time its reaction, unless a foot pedal-activated or voice-activated timer is available.

The tail-immersion test can also be used to determine nociceptive responses triggered by a very cold stimulus. In this case the test is identical, except that a fluid (e. g. ethanol) is cooled to –12 to –18 °C using an immersion-cooling device with a required accuracy of ±1 °C. *It is important to avoid lower temperatures because the skin may freeze.*

In both versions of the tail-immersion test, a temperature should be selected at which animals have a response latency of 5–15 s, *and a cut-off time of 30 s should be used.* It is usually possible to repeat this test two to three times, but some mouse strains will display latency changes if tested more often. Repeating the test is advantageous as an average response latency can be calculated for each mouse. *This is usually more accurate than single measurements and thus fewer animals may be required to obtain statistically significant results.*

Although these tests seem simple at first, they are in fact greatly influenced by the proficiency of the investigator. It is most critical to avoid any unnecessary stress during animal handling. Some investigators prefer to use specifically designed restrainers, i. e. Plexiglas tubes with a notch through which the animals' tail can be extended. In our laboratory, we mostly restrain animals manually. Independent of the restraining method, however, we find it very useful to acclimatize the animals to the handling procedure. This is especially true for inexperienced investigators. For this purpose, the animals are subjected to the entire tail-flick or tail-immersion test for three consecutive days, but without activation of the heat source, or with water warmed to 30 °C, respectively.

9.4.1.2 **Hargreaves Test**

This test is conceptually similar to the tail-flick assay. The Hargreaves apparatus consists of multiple chambers with transparent (glass) floors, under which a mobile radiant heat source can be focussed accurately on a mouse paw, thus triggering a spinal withdrawal reflex (Hargreaves et al. 1988).

The animals are first acclimatized to the apparatus for approximately 30 min before the measurement. The light source is than moved using a targeting system directly under the paw. When the light beam is turned on, a timer is activated. When the integrated motion sensor registers a movement of the paw, the timer is stopped and the light beam is turned off. The intensity of the heat source may be varied to produce the desired (4–6 s) baseline response time. *To prevent tissue injury, a cut-off time is set at three to five times the baseline latency.* The test can be repeated several (three to five) times without causing any harm to the animal or significant changes in response latencies.

We also sometimes use the Hargreaves apparatus for the determination of tail-flick latencies. In this case the procedure is identical, except that the tail instead of a paw is targeted. Although a single session in this test takes longer due to the acclimatization of the mice to the apparatus, it is less stressful for the animals. It is therefore not necessary to familiarize the animals to handling, thus making this test faster than the tail-flick assay overall.

9.4.1.3 **Hot-plate Test**

This assay was first described by Woolfe and MacDonald (Woolfe and McDonald 1944), who varied the temperature of an enclosed zinc plate to between 45 and 70 °C and observed the responses of normal mice on this hot plate. The method was later refined by using a copper plate heated with vapors of boiling solvents to produce a controlled temperature of 54–56 °C (Eddy and Leimbach 1953). A thermostatically-controlled circulating water bath or electricity can also be used to maintain a surface at the desired temperature.

Most commercially-available modern devices, however, regulate the surface temperature electronically with an accuracy of well below 1 °C. These devices are also typically equipped with a built-in timer. The movement of the animal on the metal plate is constrained by a removable Plexiglas wall.

Most mice that are placed on the hot metal will respond within a few seconds by licking or lifting a fore -or hind-paw, shaking a foot, or making an attempt to escape (jumping). The probability of each of these responses is dependent on the temperature of the plate, the strain of mice, and the previous experience of the animals. Researchers should choose the response that is seen first and has the lowest variability. Often, licking and shaking of the hind-paw is used as an endpoint. Forepaw licking and lifting are components of normal grooming behavior. Most mice will show paw-licking, -lifting, or -shaking before they jump. Indeed, jumping is mostly a suprathreshold response and thus rather an indicator of pain-tolerance than pain sensitivity. *Therefore, to avoid unnecessary pain, jumping should only be used as a behavioral endpoint when scientifically indicated.*

We typically test mice at temperatures between 48 and 55 °C. At 52 °C, which we use most commonly, mice usually respond within 20–40 s. At this temperature, we normally use a cut-off time of 60 s. *Mice should be removed from the hot-plate immediately after showing the indicative pain behavior.* Each mouse should be tested only once since repeated testing may lead to latency changes (Mogil et al. 1998). It is possible, however, to re-test animals after several days.

9.4.2 Mechanical Stimuli

Noxious mechanical stimuli are widely used with rats, but are rather difficult to apply to the much smaller and more active mice. For example, in the commonly used test paw pressure test the paw of a rat is inserted into a mechanical device that applies a constantly increasing pressure to the paw. The rat will withdraw the paw as soon as the nociceptive threshold is reached. Mice, in contrast, often lack the strength to withdraw the paw thus making accurate measurements difficult. However, mechanical stimuli are very useful for assessing pain responses in inflammatory and neuropathic pain models, where nociceptive thresholds are much lower (Randall and Selitto 1957). We describe here the two most common applications in mice, the Randall–Selitto tail- or paw-pressure, and the von Frey filament tests.

9.4.2.1 The Tail- and Paw-pressure Test

The first method used to deliver a mechanically-induced noxious stimulus was the tail pressure test (Eddy 1928), in which Eddy measured the animal's reaction to a gradual increase in pressure applied by a mechanical device. Automated tail- or paw-pressure devices (Green and Young 1951) are now available from several manufacturers (TSE, Germany; Muromachi, Japan; Ugo-basile, Italy).

When a constantly increasing pressure is applied to a small area of the tail or paw of a mouse, the animal will show a withdrawal reflex at the pain threshold. If the pressure is further increased, the animal will make attempts to withdraw the paw and, if unsuccessful, it will vocalize. The first response is clearly a spinal reflex, but the latter two responses are mediated at supraspinal sites. In practice, it may be difficult to discriminate between the withdrawal reflex and a struggle to escape.

The weight threshold at which the animal responds is determined. It is normally difficult to repeat the test more than once, especially when relatively high pressures are required to produce a nociceptive response in healthy tissues. These tests do not produce long-lasting pain and it is thus not necessary to administer analgesic drugs after the conclusion of the experiment.

9.4.2.2 Von Frey Filament Test

The von Frey filament test, developed more than 100 years ago, is still widely used today for the assessment of tactile allodynia. Von Frey monofilaments are short calibrated filaments (nylon filaments are mainly used today) inserted into a holder which allow the investigator to exert a defined pressure on a punctiform area of the mouse paw. The animal is repeatedly stimulated with increasingly strong filaments to determine the threshold where a nocifensive paw-withdrawal response is reliably elicited. A major disadvantage of this test lies in the fact that low-threshold mechanoreceptors are also stimulated, thus the stimulus is not specific. As a consequence, it is difficult to determine in healthy tissues, whether the withdrawal response is triggered by a pain sensation. It is also very difficult to apply the stimulus to a freely moving animal. If tactile allodynia is determined, animals can be placed in a small compartment with a grid or wire-mesh floor, thus permitting the application of the stimulus to the ventral surface of the paw from underneath. Electronically-controlled instruments (Ugo-Basile, Italy) now provide a more versatile alternative to von Frey filaments. These instruments consist of a mobile pressure-actuator, which can exert a user-defined force, or a continuously increasing force on the mouse paw.

For this test, mice are placed in the transparent test chamber with a wire-mesh floor and allowed acclimatize for approximately 30 min. The tip of the filaments or force actuator is applied to the middle of the plantar surface of the mice. The withdrawal threshold is measured four to eight times pro trial and expressed as the tolerance level in grams (Chaplan et al. 1994; Hofmann et al. 2003).

9.5
Tonic and Visceral Pain Models

Thermal and mechanical noxious stimuli are usually of short duration and confined to the area of the skin that has been exposed to the stimulus. However most of the clinically relevant pain states involve pain that arises from deep tissues and visceral pain. These forms of pain are often poorly localized or diffuse and may radiate over considerable distances. The neuronal processes underlying deep and visceral pain are thought to be considerably different from those associated with cutaneous pain. In the following sections, we will describe two pain models that are commonly used in mice to assess visceral and deep pain, the writhing test and the formalin test.

9.5.1
The Writhing Test

If a chemical, noxious substance is injected into the peritoneal cavity it may activate nociceptors directly and/or produce pain through inflammation of visceral (subdiaphragmatic organs) and subcutaneous (muscle wall) tissues (Siegmund et al. 1957). The most commonly used substances for the stimulation of visceral pain in mice are glacial acetic acid (0.3–0.6 %), 2-phenyl-1,4-benzoquinone (0.02 % in 5 % aqueous ethanol), and magnesium sulfate (Koster et al. 1959).

For the assay, mice are placed in a small observation chamber (e. g. an animal cage) and habituated for at least 10 min. The noxious substance is then injected into the peritoneum in a volume of 10 ml/kg. Within minutes after the injection, a typical "writhing" response, indicative of visceral pain, can be observed. Writhing is a distinct severe contraction of the abdominal musculature and the backward extension of the hind limbs (VanderWende and Margolin 1956). The number of writhing responses is counted for 15–30 min. *Immediately after the conclusion of the test, the animal should receive an injection of an analgesic drug such as buprenorphine.*

9.5.2 The Formalin Test

The formalin test was originally described by Dubuison and Dennis using rats (Dubuisson and Dennis 1977), and was later modified for use in mice (Hunskaar et al. 1985). In this test, a small volume (20–50 µl) of 5 % formalin is injected under the dorsal surface or into the plantar surface of the hind paw. This injection will produce a biphasic response: an immediate pain response is often seen within seconds after the injection and may last for 5–10 min. These responses include lifting or licking/biting the affected paw. After this early phase, the animals show relatively few pain responses for 5–10 min, until the second phase response begins, which may last for up to 30 min (Porro and Cavazzuti 1993; Fig. 9.1).

Prior to the test, the animals are habituated to the observation chamber for 10 min. The chamber (e. g. a mouse cage) should have a transparent floor, with a mirror mounted underneath to allow a clear view of the paws. Early phase responses are observed in the 1–10-min interval following the injection and late phase responses 20–40 min after the injection. The actual time may vary, however, depending on the mouse strain.

We typically evaluate the animal responses using a time sampling procedure: a simple self-designed computer program generates a tone every 5 s and the observer

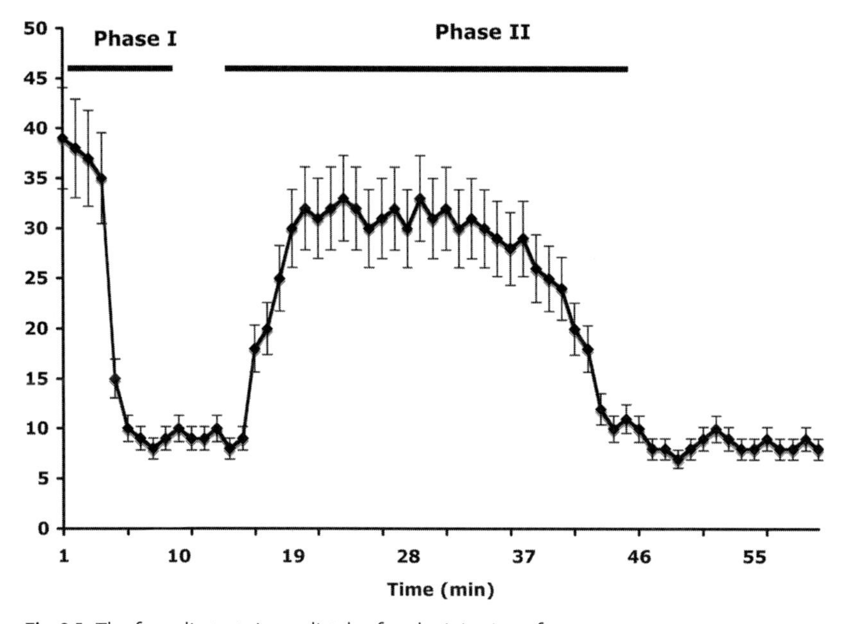

Fig. 9.1 The formalin test. Immediately after the injection of formalin under the dorsal surface of the mouse hindpaw, animals show typical pain reactions including licking, biting and lifting of the affected paw. These symptoms last for about 5 min (phase I), after which the mice are relatively free of pain, followed by a second phase of pain reactions.

scores the mouse behavior as “paw licking”, “paw lifting”, or “no pain response”. Other investigators have rated the animal responses using behavioral scales (Dubuisson and Dennis 1977). Some investigators measure the total time spent licking/biting the injected paw using a stop watch (Mogil et al. 1998; Sufka et al. 1998).

Immediately after the conclusion of the test, the animal should receive an injection of an analgesic drug such as buprenorphine. It should be noted that the formalin injection will produce a small necrotic area which may require 7–10 days to heal. All animals should be carefully monitored during this period and treated if necessary.

9.6 Hyperalgesia and Allodynia

Tissue injuries are painful, but the immediate and acute pain experienced during the injury can readily be controlled with analgesic drugs and thus does not present a clinical problem. However, following the acutely painful phase, patients may feel ongoing spontaneous aching pain, and the areas surrounding the site of the injury may become more sensitive. Disease conditions such as tissue inflammation or neuropathies can also result in decreased pain thresholds. Thus the severity of a noxious stimulus may be greatly accentuated (hyperalgesia) and normally innocuous stimuli may provoke pain (allodynia). These pain states, which can be difficult to control, involve peripheral mechanisms (e. g. activation of dormant receptors) and central sensitization (e. g. through the release of prostaglandins; Yaksh et al. 1999).

9.6.1 Hyperalgesia and Allodynia Caused by Neuropathic Pain

Neuropathic pain is the result of an injury or malfunction in the peripheral or central nervous system. The pain is often triggered by an injury, but this injury may or may not involve actual damage to the nervous system. Nerves can be infiltrated or compressed by tumors, strangulated by scar tissue, or inflamed by infection. The pain frequently has burning, lancinating, or electric shock-type qualities. Persistent allodynia, pain resulting from a non-painful stimulus such as a light touch, is also a common characteristic of neuropathic pain. The pain may persist for months or years beyond the apparent healing time of any damaged tissues (Richards 1967). Among the many causes of peripheral neuropathy, diabetes is the most common, but the condition can also be caused by chronic alcohol use, exposure to other toxins (including many chemotherapies), vitamin deficiencies, and a large variety of other medical conditions.

To study the neurobiological mechanisms and the development of neuropathic pain a number of animal models have been developed during the last decade (Martin and Eisenach 2001). These models are based on either an unilateral ligation of the sciatic or spinal nerves (i. e. the chronic constriction injury model (Bennett and Xie 1988)) or the spinal nerve ligation model (Kim and Chung 1992).

9.6.1.1 **Chronic Constriction Injury Model**

The chronic constriction injury model is the most frequently used model for the study of neuropathic pain. The model involves a loose unilateral ligation of the sciatic nerve, and shows many of the pathophysiological properties of chronic neuropathic pain in human subjects (Bennett and Xie 1988). Animals typically display tactile allodynia and thermal hyperalgesia within a few days after the ligation, and they may suffer spontaneous pain (Bridges et al. 2001). Over time, mice will also develop motor deficits in the affected leg. The clinical utility of this animal model is evidenced by the fact that compounds which are used clinically for the symptomatic treatment of chronic neuropathic pain (Sindrup and Jensen 1999) are usually effective.

Ligation of the sciatic nerve is achieved by anesthetizing mice (20–30 g) with isoflurane (3 % for induction and 1.5 % for maintenance) in air–oxygen gas. The right common sciatic nerve is exposed at the level of the middle of the thigh by blunt dissection through the biceps femoris. Proximal to the sciatic nerve trifurcation, about 10 mm of nerve is separated from adhering tissue. Three ligations are tied at 1-mm intervals around the exposed nerve using a 5–0 silk thread. The length of nerve thus affected is 6–8 mm. It is critically important to tie the ligatures correctly. If they are tied to loosely, neither hyperalgesia nor allodynia will develop. On the other hand, the nerve should not be damaged. The incision is closed in layers. In sham-operated control animals, an identical operation is performed on the same side, except that the sciatic nerve is not ligated. Although antibiotics are not necessary if sterile conditions are maintained during surgery, a systemic or local application immediately after the surgery may be useful and will not interfere with the experiment.

In a modification of this model the sciatic nerve is only partially ligated. For this purpose half of the sciatic nerve is ligated at an elevated position in the thigh. The dorsum of the nerve is carefully freed from the adhering tissues at a site near the trochanter just distal to the point at which the posterior biceps semitendinosus nerve brunches off the common sciatic nerve. The nerve is fixed in place, and an 8–0 silicon-treated silk suture is inserted into the nerve and tightly ligated so that the dorsal one-third to one-half of the nerve thickness is trapped in the ligature. This model shows many of the symptoms that characterize causalgia in humans: rapid onset, hyperalgesia, allodynia to touch, mirror image phenomena, and dependence on the sympathetic outflow (Seltzer et al. 1990). After this procedure animals show a behavior which suggests the presence of spontaneous pain. They are often hyperresponsive to non-noxious and noxious stimuli, both ipsilateral and contralateral to the operated side.

9.6.1.2 **Segmental Spinal Nerve Ligation Model**

This model is also widely used to investigate the mechanisms of neuropathic pain and for the development of analgesic drugs. The operation results in enduring behavioral signs of mechanical allodynia, heat hyperalgesia, cold allodynia, and ongoing pain (Kim and Chung 1992). Although the surgical procedure is more difficult to perform that the sciatic nerve ligation, it has the advantage that operated mice do not show signs of motor deficits beyond a mild inversion of the foot with slightly ventroreflexed toes.

Under isoflurane anesthesia, a longitudinal incision (3 cm in length, 5 mm lateral from the midline) is made at the lower lumbar sacral levels (from the caudal area of the L5 vertebra to the first sacral vertebra) to expose the paraspinal muscles on the left. The paraspinal muscle is isolated and removed from the level of the L5 spinous process to the sacrum. When the connective tissues and muscles are removed, it should be possible to see bony structures. Under a dissecting microscope, a small rongeur is used to remove the L6 transverse process, which covers the ventral rami of the L4 and L5 spinal nerves. The L5 spinal nerve is separated from adjacent tissues, and a thread of 6–0 silk is placed around the nerve which is tightly ligated. This procedure should interrupt all axons.

Before proceeding with the closure of the wound complete hemostasis should be confirmed. The muscles are sutured in layers using silk thread and the skin is closed with metal clips. In sham-operated control mice, an identical operation is performed on the same side, except that the spinal nerve is not ligated.

9.6.2
Hyperalgesia and Allodynia Caused by Tissue Inflammation

Tissue inflammation can produce persistent allodynia and hyperalgesia with a rapid onset. In experimental mouse models, tissue inflammation is usually produced by a subcutaneous injection of a small volume (20 µl) of a pro-inflammatory agent such as carrageenan (20 mg/ml), capsaicin (0.1 mg/ml), complete Freud's adjuvant containing heat-killed *Mycobacterium tuberculosis* (1 mg/ml), or zymosan A (3 mg/ml) into the hindpaw. The tissue inflammation can be monitored by the swelling of the affected paw and quantified by using a dial calliper or a plethysmometer. The inflammation produces little or no pain by itself.

9.6.2.1 Determination of Mechanical Allodynia and Thermal Hyperalgesia

Mechanical allodynia and thermal hyperalgesia are assessed at different timepoints after the surgery. We typically monitor mechanical and thermal pain responses on days 3, 6, 8, 10, and 15 after the nerve ligation and at weekly intervals thereafter. Mechanical allodynia is measured using a pressure transducer or von Frey filaments as described above. Thermal sensitivity of the hindpaws is determined with the Hargreaves test (Hargreaves et al. 1988).

The nociceptive tests are performed on the injured and control side, of operated and sham-operated animals. Operated animals typically display increased pain sensitivity and allodynia on the affected ipsilateral side, but not on the contralateral side. This should not occur in sham-operated animals. However, development of contralateral hyperalgesia and allodynia is also possible, depending on the experimental situation. Therefore, it important to compare ipsi- and contralateral responses within and between groups (Fig. 9.2).

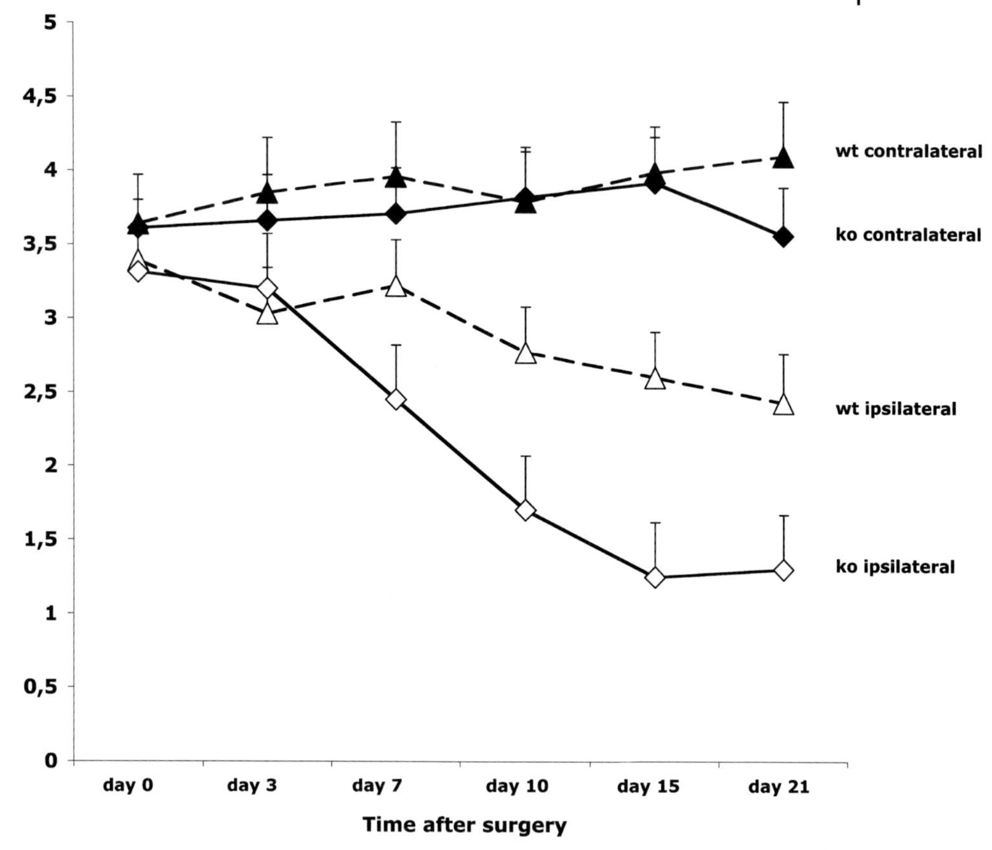

Fig. 9.2 Thermal hyperalgesia after chronic constriction injury. The response latencies of a wild-type ipsilateral (open triangle) and contralateral (closed triangle) to the operation side are shown. Also shown is a knockout mouse strain (rhombus) with an enhanced hyperalgesic response.

9.7 Stress-induced Analgesia

The perception of pain is clearly dependent on situational factors. For example, anxiety or the expectation of pain can enhance the subjective experience of pain in humans, while the stress of dangerous situations can reduce pain thresholds. The latter phenomenon is called stress-induced analgesia (SIA).

SIA can be produced in mice using a variety of stressors, including electric foot shocks (Menendez et al. 1993), social defeat stress (McLaughlin et al. 2005), restraint stress (Robinson et al. 2002; Shum et al. 2005), and forced walking stress (Furuta et al. 2003). However, a forced swim is by far the most common stressor (Konig et al. 1996). It has the advantages of being a naturally-occurring stressor, requiring no special equipment and being easy to standardize. A forced swim in warm (25–34 °C) water for 10 min produces relatively mild analgesia that can be completely reversed by opiate receptor antagonists. A brief (2 min) swim in 4 °C water results in more pronounced analgesia that is largely independent of the endogenous opioid system. For this purpose baseline pain thresholds are first determined using the hotplate or tail-flick assay. Mice are then placed in a 4-l glass beaker filled with 2 l water of the indicated temperature (±0.5 °C). After each swim, mice are patted dry and allowed to rest for 2 min (warm-water swim) or 10 min (cold-water swim), before the nociceptive test is repeated. SIA is calculated and expressed as the percentage analgesia: [(post-swim latency – baseline latency)/(cut-off time – baseline latency)] × 100.

References

Agarwal, N., S. Offermanns, et al. **2004**, Conditional gene deletion in primary nociceptive neurons of trigeminal ganglia and dorsal root ganglia, *Genesis* 38 (3), 122–129.

Bennett, G. J., Y. K. Xie **1988**, A peripheral mononeuropathy in rat that produces disorders of pain sensation like those seen in man, *Pain* 33 (1), 87–107.

Bridges, D., S. W. Thompson, et al. **2001**, Mechanisms of neuropathic pain, *Br. J. Anaesth.* 87 (1), 12–26.

Chaplan, S. R., F. W. Bach, et al. **1994**, Quantitative assessment of tactile allodynia in the rat paw, *J. Neurosci. Methods* 53 (1), 55–63.

D'Amour, F. E., D. L. Smith **1941**, A method for determining loss of pain sensation, *J. Pharmacol. Exp. Ther.* 72, 74–79.

Dirig, D. M., T. L. Yaksh **1995**, Differential right shifts in the dose–response curve for intrathecal morphine and sufentanil as a function of stimulus intensity, *Pain* 62 (3), 321–328.

Dubuisson, D., S. G. Dennis **1977**, The formalin test: a quantitative study of the analgesic effects of morphine, meperidine, and brain stem stimulation in rats and cats, *Pain* 4 (2), 161–174.

Eddy, N. B. **1928**, Studies on hypnotics of the barbituric acid series, *J. Pharmacol.* 33, 43–68.

Eddy, N. B., D. Leimbach **1953**, Synthetic analgesics. II. Dithienylbutenyl- and dithienylbutylamines, *J. Pharmacol. Exp. Ther.* 107 (3), 385–393.

Furuta, S., K. Onodera, et al. **2003**, Involvement of adenosine A1 receptors in forced walking stress-induced analgesia in mice, *Methods Find. Exp. Clin. Pharmacol.* 25 (10), 793–796.

Green, A. F., P. A. Young **1951**, A comparison of heat and pressure analgesiometric methods in rats, *Br. J. Pharmacol. Chemother.* 6 (4), 572–585.

Grotto, M., F. G. Sulman **1967**, Modified receptacle method for animal analgesimetry, *Arch. Int. Pharmacodyn. Ther.* 165 (1), 152–159.

Hargreaves, K., R. Dubner, et al. **1988**, A new and sensitive method for measuring thermal nociception in cutaneous hyperalgesia, *Pain* 32 (1), 77–88.

Harvey, R. J., U. B. Depner, et al. **2004**, GlyR alpha3: an essential target for spinal PGE2-mediated inflammatory pain sensitization, *Science* 304 (5672), 884–887.

Hofmann, H. A., J. De Vry, et al. **2003**, Pharmacological sensitivity and gene expression analysis of the tibial nerve injury model of neuropathic pain, *Eur. J. Pharmacol.* 470 (1–2), 17–25.

Hunskaar, S., O. B. Fasmer, et al. **1985**, Formalin test in mice, a useful technique for evaluating mild analgesics, *J. Neurosci. Methods* 14 (1), 69–76.

Kieffer, B. L., C. Gaveriaux-Ruff **2002**, Exploring the opioid system by gene knockout, *Prog. Neurobiol.* 66 (5), 285–306.

Kim, S. H., J. M. Chung **1992**, An experimental model for peripheral neuropathy produced by segmental spinal nerve ligation in the rat, *Pain* 50 (3), 355–363.

Konig, M., A. M. Zimmer, et al. **1996**, Pain responses, anxiety and aggression in mice deficient in pre-proenkephalin, *Nature* 383 (6600), 535–538.

Koster, R., M. Anderson, et al. 1959, Acetic acid for analgesic screening, *Fed. Proc.* 18, 412.

Lariviere, W. R., E. J. Chesler, et al. **2001**, Transgenic studies of pain and analgesia: mutation or background genotype? *J. Pharmacol. Exp. Ther.* 297 (2), 467–473.

Martin, T. J., J. C. Eisenach **2001**, Pharmacology of opioid and nonopioid analgesics in chronic pain states, *J. Pharmacol. Exp. Ther.* 299 (3), 811–817.

McLaughlin, J. P., S. Li, et al. **2005**, Social defeat stress-induced behavioral responses are mediated by the endogenous kappa opioid system, *Neuropsychopharmacology*.

Menendez, L., F. Andres-Trelles, et al. **1993**, Involvement of spinal kappa opioid receptors in a type of footshock induced analgesia in mice, *Brain Res.* 611, 264–271.

Mogil, J. S., C. A. Lichtensteiger, et al. **1998**, The effect of genotype on sensitivity to inflammatory nociception: characterization of resistant (A/J) and sensitive (C57BL/6J) inbred mouse strains, *Pain* 76 (1–2), 115–125.

Porro, C. A., M. Cavazzuti **1993**, Spatial and temporal aspects of spinal cord and brainstem activation in the formalin pain model, *Prog. Neurobiol.* 41 (5), 565–607.

Randall, L. O., J. J. Selitto **1957**, A method for measurement of analgesic activity on inflamed tissue, *Arch. Int. Pharmacodyn. Ther.* 111 (4), 409–419.

Richards, R. L. **1967**, Causalgia. A centennial review, *Arch. Neurol.* 16 (4), 339–350.

Robinson, D. A., F. Wei, et al. **2002**, Oxytocin mediates stress-induced analgesia in adult mice, *J. Physiol.* 540 (Pt2), 593–606.

Seltzer, Z., R. Dubner, et al. **1990**, A novel behavioral model of neuropathic pain disorders produced in rats by partial sciatic nerve injury, *Pain* 43 (2), 205–218.

Shum, F. W., S. W. Ko, et al. **2005**, Genetic alteration of anxiety and stress-like behavior in mice lacking CaMKIV, *Mol. Pain* 1, 22.

Siegmund, E., R. Cadmus, et al. **1957**, A method for evaluating both non-narcotic and narcotic analgesics, *Proc. Soc. Exp. Biol. Med.* 95 (4), 729–731.

Sindrup, S. H., T. S. Jensen **1999**, Efficacy of pharmacological treatments of neuropathic pain: an update and effect related to mechanism of drug action, *Pain* 83 (3), 389–400.

Sufka, K. J., G. S. Watson, et al. **1998**, Scoring the mouse formalin test: validation study, *Eur. J. Pain* 2 (4), 351–358.

Treede, R. D., R. A. Meyer, et al. **1992**, Peripheral and central mechanisms of cutaneous hyperalgesia, *Prog. Neurobiol.* 38 (4), 397–421.

Tsuruoka, M., A. Matsui, et al. **1988**, Quantitative relationship between the stimulus intensity and the response magnitude in the tail flick reflex, *Physiol. Behav.* 43 (1), 79–83.

VanderWende, C., S. Margolin **1956**, Analgesic tests based upon experimentally induced acute abdominal pain in rats, *Fed. Proc.* 15, 494.

Watanabe, S., T. Kuwaki, et al. **2005**, Persistent pain and stress activate pain-inhibitory orexin pathways, *Neuroreport* 16 (1), 5–8.

Willis, W. D., K. N. Westlund **1997**, Neuroanatomy of the pain system and of the pathways that modulate pain, *J. Clin. Neurophysiol.* 14 (1), 2–31.

Woolfe, G., A. D. McDonald **1944**, The evaluation of the analgesic action of pethidine hydrochloride (Demerol), *J. Pharmacol. Exp. Ther.* 80, 300–307.

Yaksh, T. L., X. Y. Hua, et al. **1999**, The spinal biology in humans and animals of pain states generated by persistent small afferent input, *Proc. Natl Acad. Sci. USA* 96 (14), 7680–7686.

10
Mouse Phenotyping: Immunology

Svetoslav Kalaydjiev, Tobias J. Franz, and Dirk H. Busch

10.1
Introduction

The maintenance of immunological homeostasis and immune function is attributed to complex interactions of various signaling pathways, known to be regulated by multiple genes. Although a number of traits allowing the successful control of infection, or leading to susceptibility to autoimmune diseases or cancer are now attributed to certain gene functions or genetic backgrounds, most of the genes contributing to the preservation of immunological homeostasis and regulation of specific immune functions have remained largely unknown.

A major source of information for the involvement of certain genes in immune function was the work carried out over the past 50 years through human genealogy studies of inherited immunodeficiencies. These investigations led to the identification of defective genes, as well as to the description of some intimate mechanisms of immunological homeostasis, maintenance, immune activation and regulation [1, 2]. In addition, major advances in therapy for such disease entities were first made possible by uncovering the genetic and molecular bases of disease pathology. It became clear that inherited immunodeficiency diseases were frequently caused by recessive gene defects, often linked to the X chromosome. They can affect different components of the immune system. Well-known examples are: complement system defects (e. g. paroxysmal nocturnal hemoglobinuria, a defect of DAF or CD59); impairment of phagocytic cells (e. g. leukocyte adhesion deficiency, a lack of CD18); B and T cell defects (e. g. Bruton's X-linked agammaglobulinemia due to a mutation in the *btk* gene, X-linked hyper IgM syndrome due to a mutation in the *CD40L* gene [3, 4]). In some cases different mutations lead to the same disease phenotype, and present as severe combined immunodeficiency (SCID). Examples for identified genetic alterations are: X-linked SCID (caused by mutation in the common gamma chain of several interleukin receptors), adenosine deaminase (ADA) deficiency, purine nucleotide phosphorylase (PNP) deficiency (both autosomally inherited), bare lymphocyte syndrome (lack of MHC class I or II molecules on lymphocytes and thymic epithelial cells), failure of DNA rearrangement in developing lymphocytes (caused by defects in the *RAG-1* or *RAG-2* genes), Bloom's

Standards of Mouse Model Phenotyping. Edited by Martin Hrabé de Angelis, Pierre Chambon, and Steve Brown

ISBN: 3-527-31031-2

Table 10.1 Mouse phenotyping centers measuring immunological parameters.

Mouse phenotyping centers	Website
German Mouse Clinic, Munich, Germany	http://www.mouseclinic.de
German ENU-Mouse Mutagenesis Screen Project, Munich, Germany	http://www.gsf.de/ieg/groups/genome/enu.html
Eumorphia research program	http://www.eumorphia.org
Center for Modeling Human Disease, Toronto, Canada	http://cmhd.mshri.on.ca
The Scripps Research Institute, La Jolla, USA	http://www.scripps.edu/imm/beutler/
The John Curtis School of Medical Research, Immunogenomics Laboratory, Australia	http://www.apf.edu.au
Mouse Clinic Institute, Strasbourg, France	http://www-mci.u-strasbg.fr/

syndrome caused by failure in the DNA repair system (mutation in a DNA helicase enzyme) or Wiskott–Aldrich syndrome (defect in the WAS protein presenting with impaired T cell function, reduced T cell numbers, and a failure of antibody responses to encapsulated bacteria). Another group of inherited immunological deficiencies was found to be the result of mutations in certain cytokine genes or their receptors [5, 6], for example deficiencies of IFN-gamma and IL-12 or their receptors, revealed as persistent infections with intracellular pathogens like *Mycobacteria* or *Salmonella* [7].

Until recently, the principal source of data for understanding the genetics of the immune system has been mouse models generated by targeted gene knock-out technology [8], transgenic expression of molecules of interest [9], spontaneous mutations [10], and congenic mouse strains [11, 12]. They provided useful models for studying the *in vivo* function of immunologically-relevant surface antigens, antigen receptors, histocompatibility markers, growth factors, cytokines and their receptors, as well as different intracellular signaling molecules [13].

In spite of these advances over recent decades, the information available from current human and mouse resources is by no means sufficient for understanding the complexity of genetic interactions leading to effective immunological homeostasis and immune regulation. Furthermore, there is an urgent need for new experimental models which correlate with defined human diseases of the immune system [14].

The availability of mouse and human genome sequences has marked the beginning of a new era in animal model research that greatly facilitates the attribution of new gene functions. However, whereas the generation of mouse mutant lines has been well established and even transferred to large-scale mutant production [15] via ENU mutagenesis, gene-trap, or conditional knock-in/knock-out technologies, the main problem is achieving detailed and standardized phenotypic analysis in order to identify defined alterations in the immune system. Even many existing mouse resources have never been fully phenotyped, leaving an enormous source for potential mouse models of human diseases almost untouched.

In order to face the challenge of standardized phenotypic analyses, several research centers around the world have initiated the development of generally accepted protocols for most comprehensive examinations of the mouse, which can also be applied to high-throughput technologies. The main centers and links to their websites are listed (Tab. 10.1).

10.2 Diagnostic Methods

Immunological research has always had a vast array of tools available that can be used in phenotyping studies. Looking back in history, the methods which were applied and which led to the first successful identification of hereditary immunological disorders caused by certain gene defects, can be divided into two large groups of techniques: methods for detection of antibodies and antigens, and methods for detection of cellular immunity. In addition, over recent years new developments in molecular biology and genetics have brought immunology researchers a new set of powerful tools.

10.2.1 Antibody-based Techniques

Historically, immunodiffusion techniques were one of the first assays used for quantification of serum immunoglobulin levels in the 1960 s [16, 17]. At about that time immunoelectrophoresis methods were improved and became standard for the analysis of serum proteins [18]. These techniques (and some of their modifications) were for several decades the gold standard in clinical immunology and were successfully used for the diagnosis of immunodeficiencies, monoclonal gammopathies, autoimmune and other diseases. Immunochemical and physicochemical methods such as chromatography, were frequently used for the separation and further analysis of proteins of interest. Nephelometry is another, still widely-used technique in clinical immunological analysis for the detection of antigen–antibody reactions.

The development of binder-ligand assays (also established in the 1960 s) provided a new and highly sensitive tool for the detection of proteins of interest in the pg–ng range [19]. For nearly 20 years the radioimmunoassay (RIA) has been regarded as the most sensitive assay in immunological research. But in the late 1970 s this test started to be gradually replaced by the enzyme-linked immunosorbent assay (ELISA), which provided almost the same sensitivity without the need for special radioactive chemicals. Currently ELISA, used with various modifications, is probably the most widely-used system that can be applied to phenotyping at the level of antigen–antibody interactions. It can, for example, be used for the direct measurement of immunoglobulin isotype levels, cytokine quantitation, and antigen-specific antibody responses.

In parallel to the above-mentioned techniques which detect immunological phenomena in a liquid phase, microscopic techniques have evolved that allow the detection of specific antigen/receptor molecules on the cell surface. Immunofluorescence microscopy was one of the first techniques used for detection and localization of antigens in cells and tissues [20]. This method, which enables the simultaneous detection of several antigens using antibodies with different specificities and different (color) labels, has progressively evolved over the years, for example, with the development of confocal microscopy [21] or fluorescence *in-situ* hybridization (FISH) [22]. Enzyme-linked antibodies have also become widely used in immunohistochemical [23] and immunocytochemical analyses [24]. In this context agglutination assays should not be overlooked, as they are still the gold standard for transfusion hematology, and diagnosis of hemolytic diseases of the newborn and autoimmune hemolytic anemia [25].

The complement system is a major effector mechanism in immune-complex-induced tissue damage. This aspect of the complement system has been used in the so-called hemolytic assays, where erythrocytes are lysed with different intensity depending on the amount of antibody present in the serum [26]. Since changes in serum complement activity accompany a number of diseases (including autoimmune and immunodeficiency conditions), its measurement has become a valuable diagnostic tool. Antigen–antibody reactions can also be measured indirectly by changes in complement activity in the so-called complement fixation test [26].

10.2.2 Cellular Immunity Techniques

Delayed hypersenstitivity skin testing detects cutaneous reactivity to certain antigens [27]. Although widely used in epidemiological studies, this relatively simple and crude test currently has limited potential in specialized research. The inability to mount an immune response to a defined antigen (termed as anergy) is usually associated with a number of immune deficiencies or infections. Nowadays, skin tests are mostly applied in the diagnosis of allergies [28].

A widely used test for T-cell immunocompetence, as well as in histocompatibility investigations is the mixed lymphocyte culture (MLC) [29]. In this assay changes in DNA synthesis (quantitated by the incorporation of ^{3}H-thymidine) reflect the ability of effector lymphocytes to proliferate after co-incubation with irradiated (or myomicin-treated) allogeneic stimulator lymphocytes. The MLC can be extended to evaluate the ability of effector cells to lyse ^{51}Cr-labeled target cells, providing further information regarding the function of cytotoxic T cells [30].

Assays aiming to identify phagocytic cells such as monocytes, macrophages or dendritic cells employ microscopic methods or specific staining (e. g. esterase staining for monocytes [31]); fluorescence-conjugated monoclonal antibodies against specific surface antigens). Analysis of these cell subtypes can be further extended by functional assays, such as phagocytosis of particles or antibody-coated heat-killed bacteria; *ex vivo* stimulation of thioglycolate-elicited peritoneal macrophages with LPS, lipid A, peptidoglycan, $PAM_3CysSer(Lys)_4$, resiquimod, dsRNA, CpG oligonucleotides [32, 33]. A number of specific assays have been developed to assess neutrophil function including tests to determine motility, recognition and adhesion, ingestion, degranulation, microbicidal activity [34].

Lymphocyte subpopulations such as B cells, T cells or NK cells, can be effectively detected and analyzed, with flow cytometry in particular (most commonly called "FACS", flow cytometry activated cell sorting). This is one of the most powerful tools for studying the cellular components of the immune system. This method is based on the specific binding of monoclonal antibodies labeled with fluorescence dyes to distinct cell surface antigens. It allows the simultaneous identification and quantification of various cell populations in samples of peripheral blood or from organs of the immune system, based on the differential expression of cell surface molecules. The robustness of FACS analysis is constantly increasing as a result of advances in molecular immunology, providing new sensitive detection reagents, and progress in technology that led to the invention of new machines allowing the simultaneous detection of up to 17 different fluorescence colors [35]. These new advances have provided the basis for comprehensive high-throughput immunological phenotyping, enabling an extensive analysis of the cellular populations and subpopulations of an individual in a single sample to be carried out. The more colors the FACS machine can display, the more antibodies can be included in one staining, which gives more information per staining as well as a reduction in the total number of samples, which also means that a minimum of material can be used without any loss of information. Therefore the intelligent use of distinct antibody combinations, depending on the organ of choice and cell populations that can be potentially detected within this organ, can differentiate between main cell lineages characterized by the unique expression of surface antigens such as CD3 for T cells or CD19 for B cells, in addition to providing a more detailed insight into the expression pattern of defined activation or differentiation markers within cellular subsets such as CD25 on activated/regulatory T helper cells or IgD on mature/immature B cells [36].

10.2.3 Molecular Genetic Techniques

DNA hybridization assays have been extensively used for in-depth genetic analysis of the immune system. These tests basically employ labeled nucleic acid probes complementary to certain DNA fragment of interest. The Southern blot analysis is a frequently used technique, which combines agarose electrophoresis of restrictase-fragmented DNA, blotting, and consecutive incubation with a specific probe [37]. *In-situ* hybridization is a technique where labeled probes can bind target DNA in tissue section or cell smears [22]. A major drawback of traditional hybridization techniques has been the relatively large amount of DNA required. During the last decade this has been overcome by the application of the polymerase chain reaction (PCR) technology, allowing the use of small quantities of the DNA of interest, which are amplified *in vitro* to provide quantities that can be further used in traditional hybridization analyses [38]. In addition, there are techniques for the analysis of specific RNA fragments. They have the basic advantage of being more sensitive than the above-mentioned DNA analysis methods. They provide information about the level of expression of certain genes of interest. A good example of such a method is the Northern blot, which is a modification of the Southern blot technique, but adapted for the specific characteristics of RNA [39]. Another technique known as reverse-transcriptase (RT) PCR is used for the amplification of certain DNA fragments from its corresponding RNA sequence [40].

During the last decade, a new tool for large-scale high-throughput DNA or RNA analysis has evolved – the microarray technology, which facilitates the simultaneous analysis of thousands of nucleotide sequences. Both expression monitoring and mutation detection can be accomplished using this hybridization-based technique [41]. Microarrays are being increasingly applied in immunological research and diagnostics, providing opportunities to monitor genes which are differentially expressed between samples, and to characterize new genes, or to elucidate new functions of already known genes [42].

10.3
Immunological Phenotyping at the German Mouse Clinic

At the GMC, we have initiated the development of very comprehensive protocols for the immunological analysis of the mouse. Depending on the question that is posed, several "levels" of immunological phenotyping can be outlined (Fig. 10.1).

Screening individual mouse variants within a large cohort of mice for abnormalities of the immune system requires the application of high-throughput non-mouse-consuming assays. It usually results in the subsequent establishment of a mouse mutant line. If investigations are combined with other screens (e. g. behavior), the application of challenge tests is often limited. A screen should be sensitive enough to detect minor alterations under steady-state (non-challenge) conditions, often using a limited number of samples. These assays are usually limited to a single compartment of the immune system, but nevertheless can give a quantitative overview of major cellular (lineages) and soluble components of the immune system (immunoglobulin concentrations). This screening "level" can be further extended by searching for certain other indications of immune dysfunction such as higher or lower expression of certain activation markers on particular cell subsets, or the presence of soluble factors indicative of certain diseases, such as autoantibodies for example, which should be barely detectable in healthy individuals.

A different and maybe even more "fruitful" method of screening mutants or genetically-manipulated mice for immunological phenotypes as compared to measuring only steady-state parameters is the application of challenge conditions. Although animals are usually maintained under resting conditions in a SPF facility, and some significant genetic changes will be detectable in these mice, many more genes will be activated or differentially expressed under challenge conditions. Depending on the field of interest, activation of the immune system can be accomplished in a variety of different ways to specifically examine defined immune functions or pathways, which would remain undetected under resting conditions: for example, *in vitro* stimulation of cells with different ligands and measurement of their resulting expansion or cytokine secretion. These types of assay have the

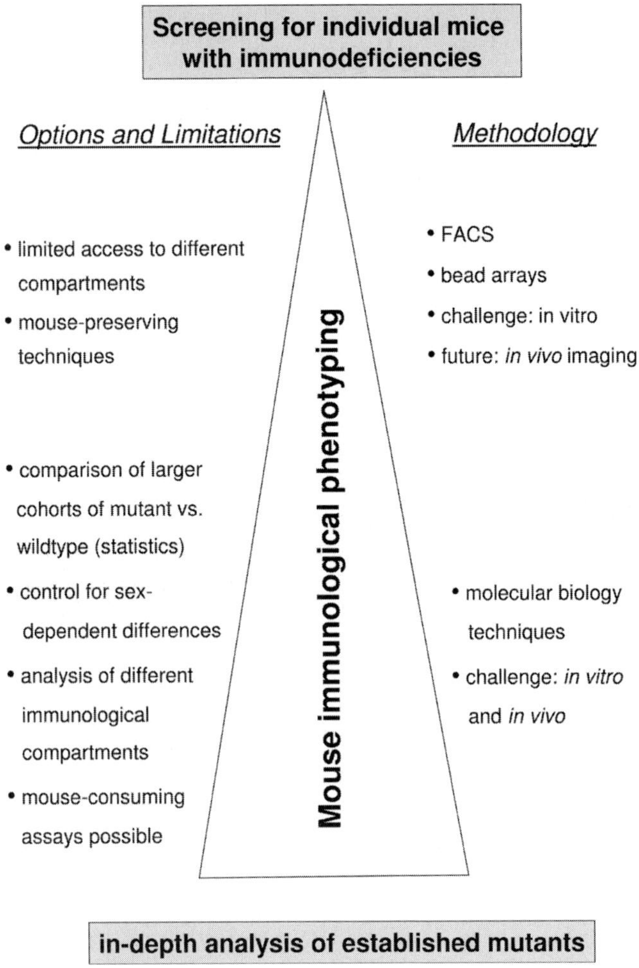

Screening for individual mice with immunodeficiencies

Options and Limitations

- limited access to different compartments
- mouse-preserving techniques

- comparison of larger cohorts of mutant vs. wildtype (statistics)
- control for sex-dependent differences
- analysis of different immunological compartments
- mouse-consuming assays possible

Mouse immunological phenotyping

Methodology

- FACS
- bead arrays
- challenge: in vitro
- future: *in vivo* imaging

- molecular biology techniques
- challenge: *in vitro* and *in vivo*

in-depth analysis of established mutants

Fig. 10.1 Screening approaches to mouse immunological phenotyping.

advantage of being repeatable in individual mice, and have been successfully applied, for example by the group of Bruce Beutler on ENU mutagenized mice at the Scripps Research Institute [32, 43].

Since the main function of the immune system is to effectively protect the host organism against pathogens, immunological phenotyping can also focus on *in vivo* challenges such as infection or immunization (see also Chapter 8 in this volume). This approach allows intensive investigation of innate and/or adaptive immune mechanisms and their capability to deal with a defined dose of antigen or infection. Immunization of ENU mice was successfully used by Chris Goodnow's group in Australia for identification of immunological variants [44, 45]. Thus, antigen-specific humoral immune responses can be monitored by means of various immunization strategies [45]. The major drawbacks of *in vivo* challenge experiments are the possible interference with other screens, and the limited possibility of repeating the challenge in the same individual mouse.

In our experience, *Listeria monocytogenes* (*L.m.*) infection fulfils most of the specific requirements needed to be successfully used in an infection screen, even in a non-mouse-consuming screen like ENU, because *L.m.* can easily be treated with antibiotics to rescue mice with an increased susceptibility to this facultative intracellular bacterium [46]. As the severity of disease and the strength of the immune response are influenced by the original infection dose and the route of infection, determination of the concentration of *L.m.* required for successful infection is crucial for the successful standardization of such an infection screen (see also Chapter 8 in this volume). An additional advantage of the *Listeria* model system is that the immune response of mice against *L.m.* is very well described [47, 48] and many parameters are suitable for monitoring the strength of disease and the status of the immune response. In addition, after primary infection mice usually develop a highly effective protective immunity against even the highest doses of *Listeria*, mainly mediated by $CD8^+$ memory T cells [49, 50]. This fact actually makes it possible to include in an immunological phenotyping screen parameters reflecting the ability to generate and maintain a functional active memory T cell population that protects the host against re-infection. Monitoring the health status and immune responses of animals within the first few days after infection can be achieved by measuring cytokine levels, mainly of IFNγ, which directly correlates with the progress and severity of disease, and can easily be determined by conventional serology techniques. A big advantage of this kind of immunological phenotyping using *L.m.* is the invention of MHC-I tetramers, which allow the detection of antigen-specific $CD8^+$ T cells [50, 51]. This technique enables direct monitoring of pathogen-specific T cells in the host organism, and their additional characterization for other differentiation and memory markers such as CD62L or CD127 [52]. For *Listeria*, many different epitopes are known in a variety of mouse inbred strains [51, 53] against which a robust $CD8^+$ T cell response is induced and detectable in different organs. As the $CD8^+$ T cell response follows distinct kinetics after primary or recall infection, the best time-points at which to study the peak burst size of expanded T cells during the effector phase is at day 7 (primary) or day 5 (recall) after infection, which also facilitates a standardized operation procedure [51].

In all instances of individual mouse phenotyping that have been described, identification of mutant animals, which differ dramatically from the rest of the test group is accomplished by calculating cut-off values for the parameter being measured, thus greatly facilitating the detection of outliers. These calculations very much depend on the variability of the measurements, and in our experience the use of mean values plus three standard deviations was satisfactory for the identification of variants which gave rise to new mutant lines.

The next "level" in mouse phenotyping is the *extensive characterization of already established mutant lines.* This approach is applied to already existing mutant lines and requires the comparison of age- and sex-matched groups of mutant and (littermate) control animals; it allows the in-depth characterization of the mutant line and identification of both profound and subtle differences between the groups studied. It involves techniques usually used for the identification of individual variants, but requires a different type of data analysis and handling, basically to search for the genotype effect. Different screens involving various non-mouse-consuming assays are applied in this approach. These include both tests performed under steady-state, baseline conditions, analyzing major leukocyte lineages and immunoglobulin isotypes in body fluids, and more sophisticated methods after *in vitro* or *in vivo* challenge, aimed at discovering the genes most likely to be involved in a certain immunological pathway or in the host immune response against pathogen challenge or genetically-altered host cells. Besides these screening techniques, further in-depth flow cytometry analysis can be carried out using cells from the different organs of the immune system, combined with "traditional" histology (HE staining), immunohistochemistry, and especially analysis of antigen-specific immune responses.

10.4 Screening Protocols

To ensure both extensive and in-depth immunological phenotyping, a combination of high-throughput screen management and appropriate sensitive techniques is required. A basic prerequisite for the successful identification of immunological mutants is the robustness of the screening assays as defined by well-established SOPs for high-throughput analysis and manageable natural variability of the measured parameters. Multi-color flow cytometry has become the standard in such immunological investigations, and has proven successful for the discovery of mutants based on the simultaneous analysis of a number of cell surface marker molecules in blood. As for the analysis of liquid components such as immunoglobulin isotypes or cytokines, bead array-based systems, such as the Lunimex X-map technology [54], have taken over from traditional methods like ELISA, mostly because of their power, as they are capable of combining the simultaneous analysis of up to 100 different analytes in a single and relatively small amount of body fluid/plasma/serum, with high sensitivity.

To improve the handling of large sample numbers and minimize the impact of human error, sample handling should be highly automated. In the case of FACS machines, computer-based compensation software is nowadays used to guarantee constant and reproducible data analysis. Sample acquisition for high-throughput FACS should be fast; this is best accomplished by combination with automated sample acquisition devices (for example, at the GMC we have implemented a Quadra 3 pipetting robot from Tomtec for sample preparation, and a high-throughput sampler (HTS) from Becton Dickinson, which automatically samples the cells from 96-well plates). To ensure careful and appropriate data analysis, automated gating and data handling are necessary (Fig. 10.2).

In the following sections we give some examples of protocols established at the German Mouse Clinic for immunological phenotyping using peripheral blood samples.

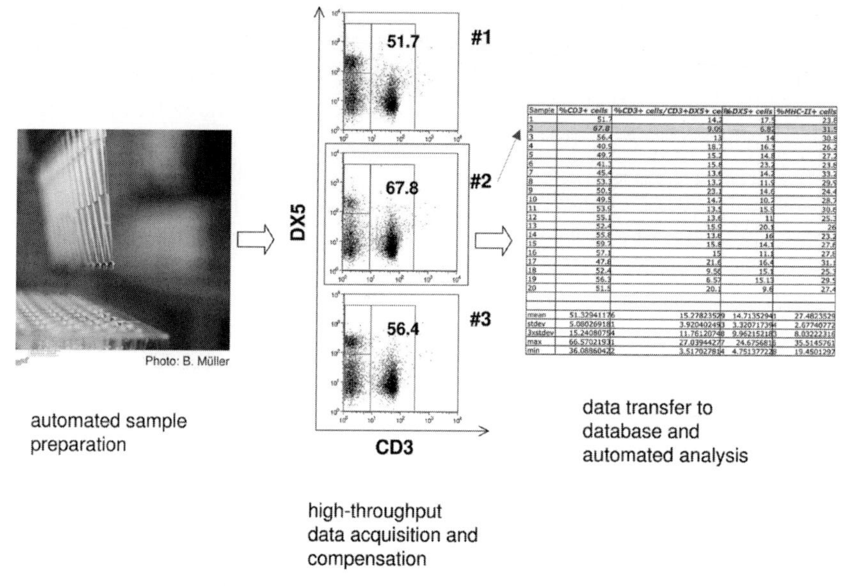

Fig. 10.2 Example sample handling for flow cytometry.

10.4.1
FACS for Leukocyte Subpopulations

10.4.1.1 **Reagents and Equipment**

Lysis buffer, NH_4Cl (0.17 M), Tris buffer (pH 7.45), PBS, FACS staining buffer (PBS, 0.5% BSA, 0.02% sodium azide, pH 7.45), 1 mM ethidium monazide bromide (EMA), Fc block, fluorescent-conjugated antibodies: mixtures 1–6 according to Tab. 10.2, 1% paraformaldehyde (PFA), 96-well U-bottomed plates, 1.5-ml tubes, nylon filter nets, pipettes/pipetting robot, centrifuge.

10.4.1.2 **Procedure**

1. Blood samples are centrifuged and the plasma is separated from the cell pellet. Resuspend the cells of one sample in 900 µl PBS (at room temperature) and filter through nylon nets over a 1.5-ml tube.
2. Distribute the cell suspension of one sample between six wells (each 150 µl) of a 96-well U-bottomed plate; the remaining cells should be pooled with other samples and used as the control for staining.
3. Centrifuge the plate (530RCF, 3 min, 4 °C).
4. Discard the supernatant and add 200 µl/well of lysis buffer. Resuspend and incubate at room temperature for 10 min.
5. Repeat step 3.
6. Resuspend cells in 200 µl/well FACS buffer and repeat step 3.
7. Add 50 µl/well EMA/Fc block solution and resuspend.
8. Incubate for 20 min in the dark at 4 °C.

Table 10.2 Example of antibody combinations used at the GMC with five-color FACS analysis of peripheral blood samples (including live/death discrimination).

	FITC	PE	PerCP	APC/Cy-5	Populations identified
1	CD5	γδ TCR	CD19	CD3	B1/B2 B cell subpopulations; γδ T cells
2	IgD	Gr-1	B220	CD11 b	B cells, granulocytes
3	DX5	MHC-II	CD3		Antigen-presenting cells, NK cells
4	CD103	CD25	CD8	CD3	Regulatory T-cell subpopulations
5	CD62L	CD45RA	CD4	CD3	Activated and memory T-cell subpopulations
6	Ly6C	CD8	CD4	CD44	Memory T-cell subpopulations

9. Add 150 µl/well FACS buffer and keep the plate exposed to light for 10 min.
10. Repeat step 3.
11. Discard the supernatant and add 50 µl of antibody mixtures 1–6 to the six wells of each sample and resuspend. Incubate for 20 min in the dark at 4 °C.
12. Add 150 µl/well FACS buffer and repeat step 3.
13. Discard the supernatant, add 200 µl/well FACS buffer and repeat step 3.
14. Fix the cells with 100 µl/well 1% PFA for 30 min in the dark at 4 °C.
15. Repeat step 3, resuspend the cells in 100 µl/well PBS and acquire the samples until a total of 25,000 cells/well is attained.

10.4.2 Bead Array for Immunoglobulin Concentrations

10.4.2.1 Reagents and Equipment

FACS buffer, Luminex beads (of five different regions, coupled with antibodies specific for mouse IgM, IgG1, IgG2 a, IgG3, and IgA, pre-mixed before each experiment in equal proportions), mouse IgM, IgG1, IgG2 a, IgG3, and IgA standards (both unlabeled and biotinylated), streptavidin-PE, Tween 20, pipettes/pipetting robot, 96-well filter plates, plate shaker, Bioplex reader.

10.4.2.2 Procedure

1. Dilute plasma samples 1:1000 in FACS buffer and add 25 µl/well. Unlabeled standards are mixed, serially diluted in FACS buffer in seven steps starting from 2.5 µg/ml, and then added to the plate in duplicate (25 µl/well).
2. Add 5 µl/well of the bead mixture.
3. Cover the plate with aluminium foil and incubate for 10 min at room temperature on a shaker (300 r.p.m.).
4. Prepare a mixture of biotinylated IgM, IgG1, IgG2 a, IgG3, and IgA standards (200 ng/ml in FACS buffer) and add 25 µl/well.

5. Cover the plate with aluminium foil and incubate for 20 min at room temperature on a shaker (300 r.p.m.).
6. Wash the samples three times with FACS buffer (100 µl/well) using a vacuum manifold (Biorad).
7. Add 1 µg/ml streptavidin-PE (50 µl/well) and incubate for 10 min on a shaker in the dark (300 r.p.m.).
8. Repeat step 6.
9. Add 100 µl/well FACS buffer/0.05 % Tween 20, shake for 30 s at 1000 r.p.m., and acquire samples into a Bioplex reader (100 beads/region/analyte).

10.4.3 ELISA for Autoantibodies

10.4.3.1 Reagents and Equipment

Coating buffer, blocking buffer, sample buffer, substrate buffer, calf thymus DNA (mixture of single-(ss) and double-stranded (ds) in equal proportions), rabbit IgG, poly-L-lysine, *p*-nitrophenyl phosphate, 96-well flat-bottomed plates, stirrer, plate washer, ELISA reader.

10.4.3.2 Procedure

1. To coat the plates, incubate plates with 50 µg/ml aqueous solution of poly-L-lysine hydrobromide (50 µl/well, 1 h, room temperature). Wash once with washing buffer (using plate washer) and coat with 50 µl/well ssDNA/dsDNA (1:1) mixture (for anti-DNA antibodies) or rabbit IgG (for rheumatoid factor). Incubate overnight at 4°C. Without removing the solution from the wells, add 100 µl/well blocking buffer and incubate for 30 min at room temperature.
2. Plasma samples: before adding the samples, wash the plate three times with 200 µl/well washing buffer. Dilute samples 1:100 (for anti-DNA antibodies) or 1 : 200 (for rheumatoid factor) in sample buffer and add 200 µl/well, two wells/sample and incubate for 90 min at room temperature. Titrate the positive control (samples from MRL/MpJ-Fas^{lpr}/J mice). The 1:10000 (anti-DNA antibodies) or 1 : 2000 (rheumatoid factor) dilutions are acceptable as the cut-off above which samples are considered to be positive.
3. Detection antibody: wash the plate three times with 200 µl/well washing buffer. Dilute anti-mouse polyvalent immunoglobulins 1 : 3000 and add 100 µl/well, incubate for 90 min at room temperature.
4. To develop the plate, wash three times with 200 µl/well washing buffer. Add 100 µl/well *p*-nitrophenyl phosphate in substrate buffer and incubate for 17 min. Read the plates on a reader.
5. All incubations should be carried out in a moist chamber.

10.5
Troubleshooting

The success of mouse immunological phenotyping is very much dependent on the following factors:

- *Appropriate housing conditions for the animals.* In this respect an SPF environment is essential. Monitoring the sanitary status of mice is really crucial for all immunological assays and should be carried out on a regular basis as infections can severely alter many immunological parameters.
- *Group size.* The cohort(s) of mice analyzed should be large enough to allow reliable identification of outliers and statistical comparisons.
- *Standardization of the assays used.* The following steps are suggested: different age- and sex-matched cohorts of mice from the same animal facility tested on different days should yield similar results; the same mouse sampled a second time (note that some changes can be expected due to a boost in hematopoesis) should also yield similar results; one sample tested twice within the same experimental setting should result in an almost identical outcome in measurements.
- Use of *internal controls* within an assay (for example, duplicate measurements) allows identification of potential technical problems.
- *Monitoring defined baseline parameters* for certain mouse lines over periods of time will reveal indications of potentially greater changes that may occur in future cohorts of mice (e. g. infections or technical changes of assays).

10.6
Outlook

Recent advances in photonics, optics, and molecular genetics have paved the way for the development of live intravital microscopy (IVM), enabling immunologists to view immune cell populations in lymphoid organs and peripheral tissues. As an adjunct to the simple technique of observation, IVM may make possible the quantification and detailed investigation of the complex behavior of immune cells in their native anatomical context [53]. Traditional technologies applied in IVM include two-dimensional imaging methods such as brightfield transillumination or epifluorescence videomicroscopy [56]. The recently developed multiphoton microscopy (MPM) features infrared pulsed laser excitation for the generation of optical sections of fluorescent signals from areas that are hundreds of micrometers below the surface of solid tissues [57]. It is hoped that the combination of MPM with IVM will enable the analysis of cell migration by time-lapse recordings of 3D tissue reconstructions [58], providing a powerful new tool with which to observe cell migration and interaction.

Over the years, various mouse imaging models have been developed to track lymphocyte activities in tissues; these models have focused on hematopoiesis, lymphopoiesis, homing to and migration within secondary lymphoid organs, inflammation, immunosurveillance by memory cells, tumors, and transplanted organs. One interesting prospect for the future application of MPM-IVM might be to focus on events leading to T-cell activation, especially the T cell–dendritic cell dialog. New models specifically developed for MPM-IVM for analysis are now emerging [55].

References

1 Fischer, A. **2001**, Primary immunodeficiency diseases: an experimental model for molecular medicine, *Lancet* 357, 1863–1869.
2 Fischer, A. **2002**, Primary immunodeficiency diseases: natural mutant models for the study of the immune system, *Scand. J. Immunol.* 55, 238–241.
3 Elder, M. E. **2000**, T-cell immunodeficiencies, *Pediatr. Clin. North Am.* 47, 1253–1274.
4 Fischer, A. **2004**, Human primary immunodeficiency diseases: a perspective, *Nat. Immunol.* 5, 23–30.
5 Kelly, J., W. J. Leonard **2003**, Immune deficiencies due to defects in cytokine signaling, *Curr. Allergy Asthma Rep.* 3, 396–401.
6 Leonard, W. J. **2001**, Cytokines and immunodeficiency diseases, *Nat. Rev. Immunol.* 1, 200–208.
7 Nichols, K. E. **2000**, X-linked lymphoproliferative disease: genetics and biochemistry, *Rev. Immunogenet.* 2, 256–266.
8 Mak, T. W., J. M. Penninger, P. S. Ohashi **2001**, Knockout mice: a paradigm shift in modern immunology, *Nat. Rev. Immunol.* 1, 11–9.
9 Bluethmann, H., P. S. Ohashi, (eds.) **1994**, *Transgenesis and Targeted Mutagenesis in Immunology*, Academic Press, San Diego.
10 Fischer, A. **2002**, Natural mutants of the immune system: a lot to learn! *Eur. J. Immunol.* 32, 1519–1523.
11 Boyse, E. A. **1977**, The increasing value of congenic mice in biomedical research, *Lab. Anim. Sci.* 27, 771–781.
12 Rogner, U. C., P. Avner **2003**, Congenic mice: cutting tools for complex immune disorders, *Nat. Rev. Immunol.* 3, 243–252.
13 Shultz, L. D. **1991**, Immunological mutants of the mouse, *Am. J. Anat.* 191, 303–311.
14 Buer, J., R. Balling **2003**, Mice, microbes and models of infection, *Nat. Rev. Genet.* 4, 195–205.
15 Brown, S. D., R. Balling 2001, Systematic approaches to mouse mutagenesis, *Curr. Opin. Genet. Dev.* 11, 268–273.
16 Ouchterlony, O. **1958**, Diffusion-in-gel methods for immunological analysis, *Prog. Allergy* 5, 1–78.
17 Ouchterlony, O. **1962**, Diffusion-in-gel methods for immunological analysis, II. *Prog. Allergy* 6, 30–154.
18 Ouchterlony, O. **1962**, Quantitative immunoelectrophoresis, *Acta Pathol. Microbiol. Scand.* 154(Suppl.), 252–254.
19 Hage, D. S. **1999**, Immunoassays, *Anal. Chem.* 71, 294R–304R.
20 Blundell, G. P. **1970**, Fluorescent antibody techniques, *Prog. Clin. Pathol.* 3, 211–225.
21 Harvath, L. **1999**, Overview of fluorescence analysis with the confocal microscope, *Methods Mol. Biol.* 115, 149–158.
22 Bartlett, J. M. **2004**, Fluorescence *in situ* hybridization: technical overview, *Methods Mol. Med.* 97, 77–87.
23 Falini, B., C. R. Taylor **1983**, New developments in immunoperoxidase techniques and their application, *Arch. Pathol. Lab. Med.* 107, 105–117.
24 Matthews, J. B. **1987**, Immunocytochemical methods: a technical overview, *J. Oral Pathol.* 16, 189–195.
25 Voak, D. **1999**, The status of new methods for the detection of red cell agglutination, *Transfusion* 39, 1037–1040.
26 Alper, C. A., F. S. Rosen **1975**, Clinical applications of complement assays, *Adv. Intern. Med.* 20, 61–88.
27 Ahmed, A. R., D. A. Blose **1983**, Delayed-type hypersensitivity skin testing: A review, *Arch. Dermatol.* 119, 934–945.
28 Devos, S. A., P. G. Van Der Valk **2002**, Epicutaneous patch testing, *Eur. J. Dermatol.* 12, 506–513.

29 Cerottini, J. C., H. R. MacDonald, K. T. Brunner **1973**, Formation of cytotoxic lymphocytes in mixed leukocyte culture systems: introduction, *Transplant. Proc.* 5, 1621–1624.

30 Jerome, K. R., D. D. Sloan, M. Aubert **2003**, Measurement of CTL-induced cytotoxicity: the caspase 3 assay, *Apoptosis* 8, 563–571.

31 Schmalzl, F., H. Braunsteiner **1970**, The cytochemistry of monocytes and macrophages, *Ser. Haematol.* 3, 93–131.

32 Hoebe, K., X. Du, P. Georgel, E. Janssen, K. Tabeta, S. O. Kim, J. Goode, P. Lin, N. Mann, S. Mudd, K. Crozat, S. Sovath, J. Han, B. Beutler **2003**, Identification of Lps2 as a key transducer of MyD88-independent TIR signalling. *Nature* 424, 743–748.

33 Lehmann, A. K., S. Sornes, A. Halstensen **2000**, Phagocytosis: measurement by flow cytometry, *J. Immunol. Methods* 243, 229–242.

34 van Eeden, S. F., M. E. Klut, B. A. Walker, J. C. Hogg **1999**, The use of flow cytometry to measure neutrophil function, *J. Immunol. Methods* 232, 23–43.

35 Perfetto, S. P., P. K. Chattopadhyay, M. Roederer **2004**, Seventeen-colour flow cytometry: unravelling the immune system, *Nat. Rev. Immunol.* 4, 648–655.

36 Baumgarth, N., M. Roederer **2000**, A practical approach to multicolor flow cytometry for immunophenotyping, *J. Immunol. Methods* 243, 77–97.

37 Southern, E. M. **1975**, Detection of specific sequences among DNA fragments separated by gel electrophoresis, *J. Mol. Biol.* 98, 503–517.

38 Eisenstein, B. I. **1990**, The polymerase chain reaction. A new method of using molecular genetics for medical diagnosis, *N. Engl. J. Med.* 322, 178–183.

39 Bartlett, J. M. **2002**, Approaches to the analysis of gene expression using mRNA: a technical overview, *Mol. Biotechnol.* 21, 149–160.

40 Ohan, N. W., J. J. Heikkila **1993**, Reverse transcription-polymerase chain reaction: an overview of the technique and its applications, *Biotechnol. Adv.* 11, 13–29.

41 Southern, E. M. **2001**, DNA microarrays. History and overview, *Methods Mol. Biol.* 170, 1–15.

42 van der Pouw Kraan, T. C., P. V. Kasperkovitz, N. Verbeet, C. L. Verweij **2004**, Genomics in the immune system, *Clin. Immunol.* 111, 175–185.

43 Hoebe, K., X. Du, J. Goode, N. Mann, B. Beutler **2003**, Lps2: a new locus required for responses to lipopolysaccharide, revealed by germline mutagenesis and phenotypic screening, *J. Endotoxin. Res.* 9, 250–255.

44 Jun, J. E., L. E. Wilson, C. G. Vinuesa, S. Lesage, M. Blery, L. A. Miosge, M. C. Cook, E. M. Kucharska, H. Hara, J. M. Penninger, H. Domashenz, N. A. Hong, R. J. Glynne, K. A. Nelms, C. C. Goodnow **2003**, Identifying the MAGUK protein Carma-1 as a central regulator of humoral immune responses and atopy by genome-wide mouse mutagenesis, *Immunity* 18, 751–762.

45 Vinuesa, C. G., C. C. Goodnow **2004**, Illuminating autoimmune regulators through controlled variation of the mouse genome sequence, *Immunity* 20, 669–679.

46 Wong, P., E. G. Pamer **2003**, Feedback regulation of pathogen-specific T cell priming, *Immunity* 18, 499–511.

47 Busch, D. H., E. G. Pamer **1999**, T lymphocyte dynamics during *Listeria monocytogenes* infection, *Immunol. Lett.* 65, 93–98.

48 Kaufmann, S. H. **1995**, Immunity to intracellular microbial pathogens, *Immunol. Today* 16, 338–342.

49 Busch, D. H., I. Pilip, E. G. Pamer **1998**, Evolution of a complex T cell receptor repertoire during primary and recall bacterial infection, *J. Exp. Med.* 188, 61–70.

50 Harty, J. T., M. J. Bevan. **1992**, CD8+ T cells specific for a single nonamer epitope of *Listeria monocytogenes* are protective *in vivo*, *J. Exp. Med.* 175, 1531–1538.

51 Busch, D. H., I. M. Pilip, S. Vijh, and E. G. Pamer, **1998**, Coordinate regulation of complex T cell populations responding to bacterial infection, *Immunity* 8, 353–362.

52 Huster, K. M., V. Busch, M. Schiemann, K. Linkemann, K. M. Kerksiek, H. Wagner, D. H. Busch **2004**, Selective expression of IL-7 receptor on memory T cells identifies early CD40L-dependent generation of distinct CD8+ memory T cell subsets, *Proc. Natl Acad. Sci. USA* 101, 5610–5615.

53 Schiemann, M., V. Busch, K. Linkemann, K. M. Huster, D. H. Busch **2003**, Differences in maintenance of CD8+ and CD4+ bacteria-specific effector-memory T cell populations, *Eur. J. Immunol.* 33, 2875–2885.

54 Fulton, R. J., R. L. McDade, P. L. Smith, L. J. Kienker, J. R. Kettman, Jr **1997**, Advanced multiplexed analysis with the FlowMetrix system, *Clin. Chem.* 43, 1749–1756.

55 Sumen, C., T. R. Mempel, I. B. Mazo, U. H. von Andrian **2004**, Intravital microscopy; visualizing immunity in context, *Immunity* 21, 315–329.

56 Mempel, T. R., M. L. Scimone, J. R. Mora, U. H. von Andrian **2004**, *In vivo* imaging of leukocyte trafficking in blood vessels and tissues, *Curr. Opin. Immunol.* 16, 406–417.

57 Denk, W., J. H. Strickler, W. W. Webb **1990**, Two-photon laser scanning fluorescence microscopy, *Science* 248, 73–76.

58 Cahalan, M. D., I. Parker, S. H. Wei, M. J. Miller **2002**, Two-photon tissue imaging: seeing the immune system in a fresh light, *Nat. Rev. Immunol.* 2, 872–880.

11
Phenotyping Allergy in the Laboratory Mouse

Thilo Jakob, Francesca Alessandrini, Jan Gutermuth, Gabriele Köllisch, Anahita Javaheri, Antonio Aguilar, Martin Mempel, Johannes Ring, Markus Ollert, Heidrun Behrendt*

* Corresponding author

11.1
Introduction

Allergy and allergic diseases represent one of the major health problems in most modern societies. They have increased worldwide in prevalence [1]. Epidemiological studies in various parts of Germany show an estimated prevalence of at least 10–20% in the general population [1, 2, 3].

Allergy can be defined as "immunologically mediated hypersensitivity leading to disease" [4]. Clinical manifestations of allergic diseases can involve various organs, but most commonly affect the skin and the mucous membranes, where the individual organism meets the environment. The most frequent allergic diseases are the IgE-mediated allergic immediate type reactions that we find in clinical conditions such as allergic rhinitis, allergic asthma, insect venom allergy or anaphylaxis, as well as the so-called delayed type of allergic reactions such as allergic contact dermatitis, which represents the most common occupational disease in many countries.

The different clinical manifestations of allergic diseases, such as allergic asthma or eczema/dermatitis have been known for a long time, however, little is known about the mechanisms that lead to these different phenotypes. It is well established that allergies have a multifactorial etiology, where genetic predisposition and environmental influences seem to interact in promoting the manifestations of the disease phenotype. However, the exact degree to which genetic and environmental factors contribute to the development of different disease manifestations is unclear. Both natural and anthropogenic environmental factors have been shown to play a role in the development and aggravation of allergic diseases [5, 6] and similarly it is well established that genetic factors play an important role in susceptibility to the development of allergic diseases. Maybe the first evidence that allergies can run in families can be found in the Julian–Claudian emperors' family with Augustus, Claudius and Britannicus all suffering from an allergy [4]. Later, the observation that asthma and hay fever often occur in families gave rise to the term "atopy" [7].

Standards of Mouse Model Phenotyping. Edited by Martin Hrabě de Angelis, Pierre Chambon, and Steve Brown

ISBN: 3-527-31031-2

With the progress in molecular genetics, it was hoped that "*the* atopy gene" would soon be discovered; however, the situation proved to be far more complex.

Several strategies have been employed to study the genetics of allergic diseases, such as linkage analysis using a candidate gene approach, genome-wide scans for linkages of positional genes and disease manifestation or identification of polymorphisms in candidate genes that lead to a structural or functional alteration associated with a disease [8]. In the meantime, many different loci on many chromosomes have been suggested to be associated with different phenotypes of allergy, e. g. the cytokine cluster on chromosome 5 q or a common locus on chromosomes 3 q and 17 q being associated with both eczema and psoriasis [9]. In addition polymorphisms of single genes such as ADAM33 have been reported to be associated with the asthma phenotype [10] and more recently novel loci have been reported that show striking associations with asthma susceptibility in population-based studies and have led to the identification of novel genes such as the G protein coupled receptor for asthma susceptibility (GPRA) [11]. Often, however, associations between genotype and phenotype cannot be explained at a functional level and do not seem to fit into current concepts of disease pathophysiology. New strategies need to be employed to learn about the function of newly detected genetic polymorphisms that show associations with disease phenotypes. These kinds of studies are commonly undertaken in animals. Model organisms such as drosophila, zebrafish or mice have mainly been used as tools for functional genome analysis. The analogy of the murine and human immune system makes the mouse the most powerful mammalian system for studying the genetics of allergy.

Several mouse mutants with characteristic gene abnormalities have greatly enhanced our understanding regarding the pathophysiology of allergic reactions. In the early 1990s, reports of interleukin (IL)-4 knockout mutants which display strongly reduced IgE serum levels together with a reduced production of Th2 cytokines, had already provided evidence for the central role of IL-4 in regulating IgE synthesis and Th2 cell differentiation [12]. Since then a wide variety of mutants such as transgenic mice, null mutants generated by targeted disruption, conditional knockout mice, knock-in mice, or mutants generated by random mutagenesis have served as valuable tools to dissect the different mechanisms involved in the manifestation of the various allergic phenotypes. Some of these mutants were generated in a hypothesis-driven approach (e. g. IL-4 null mutant, see above) and confirmed what had been expected based on existing concepts. Other mutants yielded unexpected results that opened up novel pathways that had previously not been anticipated to be involved in that particular disease manifestation (e. g. abnormal mast cell function and altered late phase passive cutaneous anaphylaxis in the Bruton tyrosinase kinase null mutant [13]). Finally there is the phenotype-driven approach in which novel phenotype/genotype relationships are identified based on a detailed phenotype analysis of mice that were generated by random mutagenesis [14].

The more that becomes known about the molecular genetics of allergic reactions, the clearer it becomes that a well-characterized phenotype is of great importance. Following the progress in "genomics" and "proteomics" it is now possible to refer to an era of "phenomics". Only when the phenotype is characterized with utmost precision, valid data from molecular genetics can be expected. With this in mind, we have summarized the most useful and reliable protocols that specifically allow the phenotyping of the most common forms of allergic reactions in the laboratory mouse.

11.2
Phenotyping Different Forms of Allergic (Hypersensitivity) Reactions

The variety of different forms of allergic reactions can be grouped according to the pathomechanisms involved. Despite recent debate [15, 18] the classification of Gell and Coombs [19] is still a valuable tool that differentiates distinct types of hypersensitivity reactions.

Type I hypersensitivity comprises the classical allergic immediate type reaction that is found in clinical conditions such as allergic rhinitis, allergic asthma, insect venom allergy or anaphylaxis. The underlying mechanism involves rapid release of vasoactive mediators from mast cells and basophil granulocytes triggered by the cross-linking of surface-bound specific IgE and the allergen.

Type II hypersensitivity describes relatively rare reactions that are generated by cytotoxic antibodies directed against cell surface antigens or haptens which lead to the direct destruction of the cell or macrophage-mediated clearance of the cell. Clinical examples include drug-induced hemolytic anemia, agranulocytosis and thrombocytopenia.

Type III hypersensitivity reactions are caused by formation of aggregates of antigen : antibody complexes. These circulating immune complexes are deposited in the local tissue and activate components of the complement system that in turn trigger release of vasoactive mediators and activate inflammatory cells (Arthus reaction). Clinical examples include conditions such as allergic vasculitis, serum sickness and allergic alveolitis (farmer‘s lung).

Unlike the immediate type hypersensitivity (type I) that is mediated by antibodies, *type IV hypersensitivity* is mediated by allergen-specific effector T lymphocytes and takes some time to develop – hence *delayed type hypersensitivity* (*DTH*). Frequent forms of type IV hypersensitivities include allergic contact dermatitis, allergic drug reactions, and tuberculin reaction.

The classification of Coombs and Gell has been extended by Roitt [16] and Ring [17] to include two additional types of hypersensitivity reaction. *Type V hypersensitivity* describes reactions that result in granuloma formation ("granulomatous hypersensitivity") such as foreign body granuloma or infectious granuloma as seen in sarcoidosis. Finally, *type VI hypersensitivity* describes reactions that are triggered by autoantibodies with affinity for receptors that exert agonistic activity ("stimulating hypersensitivity") such as autoantibodies in myasthenia gravis or autoimmune thyroiditis.

Even though all types of hypersensitivity reactions can be induced in mice, for the purpose of this chapter we will focus on the more frequent forms of allergic reactions that include type I immediate type hypersensitivity (type I), immune complex-mediated Arthus reaction (type III), delayed type hypersensitivity (type IV) and granulomatous hypersensitivity (type V).

11.2.1
Immediate Type Hypersensitivity (Type I)

11.2.1.1 Total IgE Baseline Levels in Laboratory Mice

The underlying mechanism of type I hypersensitivity involves activation of mast cells and basophils via IgE-dependent cross-linking of cell membrane Fcε receptors. A number of allergic diseases, such as allergic rhinoconjunctivitis, allergic asthma or atopic eczema, are associated with increased total and specific IgE levels. The determination of total serum IgE levels plays an essential role in the diagnosis of atopic diseases and is used to identify individuals at risk of developing manifestations of atopic disease later in life. Although there are a large number of mouse

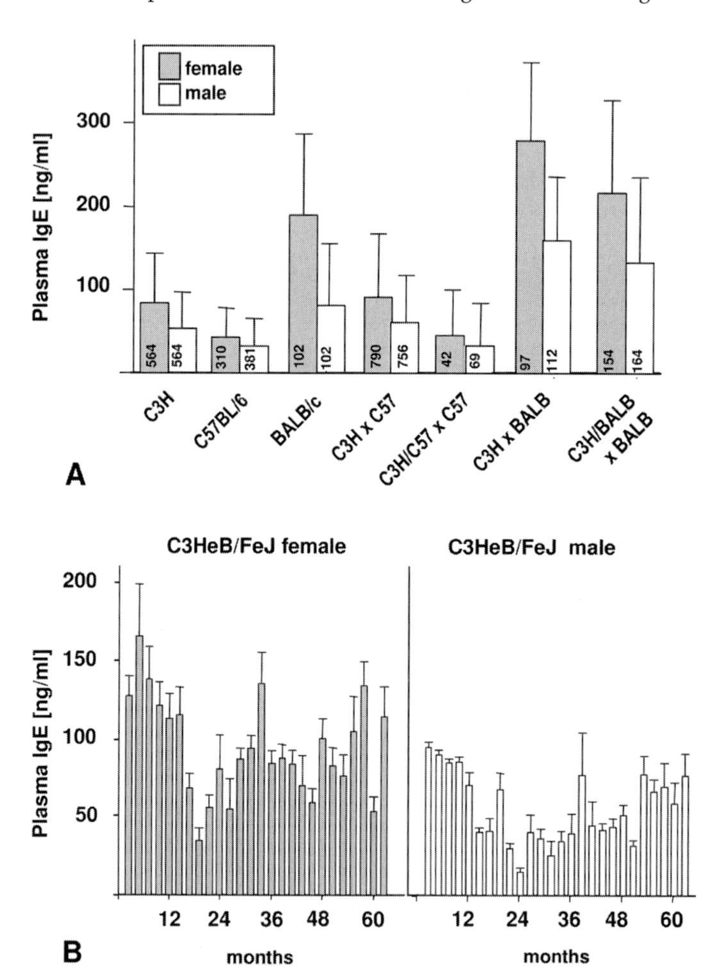

Fig. 11.1 (A) Basal plasma IgE levels in different mouse strains at the age of 12 weeks. Plasma IgE concentrations were significantly lower in male as compared to female mice in all strains tested. Plasma IgE levels were determined by ELISA (n = 42–564, as indicated by the numbers on the bars). (B) Variation of plasma IgE levels in female and male C3HeB/FeJ wild-type population tested at the age of 12 weeks over a 60-month period ($n = 20$/time interval).

models for atopic diseases [20], little data is available on baseline IgE levels in various mouse strains. As part of a large-scale ENU mouse mutagenesis screen we were able to analyze basal plasma IgE levels in large numbers of wild-type animals. Plasma was analyzed using an isotype-specific sandwich ELISA for murine IgE with a lower detection limit of 1 ng/ml as described previously [21]. In brief, 96-well microtiter plates were coated with polyclonal sheep anti-mouse IgE (The Binding Site, Birmingham, UK), loaded with purified mouse IgE standards (Clone-IgE-3, Pharmingen) and mouse plasma samples at dilutions of 1 : 4 to 1 : 40 and developed using biotinylated rat anti-mouse IgE (Clone R35–118, Pharmingen, Heidelberg, Germany), streptavidin peroxidase (Calbiochem, La Jolla, CA, USA), tetramethylbenzidine (Fluka, Deisenhofen, Germany) and analyzed with a standard microwell ELISA reader at 450 nm. As shown in Fig. 11.1 A basal IgE levels differ between male and female mice of various mouse strains. In addition, basal IgE levels very much depend on the type of mouse strain analyzed (e. g. BALB/c > C3H > C57BL/6). The outcross of C3H mice with different strains results in altered basal IgE plasma levels, depending on the strain that is used. While outcross onto a BALB/c background results in elevated basal IgE levels, outcross onto a C57BL/6 background results in lower basal IgE levels (Fig. 11.1 A). In addition it is important to note that basal plasma IgE levels within one mouse population seems to undergo some fluctuation over time as demonstrated in Fig. 11.1 B.

11.2.1.2 **Allergen-specific IgE**

In humans, the determination of total IgE has its limits as a single screening parameter for allergy, since a large number of patients can still develop allergic sensitization together with allergen-specific IgE antibodies in their serum without an increase in total serum IgE levels. This explains why other parameters in addition to total IgE are required to detect patients, and in the case of murine models, mutants with an allergy phenotype. Such additional information can be obtained when animals are sensitized to model allergens such as ovalbumin (OVA) and the allergen-specific humoral immune response (e. g. OVA-specific IgE) is analyzed using established immunoassay techniques adapted to a high-throughput platform. Depending on the kind of question that is being asked different sensitization protocols can be employed (see below). Again the genetic background is important and will influence the outcome of the specific immune response. As shown in Fig. 11.2 the magnitude of the allergen-specific IgE response in BALB/c is much more pronounced than in C3H mice, which in turn are capable of responding more vigorously than C57BL/6 mice. In addition allergen-specific IgG1 and IgG2 a can be determined to evaluate the magnitude and type of response. In this context allergen-specific IgG2 a is indicative of a Th1-dominated response, while allergen-specific IgG1 and IgE suggest a Th2-dominated immune response. The use of different adjuvants during primary sensitization determines the outcome of the type of immune response. While the absorption of allergen to aluminium hydroxide (alum) leads to Th2-dominated antibody production, addition of CpG oligonucleotide induces a Th1-dominated humoral immune response (Fig. 11.2 B).

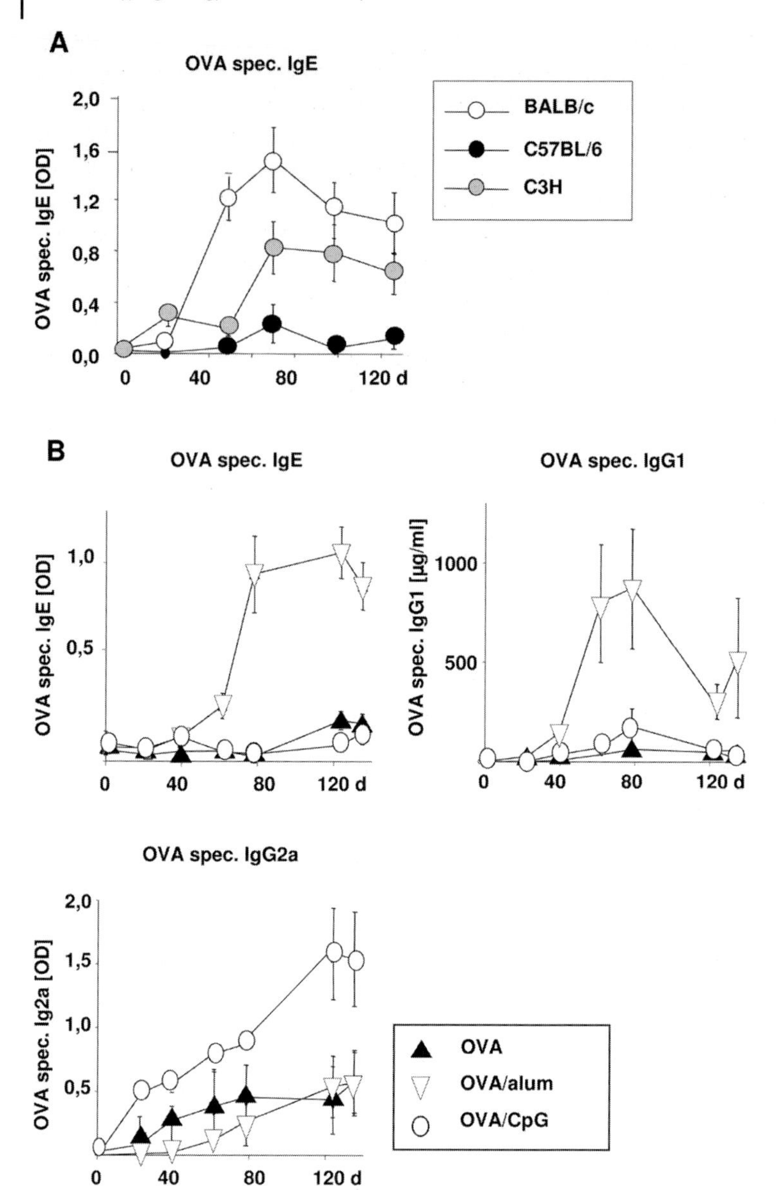

Fig. 11.2 Analysis of allergen-specific humoral immune response after OVA sensitization. (A) Different laboratory mouse strains (BALB/c, C3H and C57BL/6) were sensitized with OVA/alum (1 µg/2 mg/200 µl) on days 0, 7, 28, 56 and 84. Venous blood was obtained at the indicated time points and OVA-specific IgE was determined by ELISA. (B) Use of different adjuvants to determine the outcome of the immune response. BALB/c mice were sensitized with OVA, OVA/alum or OVA/CpG at days 0 and 7 and were subsequently boosted with OVA on days 28, 56 and 84. The outcome of the humoral immune response was determined by ELISA for OVA-specific IgE, IgG1 and IgG2 a.

11.2.1.3 Passive Cutaneous Anaphylaxis (PCA)

Type I sensitization can be passively transferred by local injection of antigen-specific IgE into the murine skin. Subsequently, a biphasic immune reaction, with an (anaphylactic) early and a late phase, is provoked by i. v. administration of the specific antigen. PCA was originally developed in the guinea pig and rat and subsequently adapted to the murine system. In this context it is important to know that mast cell degranulation in rats can only be activated by cross-linking IgE but not IgG1, while in the mouse mast cell activation can be triggered by cross-linking either allergen-specific IgE or IgG1. Since rat mast cells can also be activated by murine IgE (but not murine IgG1) allergen-specific IgE immune responses obtained in mice were frequently quantitated using PCA reaction in rats (since PCA reactions in mice did not allow the differentiation of allergen-specific IgE or IgG1 responses). Despite this limitation, PCA in mice has still proven to be an important immunological tool in the analysis of murine mast cell function [22–26] (see Fig. 11.3).

PCA Early Phase Reaction PCA has mostly been employed as a bioassay to quantitate different degrees of allergic sensitization. With the development of assay systems that allow the direct quantification of allergen-specific IgE, PCA reactions have become less common. However they are still an extremely useful tool for the analysis of the *in vivo* function of mast cells – the key effector cell of immediate type allergy. For phenotyping mast cell function *in vivo* a number of protocols have been described that follow the same basic principle. Antigen-specific IgE is injected s.c. in the dorsum or into both sides of the pinna of the ear. After 20–48 h, antigen together with a dye (e. g. Evans blue) is administered intravenously.

Seconds to minutes after antigen challenge, a wheal and flare reaction can be observed, which is characterized by an itchy red swelling at the site of the previous IgE injection [4, 25]. The underlying pathomechanism is the cross-linking of IgE molecules bound on perivascular localized mast cells by antigen, resulting in the release of preformed vasoactive biogenic amines, histamine, serotonin, which are stored in mast cell granules. The resulting wheal and flare are the consequence of vasodilatation and vascular leakage at the site of postcapillary venules, a phenomenon which is widely used by allergists in the diagnosis of IgE-mediated or IgE-associated diseases, such as allergic rhinitis, allergic asthma or atopic eczema [4, 25]. This immediate type reaction can be recorded either as swelling of the ears using an odimeter or more elegantly as extravasation of Evans blue dye measured by colorimetry at 620 nm or macroscopically recorded as the diameter of blue lesions [26].

Even though IgE of different specificity [28] can be utilized to passively sensitize mast cells, the most frequently used approach utilizes anti DNP-IgE antibodies as the "gold standard in PCA". In this context it is important to realize that some of the anti-DNP-IgE antibodies that are available display intrinsic activity leading to enhanced cytokine production, mast cell survival and spontaneous histamine release mediated by FcεRI even in the absence of the antigen [29]. Based on these studies careful selection of the DNP-specific IgE-antibody is advisable. The anti DNP-IgE clone H 1 DNP-ε-26 generated by Futong Liu, University of San Fransisco, is

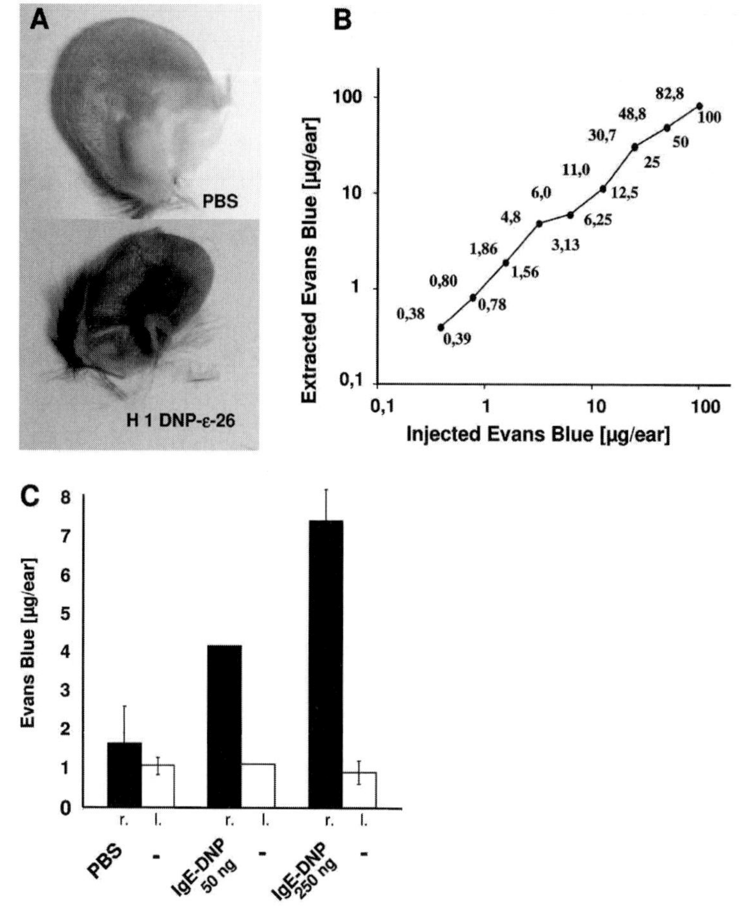

Fig. 11.3 Analysis of IgE-mediated type I hypersensitivity. (A) Passive Cutaneous Anaphylaxis (PCA). Anti DNP-IgE (H 1 DNP-ε-26, 250 ng) was injected into the dorsum of the right ear in a volume of 20 µl, the left ear was sham injected with 20 µl PBS. Twenty hours later DNP-HSA (100 µg/200 µl) was injected in 0.5 % Evans blue solution into the tail vein. One hour later, animals were sacrificed, ears removed, photo-documented and processed for quantification of extravasated Evans blue. (B) Recovery of Evans blue from mouse ear tissue: defined quantities of Evans blue were injected into the dermis of mouse ears. One hour later animals were sacrificed, ears removed and the dye was extracted from tissue by overnight incubation in 0.7 ml 1 N KOH and the subsequent addition of 0.6 N H_3PO_4 and acetone (5 : 13). After vigorous shaking and pelleting, precipitates were filtered off and Evans blue was measured colorimetrically at 630 nm. (C) Mice were injected in the dorsum of the right ear with IgE-DNP antibody at the indicated concentrations in a volume of 20 µl PBS. Left ears were injected with PBS only and served as internal controls. Twenty hours later DNP-HSA (100 µg/200 µl) was injected systemically in a 0.5 % Evans blue solution. One hour later, mice were sacrificed, ears removed and the amount of extravasated dye was measured colorimetrically at 630 nm.

frequently chosen for PCA since it displays low intrinsic activity and seems to produce reproducible Evans blue extravasation upon antigen challenge using DNP-HSA [26, 30]. Evans blue can be extracted from tissue by overnight incubation in 0.7 ml 1 N KOH with the subsequent addition of 0.6 NH_3PO_4 and acetone (5 : 13). After vigorous shaking, extracts are filtered to remove precipitates and extravasated dye can be measured colorimetrically at 630 nm.

PCA Late Phase Reaction Two to four hours after the initial wheal and flare reaction newly synthesized mediators are secreted from activated mast cells, including cytokines (IL-3, 4, 5, 6, 8, 13) chemokines (lymphotactin, MIP1α) and arachidonic acid derivatives (LTB4, LTC4; reviewed in [31]). The resulting late phase reaction is predominantly triggered by TNF-α and MIP-2, the murine analog of human IL-8 [32]. Clinically, the late phase reaction is characterized by a prolonged swelling of the ear, which is induced by recruitment of inflammatory cells in response to mast cell-derived mediators. Quantification of the late phase reaction can be achieved by measuring the ear thickness with a dial thickness gauge (oditest micrometer).

Experimental Protocol for PCA Early Phase Reaction [32]

- Antigen-specific IgE (20–500 ng) is injected intradermally at the dorsum of the pinna of the ear.
- After 20–48 h, 100–250 µg antigen is administered intravenously together with a dye (e. g. Evans blue 0.5 %).
- The antigen-triggered early phase reaction results in extravasation of Evans blue at 10–30 min, which can be monitored visually or quantified after obtaining the ear tissue.
- Evans blue can be extracted from tissue by overnight incubation in 0.7 ml 1 N KOH and the subsequent addition of 0.6 N H_3PO_4 and acetone (5 : 13).
- After vigorous shaking, extracts are centrifuged and filtered to remove precipitates.
- Extravasated dye can be measured colorimetrically at 630 nm [26].
- Recommended group size: five animals.

Experimental Protocol for PCA Late Phase Reaction [32]

- Passive sensitization of test animals with i. v. injection of H 1 DNP-ε-26.
- After 20 h 25 µl 0.75 % Dinitrofluorobenzene (DNFB), dissolved in acetone/olive oil (4 : 1) is applied.
- Ear thickness is measured with a micrometer at time points 1, 2, 4, 8, 12, 24, 48 and 72 h after antigen challenge.
- The delta in ear thickness between control groups (not passively sensitized) and sensitized mice reflects the late phase reaction triggered by mast cell-derived mediators (e. g. TNF-α, MIP-2) [13].
- Recommended group size: five animals.

For phenotyping mouse mutants, analysis of the early and late phase PCA enables different aspects of IgE-mediated mast cell activation to be studied. Alterations in the early phase reaction in mutants can help to detect defects in degranulation of preformed bioactive amines or the inhibition of their synthesis. Conversely, altered synthesis and secretion of cytokines, chemokines and arachidonic acid derivatives, most importantly of TNF-α and MIP-2, can be detected within aberrant late phase reactions. Since it is becoming increasingly evident that mast cells have a variety of functions in the orchestration of immune reactions and respond to a variety of stimuli aside from IgE-receptor cross-linking including pathogen-associated molecular patterns (e. g. LPS) or heterotopic binding of T cells, additional assays will be needed for the investigation of these mast cells effector arms.

Active Cutaneous Anaphylaxis (ACA) [27] In contrast to PCA, which is based on the adoptive transfer of sensitization by local injection of antigen-specific IgE, mice can be sensitized using standard sensitization protocols (compare 11.2.1.2). Subsequent intracutaneous injection of same antigen will trigger a localized early and late phase reaction. Intravenous injection of Evans blue prior to antigen administration allows the degree of local mast cell activation to be quantified using the same methods for measuring the Evans blue extravasation as those described above [26]. When phenotyping mouse mutants, ACA can provide additional information about the elicitation phase of allergic sensitization, however it does not differentiate between IgE- or IgG1-mediated phenomena.

11.2.1.4 Allergic Airway Inflammation, BAL, Body Plethysmography

A classical manifestation of type I hypersensitivity is the allergen-induced allergic airway inflammation, which is characterized by periodic airflow limitation, inflammation and airway hyperresponsiveness (AHR). Allergic airway inflammation and AHR are closely linked and interdependent. Lymphocytes and eosinophils are considered to be the key effector cells: T cells produce soluble factors which play a pivotal role in recruiting inflammatory cells in the lungs and eosinophil accumulation, activation and release of cationic proteins are implicated in many pathological changes which affect airway function [33]. AHR is defined as an exaggerated physiologic response of the airways to both specific (e. g. allergens in allergic individuals) and non-specific stimuli (e. g. cholinergic agonists, cold air, airborne dust, etc.). Long-standing asthma is further characterized by structural changes in the airways known as "airway remodeling", where myofibroblasts and collagen deposition are of particular importance [34].

Because the mechanisms of allergic airway inflammation and AHR cannot be exhaustively investigated in human subjects, many different animal species have been used in models of allergic airway inflammation and AHR [35]. In an effort to understand in greater detail the immunopathogenesis and susceptibility to allergic respiratory tract diseases such as bronchial asthma, we have also established an allergic airway disease model induced by exposure to an ovalbumin aerosol in allergen-sensitized mice, as a model for late-phase events such as airway eosinophilia and hyperresponsiveness.

Bronchoalveolar Lavage (Fig. 11.4) Bronchoalveolar lavage (BAL) is a widely used technique developed in order to evaluate inflammatory and immune processes in

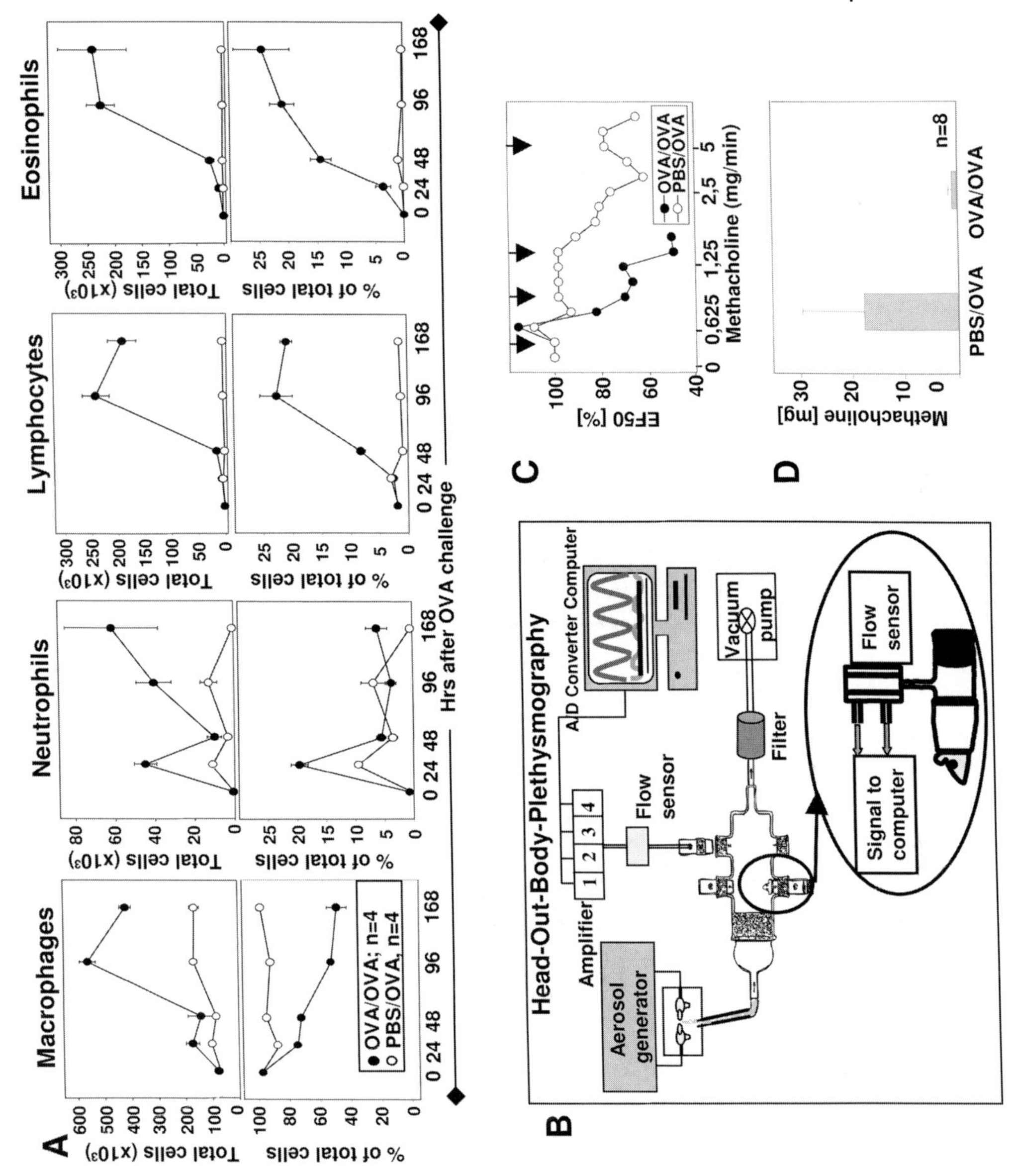

Fig. 11.4 (A–D) Analysis of bronchoalveolar lavage and airway hyperreactivity: (A) Cellular composition of BAL in BALB/c mice on consecutive days after allergen challenge, showing the 24-h peak of neutrophils and the steady increase of all inflammatory cells 48 h after allergen challenge (mean ± SD, $n = 4$). The cells are expressed both as total cell number (upper panel) and as percentages (lower panel). (B) Schematic diagram of the head-out body plethysmograph illustrating the exposure chamber and the equipment used to analyze the breathing parameters online. (C) Change of airflow during expiration at mid-tidal volume (EF_{50}) induced by increasing doses of methacholine in two representative animals, one non-sensitized/OVA-challenged (open circles) and one OVA-sensitized/OVA-challenged (filled circles). The dose of methacholine required to induce AHR in OVA-sensitized animals is considerably lower than that required in respective controls. (D) Mean cumulative dose of methacholine (± SEM, $n = 8$) required to obtain a 50% reduction in EF_{50} in non-sensitized/OVA-challenged and OVA-sensitized/OVA-challenged animals.

the human lung [36]. This technique has also been employed in screening different animal models of lung disease and has been shown to be very useful for evaluating allergic phenotypes [33]. BAL allows the quantification and differentiation of both the cellular infiltration in the airways and various inflammatory markers in the cell-free BAL fluid, i. e. the status of the allergic airway disease.

At the end of the sensitization period, the mice undergo allergen challenge via the airways by nebulization of 1% OVA in PBS for 20 min (Pari Boy, Starnberg, Germany). This procedure induces allergic inflammation in the airways, which can be evaluated by BAL. BAL cell profiles from allergic phenotypes are characterized by a peak of neutrophils 24 h after allergen challenge and by an increase in eosinophils, lymphocytes, neutrophils and alveolar macrophages 48 h after allergen challenge. A typical example in BALB/c mice is shown in Fig 11.4 A. The eosinophil count in the BAL fluid is of crucial importance in the identification of strain-related allergic response, and A/J mice seem to be the most responsive [37]. The quantification of lymphocyte subsets and the analysis of their activation status by flow cytometry can be very helpful in answering specific questions in defined mouse allergy models [38]. The Th1/Th2 phenotype in BAL fluid can be easily evaluated using commercially available bioassays for cytokine detection [39]. Furthermore, an increase in protein concentration, caused by enhanced permeability of the bronchoalveolar–capillary barrier, can be analyzed with the Coomassie protein assay. Lastly, lung histology is used to evaluate the perivascular and peribronchiolar cellular infiltrate, which in the allergic phenotype is mainly characterized by eosinophils and lymphocytes, epithelial mucus-cell hyperplasia and eventually the structural changes seen in airway remodeling [34]. For phenotyping airway remodeling, A/J mice seem to be the most suitable mouse strain [40].

Experimental Protocol for BAL and Lung Histopathology

- After euthanization and exsanguination, the animal's chest is opened and the trachea cannulated.
- BAL is performed five times with 0.8 ml PBS instilled into the lungs and the fluid harvested by gentle aspiration.
- Lavage fluid is collected, centrifuged at 600 r.p.m. for 10 min and the supernatant is stored at – 20 °C for further evaluation.
- Cells are resuspended in 1 ml PBS. After quantification of total cell viability and yield, samples of BAL fluid are centrifuged at 600 r.p.m. for 10 min, placed on slides and the different cell types present are counted. These slides are fixed and stained with Diff-Quick (Dade Behring, Newark, USA) and a total of 300 cells in each sample are counted by microscopy. Macrophages, neutrophils, lymphocytes and eosinophils are enumerated.
- Lungs obtained from separate animals are instilled with 4% buffered formalin, removed, and fixed in the same solution. After paraffin embedding, sections for microscopy are stained with hematoxylin and eosin, periodic acid Schiff (PAS), and collagen staining.
- Recommended group size: six animals.

Body Plethysmography Body plethysmography is a technique which has been traditionally employed to evaluate AHR in asthmatic patients [41]. This system has been widely used in recent years to phenotype allergic animal models, by assessing different breathing parameters in unanesthetized and unrestrained animals [42, 43]. In the strain distribution pattern for phenotyping asthma-treated AHR, A/J mice were identified as the strain showing the maximum response and C57BL/6J and C3H/HeJ mice showed the minimum response; BALB/c mice were intermediate in responsiveness [37, 39].

The system we use is the head-out body plethysmograph, according to the method of Alarie et al. [44] as shown in Fig. 11.4 B. Briefly, the system consists of a glass-made body plethysmograph which the mouse is allowed to enter; the head of the mouse is fixed by a neck collar facing into the chamber, where it is exposed to aerosolized methacholine. The pressure changes which arise in the body plethysmograph from the inspiratory and expiratory fluctuations during breathing are transformed into flow and volume signals so that automated data analysis provides tidal volumes (TV), respiratory rates (f), time of inspiration and expiration and airflow during expiration at mid-tidal volume (EF_{50}). The latter parameter is used as an index of airway-flow obstruction. Breathing patterns are measured in non-sensitized and in OVA-sensitized mice 24 h after two allergen exposures carried out on two consecutive days (1% OVA in PBS, 20 min, as described above). Baseline values are usually similar in allergic and non-allergic animals, but in the course of methacholine challenge the EF_{50} values decrease considerably in the allergic phenotypes (Fig. 11.4 C, D).

Experimental Protocol for Body Plethysmography

- After placing the animals into the chamber, they are allowed to adapt to the environment for 10 min allowing them to establish a normal breathing pattern; baseline values are then measured for a period of 5 min.
- The animals are nebulized with increasing amounts of methacholine in PBS (0, 0.625, 1.25, 2.5, 5, 10 mg per 1 min), with 4 min of rest between each nebulization; lung function during the inhalation and rest periods is monitored at regular intervals.
- Data is continuously recorded until a 50% reduction in EF_{50} is reached. At this point, the measurements are interrupted and the animals are allowed to recover.
- Recommended group size: eight animals. Note: known risk factors that could lead to incorrect measurements are possible infections and differences in body weight and size. It is recommended that the measurements are taken under SPF conditions at similar times of the day and that animals of the same sex and similar age are compared in order to reduce the variance of the data.

11.2.1.5 Impact of Different Sensitization and Challenge Protocols on Parameters of Allergen-induced Airway Inflammation

To determine the importance of the dose of allergen administered and the number of aerosolized allergen exposures in modulating the physiologic, inflammatory and immunologic features characteristic of allergen-induced airway inflammation, and

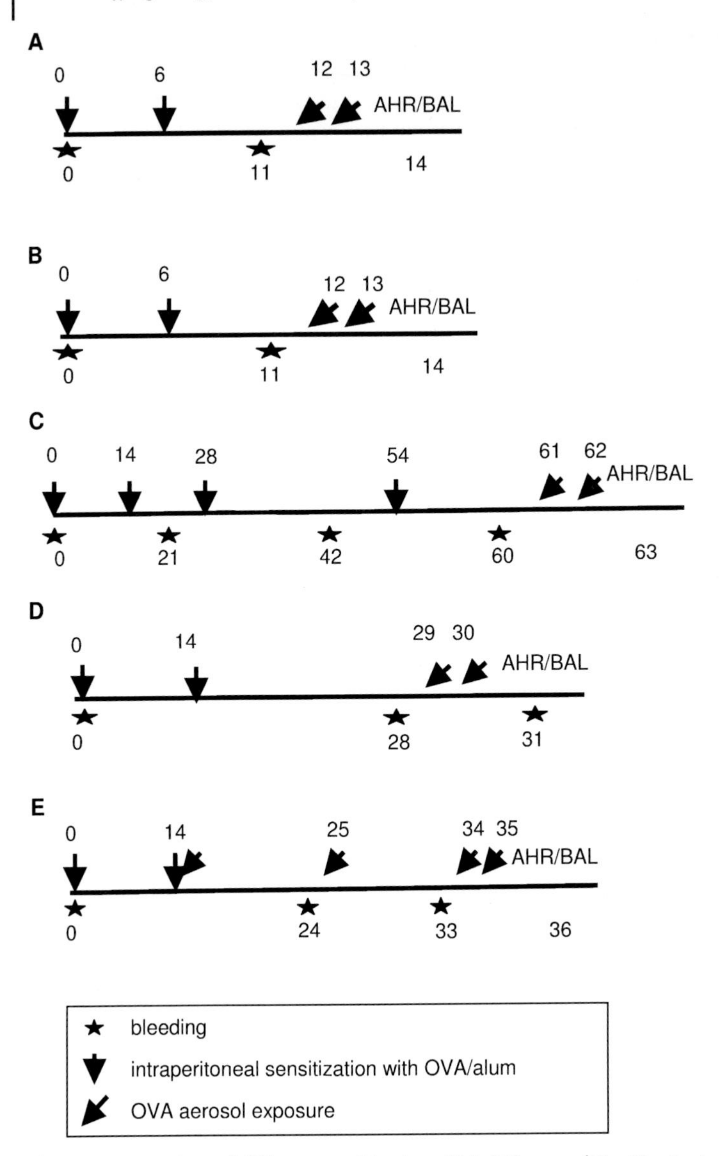

Fig. 11.5 Comparison of different sensitization protocols: Mice were sensitized by intraperitoneal injection of different doses of OVA complexed with 2 mg Alum on the days indicated. (A) OVA 2 × 200 µg i. p.; (B) OVA 2 × 50 µg i. p.; (C) OVA 4 × 1 µg i. p.; (D) OVA 2 × 20 µg i. p.; (E) OVA 2 × 20 µg i. p. plus OVA 1% aerosol 2 × 20 min. In one group (E) mice additionally received aerosolized OVA to boost the sensitization. Finally, all animals were analyzed 24 h after their last aerosol challenge which consisted of 1% OVA aerosol exposure for 20 min on two consecutive days.

Table 11.1 Different sensitization and challenge protocols (see Fig. 11.5 for details).

Protocol	Length of protocol (days)	OVA i. p. injection (times)	OVA aerosol exposure (times)	OVA-specific antibody	Eosinophilia	AHR
A	14	2	2	–	–	–
B	14	2	2	–	–	–
C	63	4	2	++	++	++
D	31	2	2	+	+	+
E	36	2	4	++	+++	++

–, no response; +, mild; ++, moderate; +++, severe.

also to optimize time requirements for the sensitization protocol, we analyzed various ovalbumin sensitization and challenge protocols based on previous studies [21, 45–48]. BALB/c mice (6- to 8-week-old females) were sensitized with OVA/alum by the intraperitoneal route and challenged with exposure to aerosolized OVA (1% in PBS) using five different protocols (see Fig. 11.5 for details). Interaperitoneal injections consisted of different OVA (grade V; Sigma, St. Louis, MO, USA) concentrations adsorbed onto aluminium hydroxide gel (alum; Pierce, Rockford, IL, USA). Twenty-four hours after the last OVA exposure, AHR, total and differential cell counts in BAL fluid, as well as levels of circulating OVA-specific antibodies were measured in the mice.

Mice immunized with very high and high doses of OVA and challenged twice in a very short period of time (Fig. 11.5 A, B) showed neither a significant increase in airway hyperresponsiveness and eosinophilia nor in OVA-specific antibody production compared to control animals that received PBS. Mice immunized with intermediate doses of OVA (Fig. 11.5 D) showed an increase in OVA-specific antibody levels, although they did not show a significant increase in airway reactivity. Mice sensitized several times with a very low dose of OVA and challenged twice (Fig. 11.5 C) showed significant OVA-specific antibody induction, bronchial eosinophilia and a solid AHR. Similar results were obtained with a shorter protocol in which mice received additional OVA aerosol exposures to boost sensitization (Fig. 11.5 E) and which reliably induced OVA-specific antibody production, a pronounced eosinophilia and airway hyperresponsiveness. Thus, with regard to high-throughput screening, this sensitization scheme appears to be more suitable for the analysis of large numbers of animals in the search for mouse mutants that display allergic phenotypes which are not obvious under baseline conditions (data summarized in Tab. 11.1).

11.2.2 Immune Complex Mediated Hypersensitivity (Type III, Arthus Reaction)

A type III hypersensitivity reaction, also known as the Arthus reaction, typically involves antigen-specific IgG antibodies and activation of complement through the formation of immune complexes. Common examples of immune complex diseases where type III reactions are of pathogenetic relevance are cutaneous and systemic vasculitis, rheumatoid arthritis and systemic lupus erythematosus. The traditional pathogenetic concept of the Arthus reaction involves the activation of the classical complement cascade by formation of immune complexes in the circulation or in local tissue through the engagement of complement-activating, antigen-specific IgM or IgG subclass antibodies followed by the subsequent activation of mast cells and endothelial cells through the binding of complement-derived anaphylatoxin C5 a to the C5 a receptor (C5 aR) [49]. These immunoglobulins can be directed either to autoantigens or to foreign antigens of environmental, medicinal or microbial origin. More recently, the pathogenetic concept of the type III inflammatory reaction was revised with the help of experimental animal models. The most common approach for experimentally modeling type III inflammation is the reverse passive Arthus reaction, which can be applied to various tissue and organ systems such as skin, lung, and joint tissue to model diseases such as cutaneous (allergic) vasculitis, immune-complex pneumonitis, and immune-complex arthritis. Using this animal model system it has been demonstrated over the last decade that in a manner similar to immediate-type allergy where IgE engages the cellular Fc receptor for IgE (FcεRI) on mast cells or basophils, the type III hypersensitivity reaction is fully dependent on the engagement of both the FcγRI (CD64) by IgG and the C5 aR on mast cells [50–52]. Animals with a genetic deficiency of the mast cells show a severe reduction in the inflammatory response in the passive Arthus reaction [51, 53]. The role of complement, which is regarded to be less important than previously anticipated, appears to be dependent on species-specific factors [51, 54]. The most complete abrogation of the type III inflammatory response can be achieved with a genetic deletion of both a functional FcγRI and the C5 aR, thereby causing the two receptors to act in a co-dominant fashion [50–52]. The relative importance of the two receptors for the overall inflammatory reaction during type III hypersensitivity is also dependent on the tissue and organ in which the reaction is elicited. Once activated through either of the receptors during type III inflammation, mast cells provide an ample source of preformed TNF-α which is important for the subsequent steps of the reaction such as recruitment of neutrophil granulocytes to local tissue sites [51]. The reverse passive Arthus reaction is the preferred method for screening and investigation of animal mutants with assumed defects in the type III inflammatory cascade because of its ease of implementation and the readily available read-out methods. In the following sections we describe the methodology and possible variations of the reverse passive cutaneous Arthus reaction in skin as a versatile tool with which to phenotype mouse mutants. The elicitation of the reverse passive Arthus reaction in other tissues such as lung or peritoneal cavity differs slightly with respect to concentrations of antibody and antigen [53].

11.2.2.1 Reverse Passive Cutaneous Arthus Reaction

The reverse passive cutaneous Arthus reaction (RPAR) is initiated by the intradermal injection of antibody followed by the immediate intravenous administration of antigen (see scheme presented in Fig. 11.6). The intradermal injection sites should be free of fur and disinfected; the abdominal region or back of the animal are the

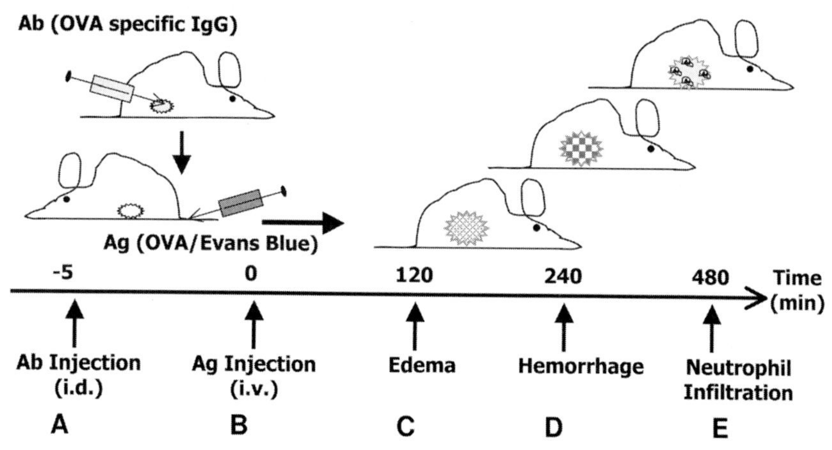

Fig. 11.6 The cutaneous reverse passive Arthus reaction: a functional phenotype assay to analyze type III hypersensitivity reactions. (A) The cutaneous reverse passive Arthus reaction is initiated by intradermal injection of monospecific polyclonal anti-OVA IgG (30–100 µg purified IgG). (B) Subsequently, soluble antigen (10–20 µg OVA per g body weight) is injected intraveneously together with the dye indicator Evans blue (1–2%). (C–E) The following sequential events occur in the reverse passive Arthus reaction, edema formation, hemorrhage (erythrocyte extravasation) and neutrophil infiltration. The time-dependent sequelae of immune complex formation in the cutaneous tissue are monitored by measuring the size of the dye extravasation (C, as an indicator for edema), the size of the purpura/hemorrhage (D, as an indicator of erythrocyte extravasation), and the extent of tissue infiltration with neutrophil granulocytes (E) as analyzed by histologic staining of tissue sections with hematoxylin and eosin and subsequent scoring by an investigator blind to the study or by using computer-based histopathologic scoring algorithms. Ag, antigen; Ab, antibody.

preferred anatomical sites. The antibody to be injected is of crucial importance: only purified antibody (IgG or IgM fraction) should be injected to avoid non-specific inflammation. Polyclonal monospecific antibodies are most commonly used. Alternatively, a mixture of complement-activating monoclonal antibodies directed to different epitopes of the same antigen is also suitable for eliciting an adequate inflammatory response. Immediately after antibody injection, the soluble antigen is injected intravenously together with a dye indicator such as Evans blue to measure edema formation. Intradermal and intravenous injections with physiological saline serve as specificity controls. The most common antigen used to elicit RPAR is chicken egg albumin (ovalbumin/OVA). Typically, 30–100 µg of purified OVA-specific polyclonal IgG are injected intradermally at various sites, followed by the intravenous injection of 10–20 µg OVA antigen per gram body weight together with 1–2% Evans blue dye [51, 53, 54]. As an alternative to Evans blue dye, 10^6 cpm of ^{125}I-labeled albumin can be used to track extravasation of macromolecules [50]. Before injection, all protein solutions should be centrifuged at 12,000 g (5 min) to remove any aggregates.

Quantification and Analysis of Reverse Passive Arthus Reaction The typical sequence of events in the RPAR include edema formation as an expression of macromolecular extravasation after 2–4 h, hemorrhage (erythrocyte extravasation) after 4–8 h, and neutrophil infiltration into the tissue after 8–12 h (Fig. 11.6). Ideally, group sizes of at least five animals per group should be used. All quantifications should be carried out in a blinded fashion. The degree of edema formation can be quantitated either by calculating the area of the dye extravasation on the everted skin using a hand-held digitizer by measuring the diameter of the dyed area directly or by extraction of Evans blue from defined skin biopsies with subsequent spectrophotometric analysis (method described in Section 11.2.1.3). Other researchers have weighed defined punch biopsies without using Evans blue dye [53]. Where ^{125}I-labeled protein markers have been used to identify extravasation, defined slices of skin are measured using a gamma-counter [50]. Extravasation of macromolecules during the inflammatory process of RPAR is followed by extravasation of erythrocytes. This can be observed after 4–8 h as a defined area of hemorrhage on the everted animal skin, the size of which can be calculated with a hand-held digitizer in a manner similar to that used for measuring the size of dye extravasation as a marker of edema formation. The third hallmark of RPAR, usually peaking as early as 8 h after the induction of the inflammatory reaction, is the massive tissue infiltration by neutrophil granulocytes. The analysis of neutrophil infiltration is commonly achieved by staining histologic tissue sections with hematoxylin and eosin, the subsequent should be carried out by a histopathologist who is blind to the study. As an alternative, histopathologic scoring algorithms can be used, that either score the density or the size of the inflammatory infiltrate [55]. Myeloperoxidase (MPO) activity can be determined in defined skin punch biopsies or other tissue biopsies as an independent quantitative parameter of neutrophil infiltration during the course of RPAR [50, 52, 55].

11.2.3
Delayed Type Hypersensitivity (Type IV)

In the pathogenesis of allergic contact dermatitis – a classical manifestation of a lymphocyte-mediated delayed type hyersensitivity (type IV) – we differentiate between a "clinically silent" sensitization phase (afferent phase) initiated by the first contact with the allergen and an elicitation phase (efferent phase) which is triggered by re-encounter of the allergen and which results in the clinical manifestation of the disease, i. e. allergic contact dermatitis. To phenotype the delayed type allergic immune responses in mouse mutants, different approaches can be employed that separately address the afferent phase and the efferent phase of allergic contact dermatitis [56]. The local lymph node assay addresses only the events which occur during the sensitization phase, while the contact hypersensitivity assay is used to investigate the sensitization and elicitation phase separately.

11.2.3.1 Local Lymph Node Assay

The local lymph node assay (LLNA) was originally developed as a rapid and sensitive method for the identification of chemical substances with contact-sensitizing potential and is approved as a testing method by the OECD and US National Institute of Health [57, 58]. For phenotyping mouse mutants, a modified approach can

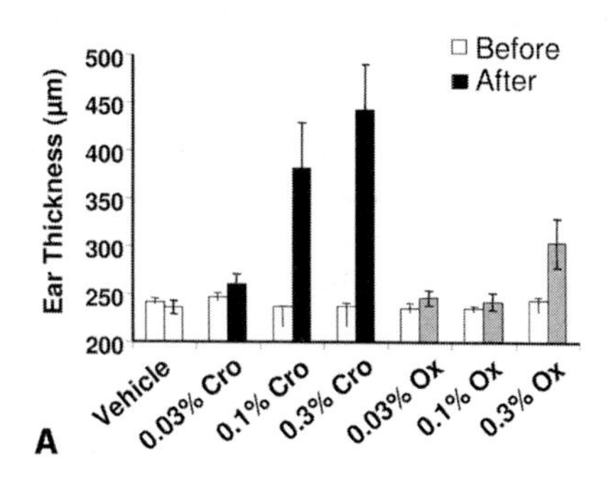

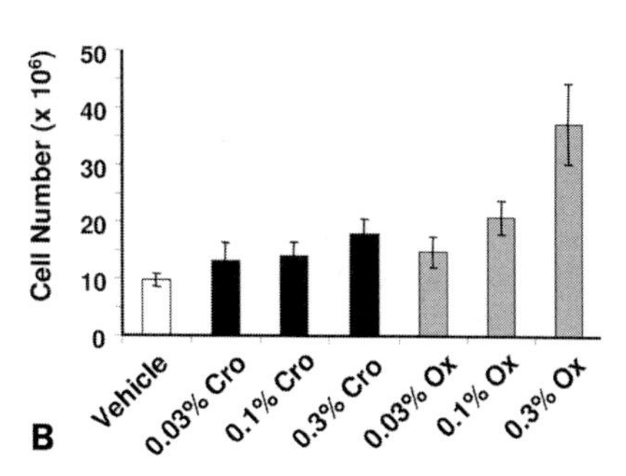

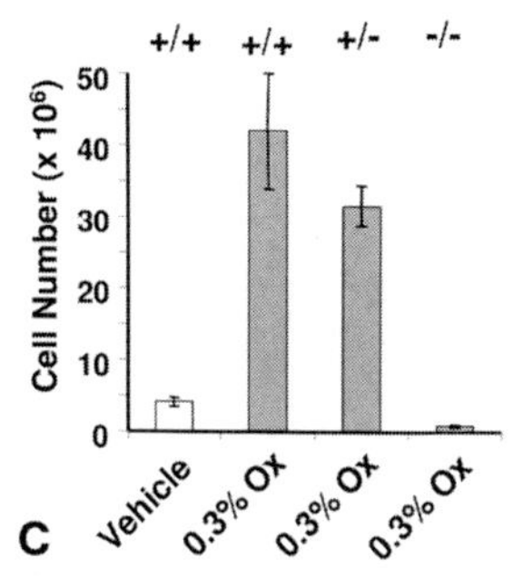

Fig. 11.7 Ear swelling response in mouse and local lymph node assay (LLNA). Mice were anaesthetized and ear thickness was recorded using a micrometer gauge. Irritant (25 µl) (crotonoil) or contact sensitizer (oxazolon) dissolved in acetone/olive oil (4 : 1) was applied to the dorsum of both ears on three consecutive days. On day 4, mice were sacrificed, the ear thickness was recorded and the draining lymph nodes of the auricular region were excised. (A) A significant dose-dependent increase of ear thickness following application of the irritant crotonoil was observed. (B) A significant dose-dependent increase in $CD45^+$ leukocytes in the draining lymph nodes following application of the contact sensitizer (oxazolon) was observed. There was also a minimal increase in the number of $CD45^+$ cells following application of the highest concentration of the irritant crotonoil. (C) An abnormal increase in $CD45^+$ leukocytes in the auricular lymph nodes of homozygous animals of a mutant with a defect in T cell development was noted. Wildtype mice and heterozygous littermate controls displayed a normal response.

be used to detect aberrant immune responses to known contact-sensitizing agents such as oxazolone or potassium dichromate [59, 60]. The rationale of this test system is based on the observation that contact-sensitizing agents (i. e. allergens) induce a strong proliferation of lymphocytes in draining lymph nodes during the sensitization phase, whereas irritants lead to local inflammation but not to significant changes of cell proliferation in regional lymph nodes. The OECD-approved LLNA has been modified and improved since then. The most recent approach is the integrated model for the differentiation of chemical-induced allergic and irritant skin reactions (IMDS). This test system integrates two observations: (1) contact sensitizers/allergens lead to lymphocyte proliferation in draining lymph nodes with almost no local inflammation at the exposure site; (2) irritants provoke local inflammation with no or very low lymphocyte proliferation in draining lymph nodes. The protocol is based on the LLNA with additional measurement of ear thickness (Mouse ear swelling test, MEST). By integrating information obtained from the ear swelling response (increase in ear thickness after exposure) and cell counts from the draining lymph node obtained by coulter counter or flow cytometric analysis, the effects of irritant substances and contact sensitizers can be differentiated. In addition the IMDS is faster than the standard LLNA and can be carried out without the use of radioactivity [59].

The immune response to contact allergens during the sensitization phase in different mouse strains or mutants can be phenotyped by measuring the increase in leukocyte number in draining lymph nodes using quantitative flow cytometry (TrueCount® beads). In addition to measuring total leukocyte numbers (CD45 staining) this allows the quantification of lymphocyte subsets (such as $CD4^+$ T cells, $CD8^+$ T cells or B cells) and the analysis of the activation status at the single-cell level (i. e. CD25, CD69 or I-A) to elucidate specific mechanisms which occur in defined mouse mutants. Deficiencies in irritation reactions are detected by differences in ear swelling in mutant compared to wild-type mice. Figure 11.7 shows an example of the ear swelling response and lymph node cell counts in BALB/c wild-type mice challenged with an irritant (croton oil), contact sensitizer (oxazolone) or vehicle control. Figure 11.7 C shows the results of a local lymphnode response (number of CD45 positive leukocytes in draining auricular nodes) in a mouse mutant identified in the ENU mutagenesis phenotyping screen which shows a defect in lymphocyte development.

Experimental Protocol of IMDS (modified from [60])

- Examples of suitable contact sensitizers are oxazolone 0.3 % (4-ethoxy-methylene-2-phenyl-3-oxazalin-5-one), DNFB 0.3 % (2,4-dinitro-1-fluorobenzene) or cinnamic aldehyde 10 %.
- Preferred irritants are croton oil 0.3 %, glutaraldehyde 3 % or SDS 40 %.
- Suitable vehicles: acetone/olive oil (4 : 1) or dimethylacetamide/ethanol/acetone (4 : 3 : 3)
- Recommended group size is five animals. Note: mice with ear marks (punches or tags) are not suitable for the ear swelling test.
- On day 1 ear thickness is measured with a micrometer and test substances are applied in appropriate dilution on the dorsum of the ear in a volume of 10 µl/ear.
- On days 2 and 3 test substances are applied as described for day 1.
- On day 4 ear thickness is recorded, mice are sacrificed and draining auricular lymph nodes excised and pooled for every animal.
- Single cell suspensions are generated by disaggregation through a 40-µm mesh, cells are washed twice with 10 % FCS and resuspended in FACS buffer containing 5 % FCS and 0.01 % w/v Na-azide.
- Absolute leukocyte numbers are obtained by FACS measurement using TrueCount® beads (BD Immunocytometry Systems) and anti-CD45 antibodies.
- Optional analysis of further cell populations (e. g. CD4/ CD8/ B220) and cell activation markers (CD25/ CD69/ I-A).

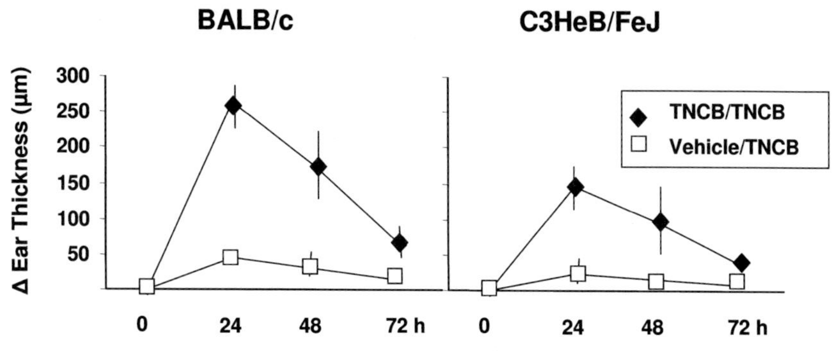

Fig. 11.8 Contact hypersensitivity response (Type IV hypersensitivity reaction). Female 8–12-week-old BALB/c or C3Heb/FeJ mice were sensitized on day 0 by painting the shaved abdominal skin with the contact sensitizer TNCB (1% w/v) in acetone/olive oil (4 : 1) or with vehicle control (acteone/olive oil 4 : 1). On day 6 ear thickness was recorded with a micrometer gauge and animals were challenged by applying an elicitation dose of 10 μl TNCB (0.3% w/v) in acetone/olive oil to the dorsum of the ear. At 24, 48 and 72 h the change in ear thickness was recorded. In non-sensitized animals (vehicle) that were challenged with TNCB (Vehicle/TNCB) only a minor ear swelling response was observed, which was due to the irritancy of the contact sensitizer. Note that depending on the mouse strain used, different magnitudes of response can be induced by the same contact allergen.

11.2.3.2 **Contact Hypersensitivity Assay**

The contact hypersensitivity (CHS) assay is a simple *in vivo* test for cell-mediated type IV hypersensitivity reactions. Exposure of epidermal cells to exogenous chemical haptens leads to a delayed type hypersensitivity reaction that can be quantified. As outlined above (see Section 11.2.3) CHS consists of a silent sensitization phase and a symptomatic elicitation phase. During the sensitization phase epidermal dendritic cells (Langerhans cells) take up the allergen, process it and transport it to the local lymphoid tissue. Here Langerhans cells present the allergen to naive T cells and prime the allergen-specific immune response. The elicitation phase is initiated when an already sensitized animal is re-exposed to the same allergen. It is characterized by localized swelling of the tissue, an inflammatory cellular infiltrate and epidermal thickening (ancanthosis) and corresponds to eczematous skin lesions that are found in allergic contact dermatitis in humans [56, 61, 62].

In phenotyping mouse mutants the CHS assay is a valuable tool with which to functionally analyze T cell-mediated immune responses *in vivo* [63]. Depending on the protocol used, this assay can be used to analyze mechanisms relevant to either the sensitization phase (e. g. dendritic cell function) or the elicitation phase (e. g. T cell function). For the CHS assay the abdomen of the mouse is shaved and exposed epicutaneously to a sensitizing dose of the contact allergen (sensitization). After 6 days the baseline ear thickness is measured and subsequently the skin on the dorsum of the ears is treated epicutaneously with an elicitation dose of the same contact allergen (initiation of the elicitation phase). Ear thickness is measured 24 h later when maximum swelling of the ear is expected. Measurement at additional time points (e. g. 12, 48 or 72 h) may lead to the detection of accelerated or prolonged response kinetics. The change in ear thickness after allergen treatment is

Tab. 11.2 Mouse ear swelling test for the analysis of type IV contact hypersensitivity: experimental conditions required and expected results.

	Sensitization (abdomen)	Elicitation (ear)	Ear swelling
Group 1	Allergen	Allergen	++++++
Group 2	Vehicle	Allergen	(+)*
Group 3	Allergen	Vehicle	– –
Group 4 (optional)	Vehicle	Vehicle	– –

*Due to irritancy of contact allergen.

used to ascertain the presence of potential defects in T cell or dendritic cell function by comparing wild-type and mutant mice. Depending on the nature of the study, alterations occurring during the sensitization phase (dendritic cell and T cell dependent) or the elicitation phase (predominantly T cell dependent) can be differentiated by comparing variously treated groups. The expected ear swelling response will depend on the type of allergen used (e. g. TNCB > DNFB > Oxazolone) and the type of mouse strain tested (BALB/c > C3H > C57BL/6; see Fig. 11.8). When comparing mutants versus wild-type mice the genetic background is crucial, i. e. mice should be backcrossed onto the same background for at least seven generations [64]. In addition, a minimum of two control groups (groups 2 and 3) should be included as outlined in Tab. 11.2. Group 2 is used to approximate the irritation effect of the contact allergen used, which may contribute to the ear swelling response observed in group 1. Group 3 is used to establish the potential irritancy of the vehicle used and group 4 is used to approximate the sensitizing potential of the vehicle. It is important to choose the right solvent system and elicitation dose for each allergen to minimize irritancy [65].

Contact Hypersensitivity Protocol (modified from [63])

- Examples of suitable contact sensitizers are oxazolone 3 % (4-ethoxy-methylene-2-phenyl-3-oxazalin-5-one), DNFB 0.5 % (2,4-dinitro-1-fluorobenzene) or TNCB 3 % (2,4,6-trinitrochlorobenzene). Note: all allergens must be handled with great care (wearing gloves, eye and mouth protection) since they are potentially sensitizing for humans.
- Suitable vehicles: acetone/olive oil (4 : 1) or dimethylacetamide/ethanol/acetone (4 : 3 : 3)
- Recommended group size is ≥ five animals. Note: mice with ear marks (punches or tags) are not suitable for the ear swelling test.
- On day 0 the abdomen of the mouse is shaved and 100 μl of contact sensitizer is applied epicutaneously in a sensitizing dose (see above).
- On day 6 baseline ear thickness is measured with a dial thickness gauge (Oditest micrometer) and recorded.

- Contact sensitizers (elicitation dose usually 1/3 of the sensitization dose) are applied epicutaneously to the dorsum of the ear in a volume of 10 μl/ear.
- On days 7, 8 and 9 ear thickness is recorded in sensitized and naive animals.
- Changes in ear thickness are calculated separately for each ear.

11.2.4 Granulomatous Hypersensitivity (Type V)

By the 1980 s granuloma formation was already considered to be a particular type of hypersensitivity reaction [17] which is mounted by an organism to prevent the spread of a potentially hazardous invading pathogen. This classification has recently been re-proposed by Rajan [18] who presented solid arguments in favor of the unique features of granuloma formation being representative of a hypersensitivity reaction.

Granulomas can be divided into various types including the classical form of infectious granulomas such as the caseating type as seen in tuberculosis, the sarcoid type as seen in sarcoidosis (typically non-caseating) and the foreign-body granuloma. A particular type of granuloma is also represented by the schistosoma egg antigen (SEA) granuloma which shows patterns of a Th2-immune reaction instead of the typical Th1-response which is seen with most other granulomas [66, 67].

The "classical" type of granuloma consists of a central antigenic stimulus which may consist of bacteria, foreign-body material and/or necrotic cell material. This central zone is surrounded by antigen-presenting cells such as macrophages and dendritic cells together with variable numbers of neutrophils. Depending on the stage of the granulomatous reaction, lymphocytes may be found at the periphery. This rather mixed composition of cells enables the function of different cell types to be analyzed in a single experimental set-up.

In order to phenotype the capacity for developing granulomatous hypersensitivity responses in mouse mutants, granuloma formation can be tested in various organs such as the liver, lung and skin. Models of infectious (mycobacterial) granulomas and foreign-body granulomas have been described for testing the individual capacity to mount a granulomatous response. The evaluation of infectious granulomas can be achieved either microscopically (by counting the granulomas in an organ of interest) or microbiologically (recovering live bacteria after granuloma disruption). Foreign-body granulomas are usually evaluated by their size and cell composition after histological staining. Therefore, inert particles (e. g. Sephadex beads) are injected subcutaneously and samples are taken from the injection site after 14 days. The material should be fixed and processed for histological evaluation or quick-frozen for molecular biology studies. However, this technique is difficult to standardize and requires experience in sophisticated histological techniques.

11.2.4.1 Experimental Protocol for Type V Hypersensitivity

For facilities that have access to mycobacteria the following protocol for infectious granuloma is proposed:

- Growth of sufficient numbers of mycobacteria (e. g. *M tuberculosis* strain Erdmann) in Middlebrook broth 7H9 containing the appropriate supplements.
- Wash bacteria in PBS and enumerate live bacteria by dilution on agar plates.
- Inject adequate numbers of live bacteria intravenously into the tail vein (e. g. 10^5–10^6 CFU) in sterile PBS containing 0.05 % Tween 80.
- Kill the infected animals ~ 7 days later and recover the lungs, liver, and spleen using aseptic techniques. Homogenate the organs in PBS-Tween and plate serially diluted samples onto 7H10 agar plates containing the appropriate supplements.
- Incubate at 37 °C for 3 weeks and enumerate the growing colonies. In parallel representative organs from infected and control mice should be recovered and stained for granuloma architecture by HE staining and for mycobacteria content by acid-fast Ziehl–Neelsen bacterial stain.
- Experiments should be carried out on groups of five animals.

For facilities that do not have access to mycobacteria the foreign-body granuloma protocol is proposed:

- 50 µl Sephadex beads (G50-medium beads, Amersham, Biosciences) are mixed with 12.5 % volume of aluminium hydroxide (Serva) and injected subcutaneously into the flank of previously shaved animals (shaving makes it easier to locate the site of injection).
- After 14 days the injection sites are sampled and either processed for histology (paraffin embedding) or quick-frozen for molecular biology studies.
- Recomended group size $n = 5$.

11.3
General Considerations, Logistics and Outlook

Phenotyping allergic reactions is of great interest for those investigators seeking to analyze specific pathways or mechanisms in an *in vivo* setting using the laboratory mouse or appropriate mouse mutants as the model organism. For this type of hypothesis-driven approach the precise phenotyping of *in vivo* reactions has enhanced our understanding of the pathomechanisms involved in allergic reactions. Normally, this involves analyzing only those specific types of reaction that seem plausible for and related to the hypothesis that is being tested. In some cases this approach is successful but in many other cases mouse mutants have not shown any obvious phenotype. This has prompted investigators at the GSF National Research Center for Environment and Health in Munich to set up a unique high-throughput screening platform for mouse mutants in which a broad spectrum of organ systems is analyzed in a hypothesis-free first-line screen covering more than 200 standardized parameters. This phenotyping platform, known as the German Mouse Clinic (GMC) (www.mouseclinic.de), includes a primary screen of genotypically-defined mouse mutants for allergic phenotypes. Newly identified mutants of the primary screen are then analyzed further and in more detail in secondary and tertiary screens.

In the primary allergy screen, mouse lines are routinely examined on the basis of their total plasma IgE levels. Future directives for primary allergy screening also include the analysis of Fc receptor expression on peripheral blood leukocytes. Primary screening of mutants is carried out in close coordination with other screens at the GMC. The decision to initiate further screening for a given mutant is based on this information, which comprises more than 200 standardized parameters. According to an established workflow of the GMC, up to 26 mutant mouse lines per year (up to 1500 mice/year) are phenotyped in the screen for allergic disorders. These include mutant mouse lines of different origin such as transgenic, knock-out, or mutants generated by ENU-mutagenesis. Within the setting of the GMC primary screen for allergic disorders a total of 1055 wild-type and mutant mice have been screened for aberrant total IgE concentrations in plasma within the first 10 months after the establishment of the GMC. These included 20 mutant mouse lines of mutagenized, transgenic or knock-out origin of which four lines with abnormal plasma IgE levels or other allergic defects have been identified to date. Using these approaches it is expected to define several new mouse models with a phenotype relevant to allergic diseases each year.

Mouse mutants that show abnormal phenotypes in the primary screen will be subjected to secondary and tertiary screens. These are undertaken in concert with the expertise from other screens of the GMC and depend on the initial observations obtained in the primary screen. In theory, all the allergy phenotyping protocols described above are suitable for this type of secondary screen. Here logistic considerations are important. Some of the test systems such as the determination of allergen specific humoral immune responses (specific IgE, IgG2a, IgG1) and allergen-induced airway hyperreactivity or analysis of inflammatory infiltrate in the bronchial alveolar lavage fluid can be conducted sequentially. The majority of the protocols described, are however, consumptive in nature and do not allow sequential analysis of the same animals in different assay systems. Thus in most cases the secondary phenotyping assay should be selected based upon the results of the primary screen. Mouse mutants with a confirmed allergy phenotype can be subjected to in-depth tertiary analysis. This type of analysis very much depends on the lead phenotypes observed in the primary and secondary test systems. Given an abnormal plasma IgE level in the primary screen and increased airway hyperreactivity in the secondary screen, it may be decided to select as the tertiary screen the analysis of mast cell function *in vivo* using the passive cutaneous anaphylaxis protocol. In contrast, if the primary screen reveals hyper IgE, and the secondary screen a lack of the allergen-specific immunoglobulin response, the tertiary phenotyping approach would rather focus on assays that analyze T-cell and B-cell function (instead of mast cell function). As exemplified above, no general recommendations can be given regarding the sequence in which the different test systems should be employed. However, given the wide variety of phenotyping assays (as described in this chapter) for the analysis of different *in vivo* functions, the chances of detecting allergy-relevant phenotypes can be considered to be very high.

In conclusion, the establishment of a successful phenotyping platform for mutant mouse lines in the search for new phenotypes with relevance to IgE-mediated allergic diseases marks an important step forward. The phenotyping approach has already proven its ability to define new disease models for allergy that can be used by the scientific community. This will ultimately lead to the discovery of new genes involved in the humoral and cellular regulation of IgE and IgE-mediated diseases and subsequently to new therapeutic strategies in the treatment of allergy.

References

1 J. Ring, U. Kramer, T. Schafer, H. Behrendt **2001**,Why are allergies increasing? *Curr. Opin. Immunol.* 13, 701–708.

2 U. Wahn, H. E. Wichmann (Hsg) **2000**, *Spezialbericht Allergien*, Metzler und Poeschel, Statistisches Bundesamt, Stuttgart.

3 J. Ring, T. Fuchs, G. Schultze-Werninghaus (eds.) **2004**, *Weißbuch Allergie in Deutschland*, Urban & Vogel Verlag, München.

4 J. Ring **2005**, *Allergy in Practice*, Springer Verlag, Berlin.

5 H. Behrendt, W. M. Becker, C. Fritzsche, W. Sliwa-Tomczok, J. Tomczok, K. H. Friedrichs, J. Ring **1997**, Air pollution and allergy: experimental studies on modulation of allergen release from pollen by air pollutants, *Int. Arch. Allergy Immunol.* 113, 69–74.

6 H. Behrendt, W. M. Becker **2001**, Localization, release and bioavailability of pollen allergens: the influence of environmental factors, *Curr. Opin. Immunol.* 13, 709–715.

7 A. F. Coca, R. A. Cook **1923**, On the classification on the phenomena of hypersensitivity, *J. Immunol.* 8, 163–182.

8 K. C. Barnes, D. G. Marsh **1998**, The genetics and complexitiy of allergies and asthma, *Immunology Today* 19, 325–332.

9 W. O. Cookson, B. Ubhi, R. Lawrence, G. R. Abecasis, A. J. Walley, H. E. Cox, R. Coleman, N. I. Leaves, R. C. Trembath, M. F. Moffatt, J. I. Harper **2001**, Genetic linkage of childhood atopic dermatitis to psoriasis susceptibility loci, *Nat. Genet.* 27, 372–373.

10 P. Van Eerdewegh, R. D. Little, J. Dupuis, R. G. Del Mastro, K. Falls, J. Simon, D. Torrey, S. Pandit, J. McKenny, K. Braunschweiger, A. Walsh, Z. Liu, B. Hayward, C. Fo, S. P. Manning, A. Bawa, L. Saracino, M. Thackston, Y. Benchekroun, N. Capparell, M. Wang, R. Adair, Y. Feng, J. Dubois, M. G. FitzGerald, H. Huang, R. Gibson, K.M. Allen, A. Pedan, M. R. Danzig, S. P. Umland, R. W. Egan, F. M. Cuss, S. Rorke, J. B. Clough, J. W. Holloway, S. T. Holgate, T. P. Keith **2002**, Association of the ADAM 33 gene with asthma and bronchial hyperresponsiveness, *Nature* 418, 426–430.

11 T. Laitinen, A. Polvi, P. Rydman, J. Vendelin, V. Pulkkinen, P. Salmikangas, S. Makela, M. Rehn, A. Pirskanen, A. Rautanen, M. Zucchelli, H. Gullsten, M. Leino, H. Alenius, T. Petays, T. Haahtela, A. Laitinen, C. Laprise, T. J. Hudson, L. A. Laitinen, J. Kere **2004**, Characterization of a common susceptibility locus for asthma-related traits, *Science* 304, 300–304.

12 R. Kuhn, K. Rajewsky, W. Muller **1991**, Generation and analysis of interleukin-4 deficient mice, *Science* 254, 707–710.

13 D. Hata, Y. Kawakami, N. Inagaki, C. S. Lantz, T. Kitamura, W. N. Khan, M. Maeda-Yamamoto, T. Miura, W. Han, S.E. Hartman, L. Yao, H. Nagai, A. E. Goldfeld, F. W. Alt, S. J. Galli, O. N. Witte, T. Kawakami **1998**, Involvement of Bruton's tyrosine kinase in FcepsilonRI-dependent mast cell degranulation and cytokine production. *J. Exp. Med.* 187, 1235–1247.

14 M. H. Hrabe de Angelis, H. Flaswinkel, H. Fuchs, B. Rathkolb, S. Soewarto, S. Marschall, S. Heffner, W. Pargent, K. Wuensch, M. Jung, A. Reis, T. Richter, F. Alessandrini, T. Jakob, E. Fuchs, H. Kolb, E. Kremmer, K. Schaeble, B. Rollinski, A. Roscher, C. Peters, T. Meitinger, T. Strom, T. Steckler, F. Holsboer, T. Klopstock, F. Gekeler, C. Schindewolf, T. Jung, K. Avraham, H. Behrendt, J. Ring, A. Zimmer, K. Schughart, K. Pfeffer, E. Wolf, R. Balling **2000**, Genome-wide, large-scale production of mutant mice by ENU mutagenesis, *Nat. Genet.* 25, 444–447.

15 J. Descotes, G. Choquet-Kastylevsky **2001**, Gell and Coombs's classification: is it still valid? *Toxicology* 158, 43–49.
16 I. M. Roitt **1984**, *Essential Immunology*, Blackwell Science, Oxford.
17 J. Ring **1988**, *Angewandte Allergologie*, J.Ring (ed.), MMV Medizin Verlag, München.
18 T. V. Rajan **2003**, The Gell–Coombs classification of hypersensitivity reactions: a re-interpretation, *Trends Immunol.* 24, 376–379.
19 P. G. H. Gell, R. R. A. Coombs **1963**, The classification of allergic reactions underlying disease. In *Clinical Aspects of Immunology*, R. R. A. Coombs, P. G. H. Gell (eds.), Blackwell Science, Oxford.
20 J. Gutermuth, M. Ollert, J. Ring, H. Behrendt, T. Jakob **2004**, Mouse models of atopic eczema critically evaluated, *Int. Arch. Allergy Immunol.* 135, 262–276.
21 F. Alessandrini, T. Jakob, A. Wolf, E. Wolf, R. Balling, M. Hrabe de Angelis, J. Ring, H. Behrendt **2001**, ENU mouse mutagenesis: Generation of mouse mutants with aberrant plasma IgE levels, *Int. Arch. Allergy Immunol.* 124, 25–28.
22 Z. Ovary **1958**, Passive cutaneous anaphylaxis in the mouse, *J. Immunol.* 81, 355–357.
23 Z. Ovary, O. G. Bier **1953**, Quantitative studies on passive cutaneous anaphylaxis in the guinea pig and its relationship to the Arthus phenomenon, *J. Immunol.* 71, 6–11.
24 H. Nagai, T. Sakurai, N. Inagaki, H. Mori **1995**, An immunopharmacological study of the biphasic allergic skin reaction in mice, *Biol. Pharm. Bull.* 18, 239–245.
25 A. K. Abbas, A. H. Lichtman **2003**, Immediate hypersensitivity. In *Cellular and Molecular Immunology*, A. K. Abbas, A. H. Lichtman (eds.), Elsevier, Philadelphia, 432–453.
26 N. Inagaki, S. Goto, H. Nagai, A. Koda **1986**, Homologous passive cutaneous anaphylaxis in various strains of mice, *Int. Arch. Allergy Appl. Immunol.* 81, 58–62.
27 N. Inagaki, T. Miura, H. Nagai, A. Koda **1992**, Active cutaneous anaphylaxis (ACA) in the mouse ear, *Jpn J. Pharmacol.* 59, 201–208.
28 K. Zhang, C. L. Kepley, T. Terada, D. Zhu, H. Perez, A. Saxon **2004**, Inhibition of allergen-specific IgE reactivity by a human Ig Fcgamma-Fcepsilon bifunctional fusion protein, *J. Allergy Clin. Immunol.* 114, 321–327.
29 J. Kitaura, J. Song, M. Tsai, K. Asai, M. Maeda-Yamamoto, A. Mocsai, Y. Kawakami, F. T. Liu, C. A. Lowell, B. G. Barisas, S. J. Galli, T. Kawakami **2003**, Evidence that IgE molecules mediate a spectrum of effects on mast cell survival and activation via aggregation of the FcepsilonRI, *Proc. Natl Acad. Sci. USA* 100, 12911–12916.
30 F. T. Liu, J. W. Bohn, E. L. Ferry, H. Yamamoto, C. A. Molinaro, L. A. Sherman, N. R. Klinman, D. H. Katz **1980**, Monoclonal dinitrophenyl-specific murine IgE antibody: preparation, isolation, and characterization, *J. Immunol.* 124, 2728–2737.
31 Y. A. Mekori **2004**, The mastocyte: the "other" inflammatory cell in immunopathogenesis, *J. Allergy Clin. Immunol.* 114, 52–57.
32 S. Klemm, J. Gutermuth, L. Hultner, T. Sparwasser, H. Behrendt, C. Peschel, T. W. Mak, T. Jakob, J. Ruland, **2006**, The Bcl10/Malt1 Complex Segregates, FcεRI-Mediated NF-κB Activation and Cytokine Production from Mast Cell Degranulation and Leukotriene Synthesis. *J. Exp. Med.* 203, 337–347.
33 E. Adelroth **1998**, How to measure airway inflammation: bronchoalveolar lavage and airway biopsies, *Can. Respir. J.* 5(Suppl. A), 18A–21A.
34 H. Tanaka, T. Masuda, S. Tokuoka, M. Komai, K. Nagao, Y. Takahashi, H. Nagai **2001**, The effect of allergen-induced airway inflammation on airway remodeling in a murine model of allergic asthma, *Inflamm. Res.* 50(12), 616–624.
35 A. Wanner **1990**, Utility of animal models in the study of human airway disease, *Chest* 98(1), 211–217.
36 G. W. Hunninghake, J. E. Gadek **1979**, Inflammatory and immune processes in the human lung in health and disease: evaluation by bronchoalveolar lavage, *Am. J. Pathol.* 97(1), 149–206.
37 S. L. Ewart, D. Kuperman, E. Schadt, C. Tankersley, A. Grupe, D. M. Shubitowski, G. Peltz, M. Wills-Karp **2000**, Quantitative trait loci controlling allergen-induced airway hyperresponsiveness in inbred

mice, *Am. J. Respir. Cell. Mol. Biol.* 23(4), 537–545.

38 L.S. van Rijt, H. Kuipers, N. Vos, D. Hijdra, H. C. Hoogsteden, B. N. Lambrecht **2004**, A rapid flow cytometric method for determining the cellular composition of bronchoalveolar lavage fluid cells in mouse models of asthma, *J. Immunol. Methods* 288(1–2), 111–121.

39 U. Herz, A. Braun, R. Rückert, H. Renz **1998**, Various immunological phenotypes are associated with increased airway responsiveness, *Clin. Exp. Allergy* 28(5), 625–634.

40 K. Shinagawa, M. Kojima **2003**, Mouse model of airway remodeling: strain differences, *Am. J. Respir. Crit. Care Med.* 168(8), 959–967.

41 L. Smith, E. R. McFadden Jr **1995**, Bronchial hyperreactivity revisited, *Ann. Allergy Asthma Immunol.* 74(6), 454–469.

42 E. Hamelmann, J. Schwarze, K. Takeda, A. Oshiba, G.L. Larsen, C. G. Irvin, E. W. Gelfand **1997**, Noninvasive measurement of airway responsiveness in allergic mice using barometric plethysmography, *Am. J. Respir. Crit. Care Med.* 156(3), 766–775.

43 U. Neuhaus-Steinmetz, T. Glaab, A. Daser, A. Braun, M. Lommatzsch, U. Herz, J. Kips, Y. Alarie, H. Renz **2000**, Sequential development of airway hyperresponsiveness and acute airway obstruction in a mouse model of allergic inflammation, *Int. Arch. Allergy Immunol.* 121(1), 57–67.

44 R. Vijayaraghavan, M. Schaper, R. Thompson, M. F. Stock, Y. Alarie **1993**, Characteristic modifications of the breathing pattern of mice to evaluate the effects of airborne chemicals on the respiratory tract, *Arch. Toxicol.* 67, 478–490.

45 T. Morakata, J. Ishikawa, T. Yamada **2000**, Antigen dose defines T helper 1 and T helper 2 responses in the lungs of C57BL/6 and BALB/c mice independently of splenic responses, *Immunol. Lett.* 72, 119–126.

46 E. Hamelmann, K. Tadeda, A. Oshiba, E. W. Gelfand **1999**, Role of IgE in the development of allergic airway inflammation and airway hyperresponsiveness-a murine model, *Allergy* 54, 297–305.

47 Y. Zhang, W. J. Lamm, R.K. Albert, E. Y. Chi, W. R. Jr. Henderson, D. B. Lewis **1997**, Influence of the route of allergen administration and genetic background on the murine allergic pulmonary response, *Am. J. Respir. Crit. Care Med.* 155, 661–669.

48 K. Sakai, A. Yokoyama, N. Kohno, H. Hamada **2001**, Prolonged antigen exposure ameliorates airway inflammation but not remodeling in a mouse model of bronchial asthma, *Int. Arch. Allergy Immunol.* 126, 126–134.

49 C. A. Janeway, P. Travers, M. Walport, M. J. Shlomchik **2005**, The immune system in health and disease. In *Immunobiology*, 6th edn, Garland Publishing, New York.

50 D.L. Sylvestre, J.V. Ravetch **1994**, Fc receptors initiate the Arthus reaction: redefining the inflammatory cascade, *Science* 265, 1095–1098.

51 D. L. Sylvestre, J. V. Ravetch **1996**, A dominant role for mast cell Fc receptors in the Arthus reaction, *Immunity* 5, 387–390.

52 U. Baumann, J. Kohl, T. Tschernig, K. Schwerter-Strumpf, J. S. Verbeek, R. E. Schmidt, J. E. Gessner **2000**, A codominant role of Fc gamma RI/III and C5aR in the reverse Arthus reaction, *J. Immunol.* 164, 1065–1070.

53 U. Baumann, N Chouchakova, B. Gewecke J. Kohl, M. C. Carroll, R. E. Schmidt, J. E. Gessner **2001**, Distinct tissue site-specific requirements of mast cells and complement components C3/C5 a receptor in IgG immune complex-induced injury of skin and lung, *J. Immunol.* 167, 1022–1027.

54 A. J. Szalai, S. B. Digerness, A. Agrawal, J. F. Kearney, R. P. Bucy, S. Niwas, J. M. Kilpatrick, Y. S. Babu, J. E. Volanakis **2000**, The Arthus reaction in rodents: species-specific requirement of complement, *J. Immunol.* 164, 463–468.

55 H. Pfister, M. Ollert, L. F. Frohlich, L. Quintanilla-Martinez, T. V. Colby, U. Specks, D. E. Jenne **2004**, Antineutrophil cytoplasmic autoantibodies against the murine homolog of proteinase 3 (Wegener autoantigen) are pathogenic *in vivo*, *Blood* 104, 1411–1418.

56 T. Jakob, M. C. Udey **1999**, Epidermal Langerhans cells: from neurons to nature's adjuvants, *Adv. Dermatol.* 14, 209–258.

57 OECD, Skin Sensitization: Local lymph node assay. *OECD Guidelines for the Testing of Chemicals – Online Edition*; ISSN: 1607310X; www.oecd.org.

58 National Institute of Environmental Health Sciences and National Institutes of Health **1999**, The Murine Local Lymph Node Assay, *NIH Publication* No. 99, 4494.

59 B. Homey, C. von Schilling, J. Blumel, H. C. Schuppe, T. Ruzicka, H. J. Ahr, P. Lehmann, H. W. Vohr **1998**, An integrated model for the differentiation of chemical-induced allergic and irritant skin reactions, *Toxicol. Appl. Pharmacol.* 153, 83–94.

60 H.W. Vohr, J. Blumel, A. Blotz, B. Homey, H. J. Ahr **2000**, An intra-laboratory validation of the Integrated Model for the Differentiation of Skin Reactions (IMDS): discrimination between (photo)allergic and (photo)irritant skin reactions in mice, *Arch. Toxicol.* 73, 501–509.

61 T. Jakob, J. Ring, M. C. Udey **2001**, Multistep navigation of Langerhans/dendritic cells in and out of the skin, *J. Allergy Clin. Immunol.* 108, 688–696.

62 T. Jakob, C. Traidl-Hoffmann, H. Behrendt **2002**, Dendritic cells-the link between innate and adaptive immunity in allergy, *Curr. Allergy Asthma Rep.* 2, 93–95.

63 A. A. Gaspari, S. I. Katz **1991**, Contact hypersensitivity. In *Current Protocols in Immunology*, J. E. Coligan, A. M. Kruisbeek, D.H. Margulies, E. M. Schevach, W. Strober (eds.), Wiley, New York, 4.2.1–4.2.5.

64 M. Furue, K. Tamaki **1985**, Induction and suppression of contact hypersensitivity, *J. Invest. Dermatol.* 85, 139–142.

65 T. Tamaki, H. Fujiwara, S. I. Katz **1981**, The role of epidermal cells in the induction and suppression of contact hypersensitivity, *J. Invest. Dermatol.* 76, 275–278.

66 S. L. Kunkel, N. W. Lukacs, R. M. Strieter, S. W. Chensue **1998**, Animal models of granulomatous inflammation, *Semin. Respir. Infect.* 13, 221–228.

67 B. C. Chiu, C. M. Freeman, V. R. Stolberg, E. Komuniecki, P. M. Lincoln, S. L. Kunkel, S. W. Chensue **2003**, Cytokine-chemokine networks in experimental mycobacterial and schistosomal pulmonary granuloma formation, *Am. J. Respir. Cell. Mol. Biol.* 29, 106–116.

12
Eye Disorders

Claudia Dalke, Oliver Puk, Angelika Neuhäuser-Klaus, Jack Favor, and Jochen Graw

12.1
Introduction

Blindness as consequence of congenital or age-related eye disorders affects more than 50 million people worldwide. In particular, congenital eye diseases such as anophthalmia, microphthalmia, colobomata or cataracts affect 1 million children and are mainly caused by mutations. On the other hand, some age-related ocular disorders such as senile cataracts and also some forms of retinal degenerations are a common product of genetic predisposition and environmental influences (e. g. UV light, nutrition). In some other disorders (e. g. glaucoma), the genetic aspect is more obvious; however, several genes seem to be involved. Because this is a difficult clinical situation, there is a need for animal models to further understand the various genetic and environmental influences. One of the prerequisites of a successful model is the detailed description of its genome, as is the case for the mouse, however it is equally important to be able to generate phenotypes in the model organism that are as similar to the clinical situation in humans as possible. Therefore, several laboratories have developed various test systems with which to determine the size of the entire mouse eye and its particular segments (e. g. cornea, anterior chamber, lens; using the optical low coherence interferometry), to analyze eye disorders in the mouse including cataracts and anterior segment dystrophies (using the slit lamp), retinal degeneration (using an ophthalmoscope and electroretinography) and glaucoma (using invasive methods to determine intraocular pressure). However, the existing test systems need to be standardized for better comparison of the findings in different mouse strains as well as in mouse mutants. Moreover, there is also an obvious need for non-invasive methods to determine the intraocular pressure and to measure the papillary reflex as indicator for the functional integrity of the entire visual system.

12.2
Medical and Biological Relevance of Eye Disorders

Blindness is a disorder that dramatically affects the lives of 50 million people worldwide and is caused by a variety of ocular diseases. The overall prevalence of blindness

Standards of Mouse Model Phenotyping. Edited by Martin Hrabé de Angelis, Pierre Chambon, and Steve Brown

ISBN: 3-527-31031-2

is a function of age, mainly affecting the older section of any population. However, there is also a remarkable number of cases of congenital or juvenile blindness (approximately 1 million children worldwide). Among the underlying causes of blindness, diseases such as microphthalmia, coloboma and cataracts are frequently responsible for congenital ocular disorders, whereas cataracts, glaucoma and retinal disorders tend to be age-related conditions. Among these, cataracts are responsible for half of all cases of blindness [1]. There is also a difference in the etiology of blindness between congenital or juvenile cases and age-related disorders: congenital cases are mainly of genetic origin (or in some instances also caused by teratogenic effects), whereas in age-related blindness the relative impact of genetics decreases and the impact of environmental influences (e. g. UV light or nutritional effects) increases correspondingly.

Mouse models are a major tool in the elucidation of the genetic and biochemical mechanisms of ocular disorders. A rapidly growing number of mouse mutants suffering from various types of eye diseases is now available (for a recent review see [2]). However, the quality of an animal model depends on its close similarity to the clinical situation in patients. This chapter will provide an overview of the similarity of ocular disorders between mouse and man, mainly from a genetic point of view, and will review the methods used for the detection of eye disorders in the mouse.

12.3 Eye Disorders in Mouse and Man

12.3.1 Mutations Affecting Early Eye Development Leading to Anophthalmia or Microphthalmia

A variety of syndromes which are associated with congenital anophthalmia or microphthalmia have been described in both mouse and man. These disorders are mainly caused by mutations in genes affecting early eye development. One of the most important genes is *Pax6*; the encoded protein belongs to a family of transcription factors, which are characterized by a paired-type homeodomain. Mouse *Pax6* heterozygous mutants have small eyes [3–5], but homozygous mutants have only the remnants of ocular tissues and die shortly after birth as a result of nasal dysfunction [6]. In humans, *PAX6* mutations mainly cause aniridia, a panocular disorder, and less commonly, isolated cataracts, macular hypoplasia, keratitis and Peters' anomaly (for an overview see [7]). As in the mouse, homozygous loss of *PAX6* function in humans is lethal to the neonate. Genotype and phenotype information regarding human *PAX6* mutations has been assembled in a freely accessible database (http://pax6.hgu.mrc.ac.uk).

A second important group of transcription factors in the eye is the *Pitx*-gene family, which also encodes a paired-like class of homeobox transcription factors. Several mutations in *Pitx2* have been shown to be associated with Axenfeld–Rieger Syndrome (a group of dominant disorders that affect the anterior segment of the eye). In mouse, deletions within the *Pitx3* promoter have been shown to cause the phenotype of the *aphakia* (*ak*) mouse mutant which lacks lenses and pupils [8, 9].

A third group of transcription factors are encoded by the *Fox* genes ("forkhead-box"; the name is based on the *Drosophila* gene *forkhead* [10]). Actually, two members of this family are responsible for severe congenital ocular disorders: *FoxC1* and *FoxE3*. Patients with Axenfeld–Rieger syndrome or iris hypoplasia harbored mutations in *FOXC1* [11], and *FoxC1* knockout mice also have anterior segment abnormalities that are similar to those reported in humans [12]. The penetrance of the clinical phenotype varies with the genetic background, indicating the influence of modulator(s). Additionally, two mutations within the DNA-binding domain of *FoxE3* were identified in *dyl* (*dysgenic lens*) mice [13], in which the lens vesicle fails to separate from the ectoderm causing the lens and cornea to fuse. Mutations in the human *FOXE3* gene are responsible for anterior segment ocular dystrophies and Peters' anomaly [14].

The "classical" microphthalmia phenotype was first described in the mouse by Hertwig [15]. Since that time, numerous mutant alleles have been identified and genetically characterized by various mutations in the gene coding for the microphthalmia-associated transcription factor Mitf, a basic helix–loop–helix leucine zipper (bHLHzip) transcription factor. The different alleles detected in the mouse are mainly recessive, but semi-dominant or dominant phenotypes have also been reported. Since the mutations affect particular cell types, which are derived from neural-crest melanocytes, the affected mice frequently also develop deafness owing to the lack of inner ear melanocytes [5, 16–18]. Mutations within the human *MITF* were estimated in about 20% of patients suffering from Waardenburg syndrome type II [19].

12.3.2 Cataracts

As outlined above, cataracts (i. e. lens opacities) are responsible for 50% of all cases of blindness worldwide. However, this is mainly true for blindness at an older age, but congenital cataracts also occur frequently. Since this phenotype can easily be observed in the mouse, a broad variety of mouse mutants have been identified and characterized at the molecular level within the last 10–20 years. Concurrently, increasing numbers of human mutations are also being characterized at the molecular level. According to experience with other age-related disorders (diabetes, Alzheimer etc), we can expect to understand the basic features of any disorder by looking at the more severe early-onset forms, in the case of eye disorders, congenital cataracts.

The analysis of congenital (mainly dominant) cataracts led to the discovery of an unexpectedly high number of mutations in genes coding for γ-crystallins, and to a lesser extent in those genes coding for α- and β-crystallins. Crystallins act as structural proteins mainly in the lens, however, they have also been detected in other ocular tissues and other organs (for a review see [2]). Although the crystallins are recognized as highly conserved proteins among vertebrate species, there are a large number of polymorphisms present not only in the human population [20, 21], but also in mouse strains [22]. However, these polymorphic sites are not distributed randomly, but are restricted to particular sites. It is tempting to speculate that these polymorphic sites will be correlated in the future to certain features of age-related cataracts.

Other structural proteins in the lens are represented by membrane proteins. The most prominent being the main intrinsic protein (MIP) which belongs to the family of aquaporins, and the gap-junction proteins which belong to the connexin families. Mutations in the corresponding genes *Mip*, *Gja3* and *Gja8* have been shown to be the cause of a wide variety of congenital, dominant cataracts in mouse and man (for a review see [2]).

12.3.3
Retinal Dysfunction and Degeneration

Retinal degeneration is a major cause of visual impairment and blindness in humans; the two most important forms of retinal degeneration are retinitis pigmentosa and age-related macular degeneration. In Germany, approximately 10 and 15% of recognized blindness is caused by these two forms [23]. Worldwide, the prevalence of retinitis pigmentosa is approximately 1 in 4000 of the population. The reasons for the progressive loss of photoreceptors are genetic abnormalities occurring in more than 100 genes affecting various ocular tissues (for reviews see [24, 24a] and corresponding web sites at http://eyegene.meei.harvard.edu and http://www.sph.uth.tmc.edu/RetNet). To understand the mechanisms of these disorders, animal models with malformed retinas or inherited retinal degeneration are increasingly being used.

The recessive mouse mutation *ocular retardation* (*or*) is one reason for a defect in retinal development. When this recessive mutation is homozygous it is characterized by blindness with obvious microphthalmia, a cataract lens, a thin retina that is morphologically poorly differentiated and the lack of an optic nerve. The *or* phenotype is caused by a mutation in *Chx10*—a gene that encodes a homeobox transcription factor [25]. A similar phenotype (microphthalmia, cataracts and severe abnormalities of the iris) has also been reported in families suffering from recessive mutations in the *CHX10* gene [26].

Another example of an eye disorder in the mouse that occurs very frequently and is present in a broad variety of laboratory strains of mice is characterized by early retinal degeneration (previous gene symbol *rd1*). It was characterized at the molecular level by a nonsense mutation in the *Pde6b* gene [27]. Detailed strain comparisons in our own laboratory showed that it is not only present in the C3H strain, but also at least in part in other strains such as 129/SvJ, SWR and FVB [28] (Dalke unpublished observations). Mutations in the human *PDE6B* gene have been found in patients suffering from recessive retinitis pigmentosa (OMIM 180072).

12.3.4
Glaucoma

Glaucoma can be regarded as a group of diseases that have a characteristic optic neuropathy as a common end-point, which is defined by both structural changes (appearance of the optic disc) and functional deficits (measured by visual field change), and is frequently accompanied by an abnormally high intraocular pressure (IOP). One reason for the high IOP are changes in the trabecular meshwork leading to an obstruction of the outflow of aqueous humor; this is particularly true for primary open-angle glaucoma. After cataracts, glaucoma is the second-most frequent ocular disorder affecting 6.7 million people worldwide (13% of blind people). There are some congenital or juvenile forms, but glaucoma mainly affects older people [1].

From a genetic point of view, only a few genes can be demonstrated to be involved in the formation of glaucoma. If glaucoma is associated with the Rieger syndrome or with an Axenfeld–Rieger anomaly, variations in the developmental control genes *PITX2* or *FOXC1* have also been shown to play causative role in this condition. In other families suffering from glaucoma, mutations in the *CYP1B1* or *GST* genes have been described to co-segregate with the occurrence of the disease. The most intensively studied gene in glaucoma research codes for Myocilin (gene symbol *MYOC*); *MYOC* is expressed both in the trabecular meshwork as well as in the optic nerve head (for a recent review see [29]). Mice with knockout mutations of *Myoc* show a similar phenotype to that seen in human patients [30]; in contrast, creation of homozygous $Cyp1b1^{-/-}$ mice did not lead to a phenotype with an equivalent among human patients [31].

12.4 Diagnostic Methods

12.4.1 History

The first mouse mutant suffering from an ocular disorder was identified by Chase in 1941 [32] as an anophthalmic mouse. This phenotype can be recognized by the naked eye and does not need the help of any instrumentation (recently, it has been shown that this phenotype is caused by a mutation in the developmental control gene *rax* [33]). In the following years, a few other eye mutants with severe phenotypes (anophthalmia, microphthalmia, cataract) were found by careful investigation.

However, the first *systematic* screening for mutations affecting the eye was initiated in the late 1970s, when Kratochvilova and Ehling [34] applied the slit lamp examination (Fig. 12.1 a) to the mouse for the first time. At that time, the assessment of genetic risk due to dominant mutations was the guiding force of interest. These mutations were first induced by ionizing radiation, but later also by a variety of chemicals culminating in the discovery that ethylnitrosourea (ENU) was the most powerful mutagen in the mouse [35]. Since that time, more than 200 independent mouse lines with ocular disorders have been established at the GSF Neuherberg [36] and are available for functional investigations into hereditary ocular disorders. Moreover, these numbers are still growing because of an ongoing ENU mutagenesis screen [37]. This approach will be complemented by gene-driven investigations in the search for insertional mutations via exon trap methods [38].

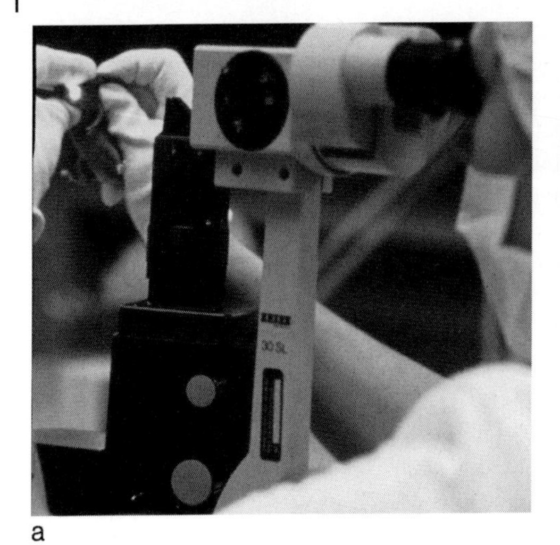
a

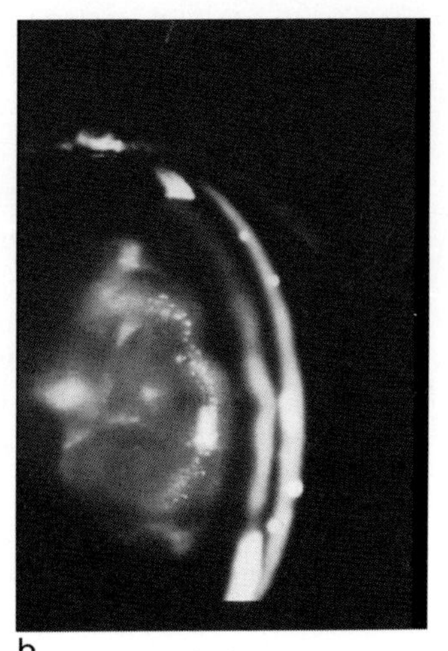
b

Fig. 12.1 Inspection of mice with a slit lamp. (a) After dilatation of the pupil with a drop of atropine, the mouse eye can be inspected using a slit lamp. The mouse is held in the beam of light and moved forward and backward until it focuses through the cornea and lens. (b) Photograph for documentation. The example shows an ENU-413 mouse, which was demonstrated to be allelic to mutations affecting the *Cryg* genes (*Cat2*^{4Neu}).

In the meantime, other centers for mouse genetics have also adopted this approach, first the MRC at Harwell in the UK [39] and later the Oak Ridge Center in Tennessee, USA [40]. Actually, mutagenesis screens for ocular disorders are a worldwide state-of-the-art concept in mouse phenotyping, however, as a variety of methods is used this leads to variations in the phenotypes discovered.

12.4.2
Routine Methods

12.4.2.1 Slit Lamp Analysis

The *slit lamp* is widely used in clinical ophthalmology to detect dysmorphologies in the anterior segment of the eye including the cornea, iris and lens. The slit-lamp examination allows the detection of even minor structural changes in the lens, such as anterior or posterior suture opacities and small areas of nuclear, cortical or lamellar opacity (Fig. 12.1 b; for an overview see [36]). Therefore, it is possible to compare different types of cataracts among different mutant lines and the development and changes in opacities during the aging process. Before or during inspection of the eye using the slit lamp, other features of ophthalmic disorders can also become apparent, particularly the presence of microphthalmia varying degree (however, the size of the eye cannot be quantified in this way). Using this system, a large number of small-eye mutants have been discovered, which were eventually identified as *Pax6* mutations [4, 5, 41].

12.4.2.2 **Length of the Axis**

Determination of the axis length in the mouse is very important. However, since the mouse eye is very small compared to a human eye (or even compared to a rat eye), it is very difficult to measure the size of a mouse eye *in vivo*. The standard technique for *in-vivo* biometry in vertebrate eyes is ultrasonography. However, due to the steep curvature of the mouse cornea, the contact area of the ultrasound transducer is too small to transmit sufficient sound energy into the eye for a satisfactory signal-to-noise ratio.

Recently, a novel non-contact pachymetric technique was described, which is based on optical low coherence interferometry. ("OLCI" – implemented in the "AC-Master", Carl Zeiss Meditec, Jena, Germany). It was originally developed for measurement of the anterior depth and of lens thickness in living human eyes prior to surgery or for follow-up studies after surgery. The principle of OLCI is based on a Michelson interferometer. The light source is a low coherence superluminescent laser diode emitting infrared light with peak emission at 850 nm and a half-band width of 10 nm. The laser beam is split using several mirrors, and finally two coaxial beams of about 50 μm in diameter propagate to the eye where they are reflected off from the cornea, the lens and the pigmented retinal epithelium. Interference between the beams occurs only if their optical path lengths are matched within the coherence length. Any interference of this type is detected by a photo cell and recorded as a function of the displacement of a movable mirror. Due to the use of coaxial beams, the measurements are largely insensitive to longitudinal eye movements. The scanning time of the moving mirror is about 0.3 s, and the resolution of the system is limited both by the coherence length and by the precision with which the position of the moving mirror can be controlled (for further technical details see http://www.meditec.zeiss.de).

This system has recently been applied to anaesthetized mice (Fig. 12.2) [55]. The measurement in our laboratory revealed some significant differences among various laboratory strains. For mutagenesis studies, mainly C3H or C57BL/6 mice (or corresponding sub-strains) have been used. Because of the absence of the *Pde6b*^{rd1} allele, we prefer the strain C57BL/6 is and determined the mean ± 3SD-interval for the four eye-size parameters. Assuming normal distribution, 99.7% of 11-week old C57BL/6 wild-type males should exhibit axial lengths between 3.430–3.622 mm, corneal thickness 72.47–130.87 μm, anterior chamber depths between 0.473–0.581 mm, and lens thickness 1.999–2.149 mm. Concerning wild-type females, 99.7% of axial lengths, corneal thickness, anterior chamber depths and lens thickness and should be ranged between 3.377–3.563 mm, 79.63–122.71 μm, 0.468–0.552 mm, and 1.999–2.119 mm, respectively. Since this system gives high resolution biometrical data of mouse eyes in vivo, it is currently used in the German Mouse Clinic (http://www.mouseclinic.de) as one of the initial methods for phenotyping mouse vision.

12.4.2.3 **Funduscopy**

Funduscopy is a widely used method for the visual inspection of the back of the eyeball (fundus) including evaluation of the condition of the retina, optic disc, choroids

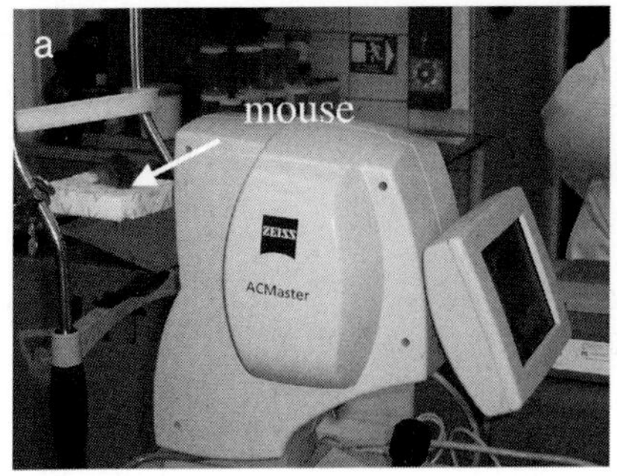

Fig. 12.2 Set-up for measuring eye size. (a) The "ACMaster" during measurement of a mouse eye. The anesthetized mouse (arrow) it is positioned on an adjustable platform, which was attached to the chin-rest of the device. (b) In this low coherence interferogram of the mouse eye, the intensity of the peaks is plotted versus the optical path length. The origin of the reflections of the cornea layers, the lens and the pigmented retinal epithelium are shown. The mouse was 13 weeks old and of a mixed genetic background (CD1 and C57BL/6J).

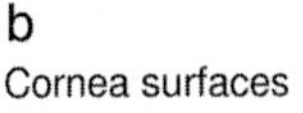

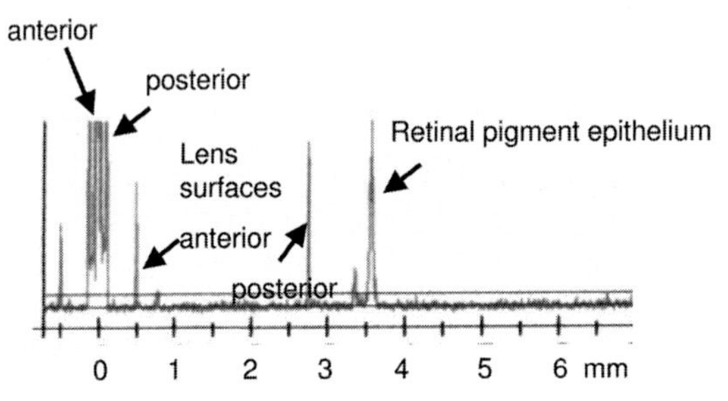

and blood vessels. Retinal malformations such as for example, retinal vessel constriction, disturbance of the retinal pigment epithelial, drusen or deposits, anomalies of the optic nerve head or abnormal vascularization can easily be recognized. Vessel attenuation and optic atrophy are common signs of retinal disorders (Fig. 12.3; [42]).

12.4.2.4 **Electroretinography (ERG)**

Electroretinography (ERG) is a non-invasive electrophysiological method for detecting the collective response of large numbers of retinal cells due to visual stimulation. The basic principle uses an electrode at the corneal surface, one reference electrode and a ground electrode. This method measures the electric potential produced by retinal cells during transmission of electric signals after perception of light. In general, the recorded responses consist of three major deflections known as a-, b- and c-waves.

The first (negative) peak, referred to as the a-wave, primarily reflects light-induced electrical changes in the rod and cone photoreceptor cells. The second, positive b-wave is believed to be generated by the activity of the neural cells of the retina; high frequency oscillatory responses can also be detected during the course of the b-wave, possibly reflecting the depolarizing responses of amacrine cells. The c-wave is generated predominantly in the pigment epithelium. Standard values for the a- and b-waves in eight different mouse strains have been reported recently [28]; two examples are given in Fig. 12.4.

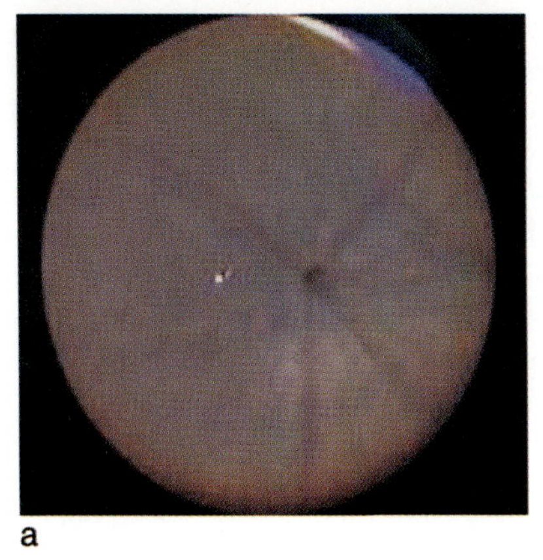

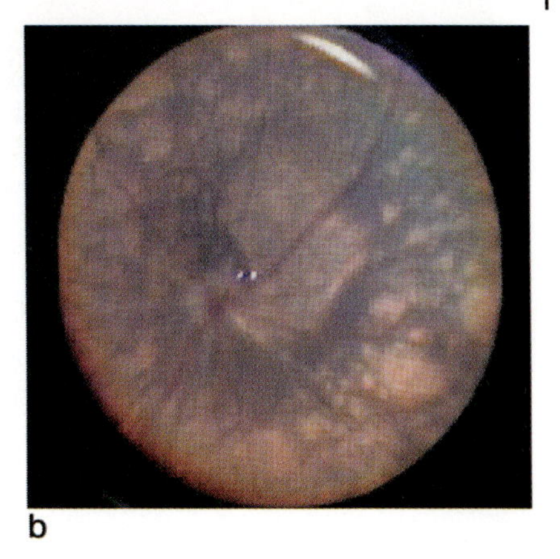

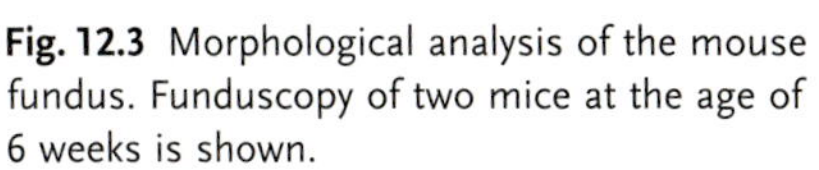

Fig. 12.3 Morphological analysis of the mouse fundus. Funduscopy of two mice at the age of 6 weeks is shown.
(a) In the wild-type C57BL/6 mouse, only faint focal patches of lighter pigment are evident.
(b) The C3H mouse suffers from retinal degeneration caused by a nonsense mutation in the *Pde6b* gene. Pigment patches and vessel attenuation are visible.

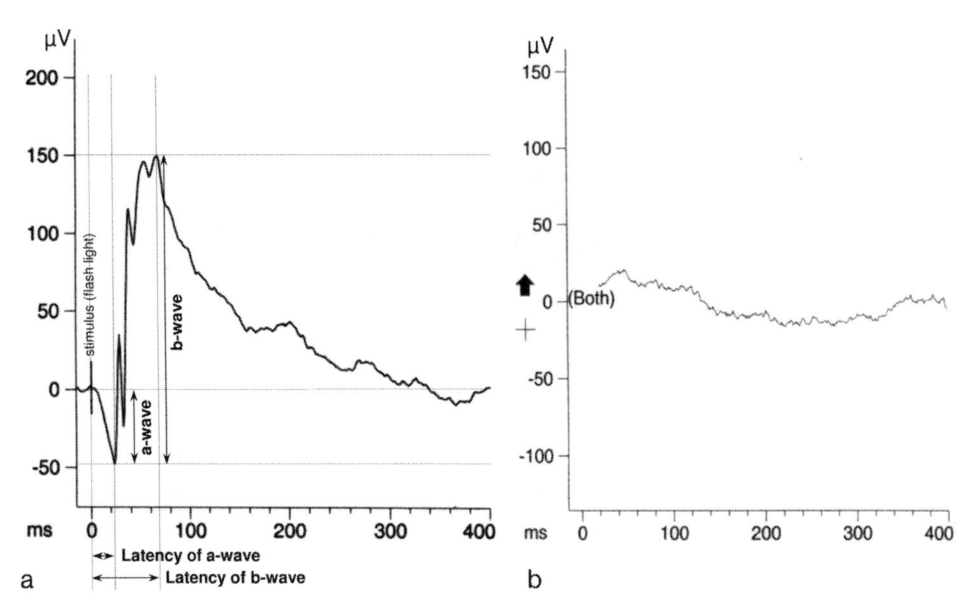

Fig. 12.4 Functional analysis of the mouse retina by electroretinography (ERG). Typical electroretinography is demonstrated for a wild-type C57BL/6 mouse (a) and a C3H mouse suffering from retinal degeneration (b) at the age of 6 weeks using light flashes of 12,500 cd/m^2. The latency period (in ms) is defined from the start point to the maximum peak of the a-wave or b-wave; the a-wave is defined as the maximal negative voltage, and the b-wave ranges from the maximal negative voltage to the maximal positive voltage.

The classical paradigm for a mouse mutant suffering from retinal degeneration is the mouse strain C3H (see Section 12.3.3). This strain is characterized by a complete loss of any ERG response at the age of 6 weeks (Fig. 12.4 b). However, in human patients retinal disorders more frequently occur at an older age. Therefore, the choice of an appropriate age for mouse screening by ERG is a compromise between the particular phenotypes that are of interest to the observer and the size of the animal facility.

ERG is a powerful tool which can be used not only in simple functional screens for retinal dystrophies, but can also be used for more detailed characterization of already-established mutant lines. ERG can be used in dark-adapted (scotopic) or light-adapted mice (photopic). The scotopic measurement mainly detects rod function, whereas the photopic condition also provides information regarding cone function. However, the contribution of the cones to the ERG components in mouse is much smaller than that in human. In humans, the majority of cones are concentrated in specialized regions (the fovea) in which they are packed together as tightly as physically possible to allow good daylight vision. In the modern mouse, which is a nocturnal animal, the retina is dominated by rods; the cones are small in size and represent just 3–5 % of the photoreceptors [43]. Therefore, under dark-adapted conditions the a-wave will result mostly from the activity of rods [44, 45]. However, mouse mutants suffering from cone defects are also available; this particular dysfunction can be detected under photopic conditions only (e. g. in *Cng3*$^{-/-}$ [46]). In addition to the analysis of a-, b- and c- waves under scotopic and photopic conditions, the intensity (e. g. from 0.0125 to 500 cd s/m^2) and frequency (e. g. in the flicker-ERG with quick photo stimuli up to a pulse rate of up to 70 Hz) of the light stimulus can also be varied. The power of such a detailed characterization has been demonstrated recently by an allelic series of five *Mitf* mutants exhibiting significant differences in their various ERG responses [47].

12.4.2.5 **The Visual Tracking Drum or Optomotor Drum**

The visual tracking drum (Fig 12.5) [48] or the optomotor drum [49] is a simple device for differentiating between mice which can see normally, and mice suffering from a severe loss of visual function. It consists of a motorized, rotating drum, whose inner walls can be lined with different cards on which black and white vertical stripes of known frequency are printed. The mouse is placed on a stationary platform in the center of the drum and observed from above while the drum is moving. In this way, any movements made by the mouse is recorded by a video tracking system. With this method the vision of the mouse can be quantified by varying the grade of stimulus and in addition, the test is not time consuming and the mouse does not need training. In a proof-of principle experiment, Thaung et al. [48] investigated the responses of 204 mice in an optomotor drum.

Head tracking of the mouse correlates with the phenomenon of nystagmus, which refers to the rhythmic, oscillating movement of the eyes. The to-and-from motion is generally involuntary; it can be a normal physiological response or a result of a pathologic problem. Railway nystagmus (also called optokinetic nystagmus) is a physiological type of nystagmus. It occurs when a person is on a moving

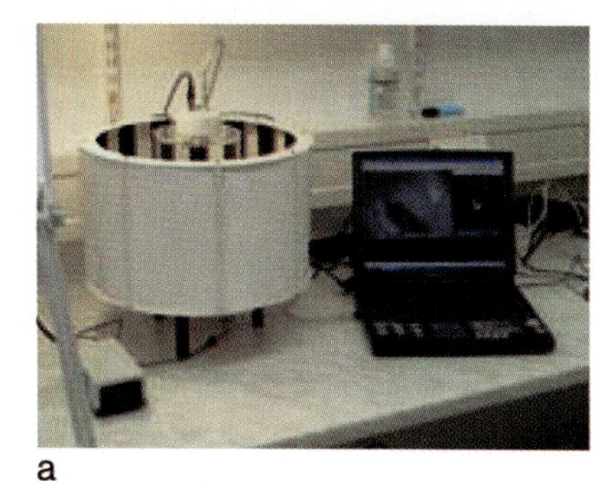
a

b

Fig. 12.5 The optokinetic drum.
(a) The drum is connected with a computer to analyze the movements.
(b) A mouse sits on the platform and follows the moving black-and white bars.

train (hence the term "railway") and is watching a stationary object which appears to be going by. The eyes slowly follow the object and then quickly jerk back to start the process again. In clinic, this phenomenon is widely used to check vision in infants.

12.4.2.6 Measurement of Intraocular Pressure

The measurement of intraocular pressure is important in the development of mouse models for glaucoma. Since glaucoma is often associated with elevated intraocular pressure (IOP), enhanced IOP is recognized as a leading symptom of this condition in clinic. Because of the large size of the human eye, non-invasive tonometric methods can be routinely used in ophthalmology. In contrast, measurement of IOP in the small mouse eye is not as easy; an invasive method has been established in the Jackson Laboratory only [49]. Based on these studies, Savinova et al. [50] carried out an extended survey of IOP in 30 mouse strains: the average IOP ranged from ~ 10 to 20 mmHg (RIIIS/J and BALB/c mice had low IOP values, and CBA and AKR mice high values; C3H, 129 and C57BL/6 mice were in the middle range). Gender does not typically affect IOP, and aging results in a decrease of IOP in some strains. Most strains tested exhibited a diurnal rhythm with IOP being highest during the dark period of the day.

One of these strain differences could be attributed to a genetic modification, since albino C57BL/6J mice homozygous for a tyrosinase mutation ($Tyr^{c\text{-}2J}$) have higher IOPs than their pigmented counterparts. Additionally, some mouse mutants with elevated IOP have been already been characterized at the molecular level (e. g. $Lepr^{db}$, $Gpnmb^{R150X}$) [50, 51]. Moreover, knockout mutants have been characterized by increased IOP (e. g. $BMP4^{-/-}$ on the C57BL/6J background [52]). These studies demonstrated that mice are indeed a practical and powerful experimental system with which to study the genetics of IOP regulation.

12.4.2.7 Histological Analysis

Phenotypic characterization of the visual system is not complete without a detailed histological analysis. If a mutant line has been established or all other *in-vivo* parameters have been tested, histology will help to explain the results obtained in other test systems. Examples are given in Fig. 12.6. Moreover, the rapid histological examination of the mouse eye has also been suggested, including the removal of one eye from an anesthetized mouse and the preparation of frozen sections; in this case the mouse can be still used for genetic investigations [53].

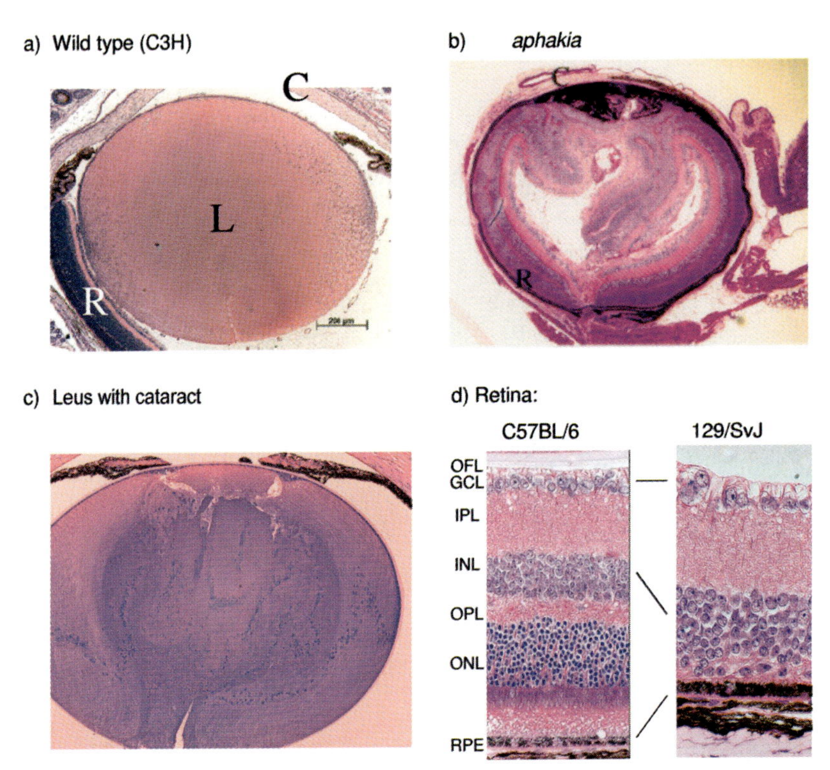

Fig. 12.6 Eye histology. Histological sections of mouse eyes are shown. (a) Wild-type C3H at the age of 7 days. (b) A general overview of the eyeball of the recessive mutant *aphakia* at the age of 2 weeks. In this particular mutant the lens is missing as a result of two mutations in the *Pitx3* promoter, and the entire eyeball is filled with badly differentiated retinal tissue. (c) A cataract mutant (*Crygc*^{MNU8}) exhibits clear clefts and vacuoles in the lens together with the presence of the usually degraded fiber cell nuclei. (d) The degenerated retina of a 129/SvJ mouse suffering from the same retinal degeneration as C3H mice is compared to a retina from a wild-type C57BL/6 mouse. C, cornea; L, lens; R, retina. GCL, ganglion cell layer; INL, inner nuclear layer; IPL, inner plexiform layer; OFL, optic fiber layer; ONL, outer nuclear layer; OPL, outer plexiform layer; RPE, retinal pigment epithelium.

12.4.3
Methods in Development

12.4.3.1 The Scheimpflug Camera

The Scheimpflug camera (e. g. Topcon SL45 (Tokyo, Japan) equipped with a black-and-white camera) can be used to quantify lens opacities and to track the progression of opacity with age in a non-invasive manner. Prior to Scheimpflug photography, whether using the vertical or horizontal slit position, a drop of atropine (1 %) should be instilled into the eyes. The optical sections thus obtained are then analyzed by linear densitometry in multiple scans parallel to the optical axis of the eye. Densitometric data will reveal characteristic light scattering patterns in distinct layers of the lenses showing not only the development of distinct types of cataracts but more importantly changes in light scattering at the pre-cataract stages. The density profile of the lenses proved to be a suitable marker for the stability and complete expression of a mutation affecting the transparency of the anterior eye segment [54]. An experienced investigator together with two technicians can screen ~ 80 mice per day (A. Wegener, personal communication).

12.4.3.2 Measurement of Intraocular Pressure

Methods of measuring the intraocular pressure by non-invasive means are currently being developed, since the invasive method is highly stressful for the mice and requires extremely sophisticated instrumentation as well as very experienced investigators. At least two non-invasive methods for the determination of IOP in the mouse have been described in the past few years. The first is *induction-impact (I/I) tonometry* based upon the rebound principle, i. e. bouncing a magnetized probe onto the eye and using its motion parameters to determine IOP [56]. The measurement of IOP using this concept is based on contacting the eye with a probe and detecting the motion as the probe collides with the eye and bounces back. The motion parameters of the probe vary according to the eye pressure and are used to calculate IOP. A prototype instrument has been constructed for the measurement of IOP in the mouse eye, and its ability to accurately and reliable measure IOP has been tested by comparing the measurements with those obtained using manometric IOP determinations in cannulated mouse eyes *ex vivo*. Using this method, an IOP value of 9.8 ± 3.9 mmHg (mean ± SD) was determined in eight ketamine-anesthetized adult C57BL/6 mice (3 to 6 months old); there was a reduction in IOP (7.6 ± 1.9 mm Hg) when ketamine–xylazine–acepromazine was used for anesthesia [57]. Comparing these data to the values observed with the invasive method, I/I-tonometry gives lower IOP values.

The second method described for the mouse uses the *Tonopen XL* (Medtronic Solan, Jacksonville, USA), which is also frequently used in clinic as a simple and rapid method to determine the IOP in humans. It is based upon an aplanation technique: the front end of the apparatus consists of two concentric cylinders, whereby the inner cylinder is connected to a non-moving force transducer and the outer cylinder defines the cornea surface that contacts the inner cylinder. In order to achieve a correct measurement, the complete area of the inner cylinder (1 mm diameter) should cover the cornea and should at least touch the inner rim of its outer cylinder (1.2 mm diameter, outer diameter of the outer cylinder is 3 mm). To access a plane surface of 1.2 mm in diameter, the cornea of an average adult mouse eye (3 mm diameter) should be compressed by 125 μm. This is possible without forcing the cornea (~ 75–100 μm thickness) against the lens, because the depth of the 3-mm diameter anterior chamber of the mouse eye is ~ 350 μm [58, 59].

Using this method, the mean IOP (± SEM) of eight ketamine/xylazine-anesthetized C57BL/6 mice was reported to be 12.36 ± 0.32 mmHg (10–27 individual readings for each mouse; [58]). This value is closer to that for adult C57BL/6J mice than the value obtained using I/I-tonometry, however, it is also lower than those values cited above for the cannulated method. To reduce the number of spurious readings (< 6 mmHg or > 80 mmHg), which were not included in the data set, the Tonopen method was used without the protective latex cup. However, the small variation reported by Reitsamer and co-workers [58] could not be reproduced [60]. Therefore, a rapid and reproducible, non-invasive method for IOP measurement in the mouse still remains to be developed.

12.4.3.3 **Pupillary Reflex**

The pupillary reflex is the reduction of pupil size in response to light. Lack of the pupillary reflex or an abnormal pupillary reflex can be caused by optic nerve damage, oculomotor nerve damage or other changes in the visual reflex pathway. For a variety of ophthalmological investigations, it is necessary to prevent the papillary reflex by atropine. In total, this reflex pathway consists of retinal ganglion cells, which convey information from the photoreceptors to the optic nerve which connects to the pretectal nucleus of the high midbrain. It bypasses the lateral geniculate nucleus and the primary visual cortex. From the pretectal nucleus neurons send axons to neurons of the Edinger-Westphal nucleus whose axons run along both the left and right oculomotor nerves. Oculomotor nerve axons synapse on ciliary ganglion neurons whose axons innervate the constrictor muscle of the iris.

The pupillary reflex has been studied in several strains of the mouse with infrared observations of pupil size (for review see [53]). This reflex is easily measured and occurs in response to a very dim stimulus. Recently, a novel device was reported consisting of a video image-processing program detecting the pupil and measuring its diameter at 25 Hz sampling rate. To stimulate, an arrangement of green LEDs, which was attached to the recording video camera, could be flashed for 40 ms [60a]. Since the pupillary reflex is dependent on a variety of parameters in the eye as well as in the central nervous system, it might be used only as a part in the entire set of visual screening methods.

12.4.3.4 **Scanning-laser Ophthalmoscopy**

Scanning-laser ophthalmoscopy is a technique for confocal imaging of the mouse eye *in vivo*. The use of lasers of different wavelengths allows obtaining information about specific tissues and layers due to their reflection and transmission characteristics. In particular, short wavelengths (blue-green) are more absorbed by the retinal pigment epithelium (RPE) and the choroid than near infrared. Therefore, blue-green laser can be used to analyse retinal structures, and the near infrared lasers to look for details in the RPE and the choroid. In addition, fluorescent dyes excitable in the blue and infrared range offer a unique access to the vascular structures associated with each layer. The major advantage of this method is to analyse retinal vessels to detect changes in the formation of the capillary network (like in mouse models for the Norrie disease) or to look for neovascularization in the retina and the choroids (like in *Crumbs-1* knockout mice). Additionally, morphological changes in the RPE can also be observed (like in deficiency for the gene *Rpe65*). For a recent overview for this method and its application see [60b]); it is an important new tool to study retinal alterations in the mouse.

12.4.4
Future Combinations of First and Secondary Screens for Vision Phenotyping

First-line screening methods need to be rapid, but should also be detailed enough to reveal all possible pathological features; in the secondary screen the results obtained should be refined and described in more detail. In the visual system, this approach is difficult to fulfill, because with all the rapid test systems currently available only particular aspects of vision can be checked. Therefore, it is necessary, even at the primary level, to combine a wide variety of test systems. Of course, analyses by slit lamp and ophthalmoscope are very rapid methods of screening for morphological alterations in the anterior or posterior of the eye, respectively; however, documentation is possible by photography only and quantification is not possible. Moreover, minor quantitative differences in the eye size obviously cannot be recognized without an appropriate technique. Therefore, the OLCI measurement was optimized for use in a primary screen. The optokinetic drum is a fast functional test of vision; however, the number of false-positive results seems to be quite high, which frequently makes repetition necessary [48]. As a functional test, ERG measurements overcome these problems; however, this method is more time consuming and requires expensive equipment. Finally, none of these test systems are able to detect changes in IOP prior to the resulting damage of the optic nerve head as observed in glaucoma. Therefore, a rapid method for IOP measurement in the mouse is urgently needed.

12.5
Screening Protocols

Despite the fact that examination of the mouse eye began some time ago, there are only a few well-established methods available in this field, in particular the slit-lamp examination size measurements, funduscopy, electroretinography and the optokinetic drum. Although screening protocols are available for these methods, slight differences in results between laboratories may still occur. For the measurement of IOP only one invasive method has been established, however, this method cannot be recommended for screening purposes.

12.5.1 Slit Lamp

There are two types of slit lamps available on the market for use in mouse phenotyping. They are manufactured by Topcon (Tokyo, Japan) and Zeiss (Jena, Germany) and various models are available. The pupils of the eye are dilated with a drop of 1 % atropine before examination. To avoid soiling the device with urine or faeces during inspection of the mice, the appropriate parts of the device should be covered with a plastic bag. The elbows and hands of the examiner should be held in a fixed position during inspection. The mouse is held by one hand while two fingers of the other hand are used to retract the eyelids, focusing is achieved by moving the mouse; anesthesia is not necessary. The light beam should be vertical and as narrow as possible and should be fixed at an angle of 30 ° to the eye. A broad slit beam is recommended for overall inspection of the eye and photography. A narrow slit produces an optic section of the eye. The light intensity should be at maximum and magnification may vary between 20 and 30-fold: a 20 × magnification is recommended for an overall inspection of the eye, whereas a 30 × magnification is recommended for detailed inspection. The optical contrast can be enhanced if the examination is carried out in dimmed light. Using an appropriate digital camera, the opacities can be easily documented (an example is given in Fig. 12.1 b).

12.5.2 Funduscopy Using an Ophthalmoscope

Funduscopy is classed as a rapid screening method. The fundus is inspected with the help of an ophthalmoscope by directing a bright light into the mouse eye. To avoid pupillary reflex, the eyes are dilated with 1 % tropicamide, 1 % cyclopentolate or 1 % atropine. In addition, a double aspheric lens (60, 78 or 90 diopter) held between the ophthalmoscope and eye is used to achieve optimal magnification and detection of fundal anomalies. For indirect ophthalmoscopy without photo-documentation, a full head brace or spectacle mounted ophthalmoscope (e. g. Sigma 150 or Omega 180; Heine, Herrsching, Germany) can be used (there are many other companies offering indirect ophthalmoscopes, e. g. American Optical, Rochester, NY, USA). Anesthesia of the mouse is only necessary for aggressive mice (in these cases a mixture of ketamine and xylazine in physiological saline can be used for i. p. administration; e. g. 137 mg/kg ketamine and 6.6 mg/kg xylazine). However, the use of anesthesia may cloud the ocular media, and in addition some mice have poor tolerance to anesthesia.

In the past, the status of the fundus was documented by photography using the Kowa Genesis small animal fundus camera (Tokyo, Japan; other possible instruments are a Nikon photo slit lamp (Tokyo, Japan), or a Kowa RC-2 human fundus camera). These instruments were used in conjunction with a condensing lens mounted between the camera and the eye. The mouse vibrissae should be trimmed with fine scissors to prevent them from obscuring the photograph.

However, the small animal fundus camera is no more available and at present no successor is offered by Kowa or another company. Therefore, other instruments or human fundus cameras need to be adapted to take fundus photographs of the mouse. One possibility is to use a surgical microscope with a 2.5 magnification, a 90D lens, and a digital camera mounted to the microscope (method modified from [61]). The angle of light entering the eye is critical to obtain good fundus images. Another critical point is to focus the microscope on the image in the lens. To keep the mouse quiet during this procedure it is recommended to anaesthetize the mouse with a Ketamine/Xylazine injection. The eyes can be coated with methylcellulose to prevent them from drying during anaesthesia. Nevertheless, the methylcellulose may cause irritating reflections.

The fundus cameras constructed for human usually need a pupil diameter of at least 4 mm and cannot be used for mice. However, there is a head-worn indirect ophthalmoscope, the Heine Video Omega 2C (Dieter Mann GmbH, Mainaschaff, Germany), that works with a pupil diameter of 1–2 mm and can be adapted to mice. The video ophthalmoscope with a built-in digital camera can be connected to a computer equipped with a video grabber (e. g. VRmAVC). Using a 40 or 60D lens held between the ophthalmoscope and the eye, digital snapshots or video sequences can be recorded (Fig. 12.2).

Details of the retinal capillary bed can be further visualized by means of *fluorescein angiography*. For this purpose, the same general procedure as that used for photographing the fundus can be used except that the standard camera is replaced with a camera specifically designed for fluorescein angiography containing a barrier filter. For observational purposes, the eyepiece is fitted with a barrier filter supplied by the manufacturer. Mice are injected i. p. with 25 % sodium fluorescein at a dose of 0.01 ml per 5–6 g body weight. The retinal vessels begin filling about 30 s after fluorescein administration. Although the timing varies due to the variable rates of intraperitoneal absorption, capillary washout usually occurs 5 min after administration of the dye [61].

12.5.3 Electroretinography (ERG)

ERG is a non-invasive method of examining the function of the entire retina. For scotopic measurement, mice are adapted to the dark in a completely darkened room for at least 12 h prior to examination. During the experiment mice are handled under dim red light. At first, mice are anesthetized with an intraperitoneal injection of a mixture of ketamine, xylazine and physiological salt solution (e. g. 137 mg/kg ketamine and 6.6 mg/kg xylazine). The pupils are dilated with one drop 1 % atropine (or 0.5 % tropicamide) applied to the cornea which may also sometimes be anesthetized with 0.5 % oxybuprocaine. To keep the body temperature at normal levels, the mice are placed on a warming pad. ERG is carried out when the mouse is fixed on a sledge (Steinbeis Transfer Center for Biomedical Optics and Function Testing, Tübingen, Germany). The reference and ground electrodes are placed subcutaneously under the scalp and the tail, respectively. Gold wires treated with one drop of methylcellulose are used as active electrodes which are placed on the cornea (Fig. 12.8). The impedance is checked before and after each measurement and should be less than 10 kΩ at 50 Hz. Each eye is exposed to light pulses of

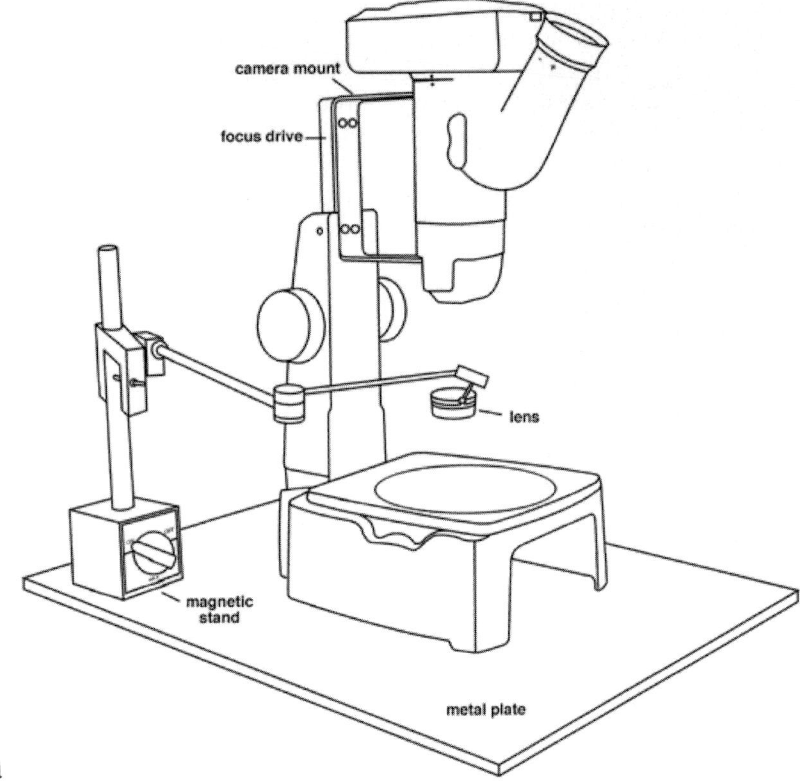

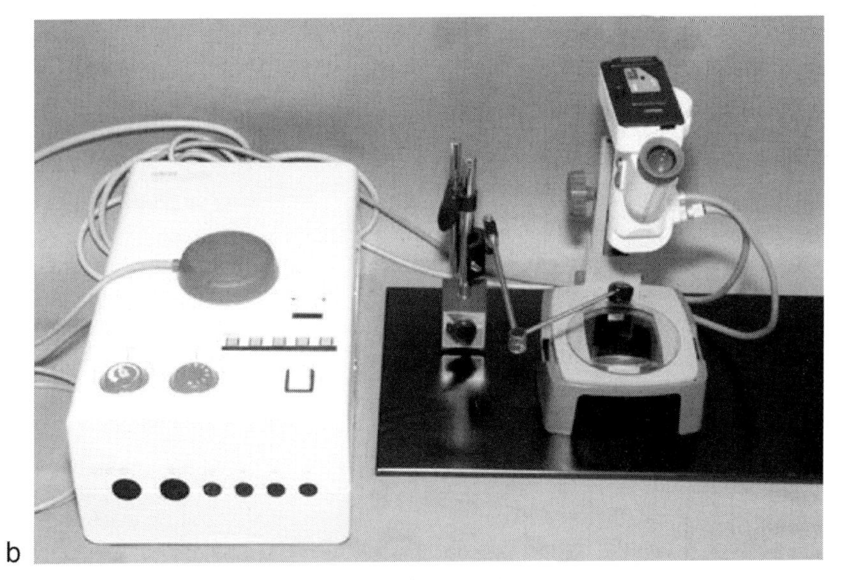

Fig. 12.7 Set-up for ophthalmoscopy. (a) Basic features of the equipment for photographing the fundus. (b) Example of actual equipment: the large white box on the left is the power supply. The foot pedal, which is normally placed on the floor, is resting on the top of the power supply [61].

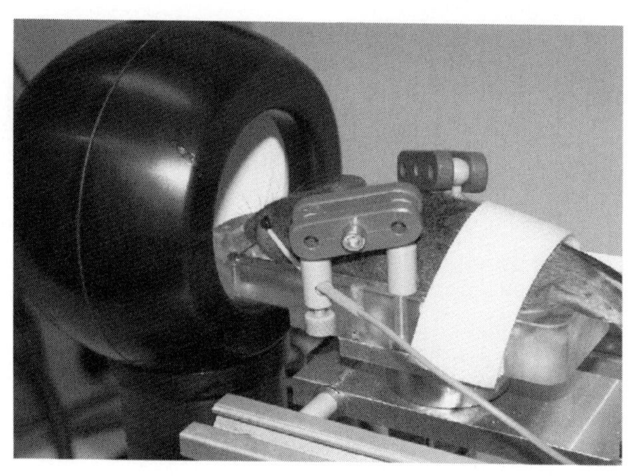

Fig. 12.8 Set-up for electroretinography. Electroretinography on a fixed, anesthetized mouse. The handheld Ganzfeld stimulator is fixed at the bottom, and the mouse is moved on its sledge into position within the Ganzfeld bowl for examination of both eyes.

different intensities using a Ganzfeld stimulator (e. g. Espion Console, Ort, UK, or Multiliner, Toennies, Höchberg, Germany).

For screening purposes, under scotopic conditions light pulses (10 ms) of 500 and 12,500 cd/m^2 are recommended [28]. Additionally, if time permits, ERG measurements can be repeated after 10 min under light (photopic) conditions in order to determine what proportion of the resulting effect is due solely to the 3 % of cones among the photoreceptor cells in the mouse (this approximately doubles the time of investigation per mouse). From 10 averaged ERG recordings the amplitude at the onset of the stimulus is set as "zero", the following a-wave is defined as the lowest negative peak amplitude. The b-wave is measured from this negative peak to the (next) maximum peak (Fig. 12.3). Additionally, the implicit time (latency) of the a- and b-waves can be determined from the onset of the a- and b-waves (Fig. 12.3). The a- and b-waves peak within the first 100–200 ms after stimulus onset, whereas the c-wave has a slow potential that lasts for several seconds. Using the filter function on the ERG device will suppress further noise effects.

It should be noted that the *Pde6b*^{rd1} mutation occurs not only in C3H mice, but also in some 129, SWR and FVB mice. Since there is an easy PCR-based test for this mutation [62], it is strongly recommended that mice are checked for the absence of this mutation before establishing a transgenic line or performing large-scale mutagenesis experiments.

12.5.4
The Optokinetic Drum

Since only a few groups in Edinburgh [48], Strassbourg (http://www-mci.u-strasbg.fr) and Tübingen [49] use an optokinetic drum, the common parameters can only be summarized. The optokinetic drum is used to assess visual perception and visual acuity in the mouse. A geared motor forms the basis of the unit and rotates the drum both clockwise and anticlockwise at a speed of approximately 2 r.p.m. The inner walls of the drum (e. g. 29–31 cm in diameter and 42–60 cm in height) can be lined with different laminated cards on which are printed black and white vertical stripes of known frequency, e. g. two, four, eight etc. The mouse is placed on a stationary platform (8–11.5 cm in diameter, 19–20 cm above the bottom of the drum) in the center of the drum and any movements are recorded using the video camera (Fig. 12.5). The lights and the camera are positioned at 30–70 cm above the drum depending on its height. This ensures that any head movements by the mouse can be seen, recorded and analyzed following the test. The mouse is not restrained and can move freely around the platform.

According to the EMPRESS protocol (http://www.empress.har.mrc.ae.uk/EMPRESS/serrlet/EMPRESS.Frameset), the test can be performed under various lighting conditions. Photopic conditions can be achieved with two or three lights equipped with 60-W bulbs such that they produce approximately 400 lx on the animal platform. The overhead spotlight should minimize the amount of shadow inside the drum. The use of a luxmeter for measuring luminance (e. g. Bioblock Scientific) is recommended.

The mouse is placed on the platform within the drum and time should be allowed for habituation prior to rotation of the drum (e. g. 10 min under a low level light (scotopic conditions), and 1–5 min under a more intense light (photopic conditions)). The mouse is tested for 2 min in both orientations of the drum (clockwise and counterclockwise). If the animal being tested has sufficient visual acuity to resolve the pattern presented, it will show movement of its head ("head tracking") to maintain fixation of the pattern. If its visual acuity is too poor to resolve the pattern, it will not show fixation maintaining behavior. This is a reflex response and not dependent on the training of the animal. A mouse with normal visual acuity can resolve an image subtending 1 ° [63].

A video-camera (e. g. Sony Digital Handycam, DCR, TRV16E) is required for recording the data. The Sony camera is recommended because it is suitable for recording under the low light conditions required for experimentation. Television equipment or a computer with the appropriate software for transferring the data, is required for viewing and quantifying the data; digital data stores are also required. For further progress, there is an obvious need for standardization of these drums with respect to their size, geometry, speed, illuminance and duration of the test.

12.6 Logistics

12.6.1 Slit Lamp

Training of the investigator is required to identify subtle opacities (e. g. suture opacities). Screening using a slit lamp is very rapid and efficient: one experienced observer can screen approximately 200 mice per day. Documentation by photography will take extra time.

12.6.2 Ophthalmoscope

Screening using an ophthalmoscope is very rapid and efficient procedure: one experienced observer will need 10 min per mouse to complete the screen if the mice are anaesthetized, otherwise the procedure can be completed in the same time as the slit-lamp examination. Documentation by photography requires extra time.

12.6.3 ERG

Using this approach, a team of two experienced people can screen 25–40 mice per day [28].

12.6.4 The Optokinetic Drum

Training is required to identify subtle movements. The optokinetic drum differentiates reliably between mice with normal vision and mice with severely decreased vision and can be applied to high-throughput screening systems since it takes 20 min per mouse [48].

12.7 Troubleshooting

12.7.1 Slit Lamp

- If the investigator wears a facemask, the ocular lenses may become clouded with her/his breath.
- If the slit is not illuminated when the instrument is switched on and the lamp is burning, it should be checked whether the slit is open and the knobs are in their correct positions (follow the instructions for the instrument).
- If the two reflection lines of the cornea and the lens cannot be seen by moving the mouse, the angle of the light beam (30 °) should be checked.
- If the pupil is not dilated, another drop of atropine (1 %) should be instilled. Care should be taken to avoid intoxicating the mouse with a large dose of atropine.
- If examination of the eye is prolonged the cornea may become dry, which adversely affects the examination of the lens. Therefore, the eyelid should be retracted for a minimum period of time.

12.7.2
Ophthalmoscope

- If the pupil is not dilated, another drop of atropine (1 %) or tropicamide should be instilled. Care should be taken to avoid intoxicating the mouse with a large dose of atropine.
- If there is an obstruction in the viewing area, ensure that the lens disc and the filter assembly are clean; if they are dirty, clean or replace them.
- If there is too much reflection, the beam can be swivelled to reduce the reflex response and improve illumination of the retina. Moreover, reflections might also be caused by dirty mirrors or scratches on the mirror, filter or objective lens. Any dirt should be removed and scratched parts replaced.
- Shadows on the light spot might be the result of dirty optics/dirty lamp envelope or loose condensing/objective lens.
- If focusing the ophthalmoscope is problematic, make sure that the lens is not damaged (for example by alcohol treatment) and that the objective lens is tightened.
- If no light spot appears, check the connection to the power supply and whether the dustproof cap has been removed.
- If the light spot cannot be fully seen by both eyes, the optics need to be adjusted according to the instructions for use.
- If the light output is intermittent (even though the power source is connected), check that the condensing/objective lens has been tightened.

12.7.3
ERG

- If the electrode impedance is not correct, the active gold-wire electrodes should be removed, cleaned and replaced on the cornea.
- If no response can be detected, all connections of wires and electrodes should be checked.
- If the animal's heartbeat interferes with the recording, the reference electrode needs to be replaced.
- A decrease in the mouse body temperature of just a few degrees is associated with a virtually undetectable ERG.
- Repeated flashing, commonly used in signal averaging, reduces rod-mediated (but not cone-mediated) ERG response amplitudes by about 20 % at high flash intensities unless the flash presentation rate is slowed to allow sufficient recovery of rod function.
- High doses of anesthetics may lead to clouding of the lens which can reduce ERG amplitudes.

12.7.4
The Optokinetic Drum

- If the mouse does not react, repeat the test again 1 h later.
- Since mice are less active in the afternoon, ideally tests should be carried out only in the morning.

12.8
Outlook

The visual system is one of the central perception systems in humans. Therefore, ocular disorders severely affect social behavior and quality of life. Since the human population (at least in the industrialized countries) is living longer, and since the frequency and severity of ocular disorders increases with age, the need for medical diagnostics and treatment will increase. Therefore, it is necessary to develop mouse models to understand in more detail the pathological mechanisms of ocular disorders. However, because of the small size of the mouse eye, diagnostic systems for the mouse are poorly developed as compared to those used in clinic. Unfortunately, only a few methods such as the slit lamp, ophthalmoscope, ERG or the optokinetic drum can be used as rapid first-line screening methods. However, for clinical purposes high-resolution methods (based upon optical coherence tomography) have recently been developed, which provide a microscopic view of the retina and optic nerve head and enable their sizes and shapes to be measured thus providing early indications of retinal degeneration [64] and glaucoma [65]. Funduscopy can be carried out in parallel [66]. Since this investigation can be undertaken in the absence of contact with the eye, it is expected that it can be easily adapted to the mouse. Since devices similar to that used to observe the posterior segment of the eye are also under development for the anterior segment, the number of diverse primary screens will be reduced in the near future to just two which will increase not only the number of mice that can be screened but also the quality of data. This will provide both qualitative and quantative data, which can be used to compile databases from which standardized data sets can be established.

Acknowledgments

Part of this work was carried out within the framework of the German Mouse Clinic (GMC), supported by the German National Genome Research Network (NGFN; R1GR0103) and the European Initiative for Standardized Mouse Phenotyping (EUMORPHIA).

References

1 G. J. Johnson, A. Foster **2003**, Prevalence, incidence and distribution of visual impairment. In *The Epidemiology of Eye Disease*, G. J. Johnson, D. C. Minassian, R. A. Weale, S. K. West (eds.), Arnold, London, UK, 3–28.

2 J. Graw **2003**, The genetic and molecular basis of congenital eye defects, *Nat. Rev. Genet.* 4, 876–888.

3 R. E. Hill, J. Favor, B. L.M., Hogan, C. C.T. Ton, G. F. Saunders, I. M. Hanson, J. Posser, T. Jordan, N. D. Hastie, V. van Heyningen **1991**, *Small eye* results from mutations in a paired-like homeobox-containing gene, *Nature* 354, 522–525.

4 J. Favor, H. Peters, T. Hermann, W. Schmahl, B. Chatterjee, A. Neuhäuser-Klaus, R. Sandulache **2001**, Molecular characterization of *Pax62Neu* through *Pax610Neu*: an extension of the Pax6 allelic series and the identification of two possible hypomorph alleles in the mouse *Mus musculus*, *Genetics* 159, 1689–1700.

5 C. Thaung, K. West, B. J. Clark, L. McKie, J. E. Morgan, K. Arnold, P. M. Nolan, J. Peters, A. J. Hunter, S. D.M. Brown, I. J. Jackson, S. H. Cross **2002**, Novel ENU-induced eye mutations in the mouse: models for human eye disease, *Hum. Mol. Genet.* 11, 755–767.

6 J. C. Grindley, D. R. Davidson, R. E. Hill **1995**, The role of Pax-6 in eye and nasal development, *Development* 121, 1433–1442.

7 D. Morrison, D. FitzPatrick, I. Hanson, K. Williamson, V. van Heyningen, B. Fleck, I. Jones, J. Chalmers, H. Campbell **2002**, National study of microphthalmia, anophthalmia, and coloboma (MAC) in Scotland: investigation of genetic aetiology, *J. Med. Genet.* 39, 16–22.

8 E. V. Semina, J. C. Murray, R. Reiter, R. F. Hrstka, J. Graw **2000**, Deletion in the promoter region and altered expression of *Pitx3* homeobox gene in *aphakia* mice, *Hum. Mol. Genet.* 9, 1575–1585.

9 D. K. Rieger, E. Reichenberger, W. McLean, A. Sidow, B. R. Olsen **2001**, A double-deletion mutation in the Pitx3-gene causes arrested lens development in aphakia mice, *Genomics* 72, 61–72.

10 D. Weigel, G. Jürgens, F. Kuttner, E. Seifert, H. Jäckle **1989**, The homeotic gene *fork head* encodes a nuclear protein and is expressed in the terminal regions of the *Drosophila* embryo, *Cell* 19, 645–658.

11 S. Komatireddy, S. Chakrabarti, A. K. Mandal, A. B.M. Reddy, S. Sampath, S. G. Panicker, D. Balasubramanian **2003**, Mutation spectrum of FOXC1 and clinical heterogeneity of Axenfeld–Rieger anomaly in India, *Mol. Vis.* 9, 43–48.

12 R. S. Smith, A. Zabaleta, T. Kume, O. V. Savinova, S. H. Kidson, J. E. Martin, D. Y. Nishimura, W. L.M. Alward, B. L.M. Hogan, S. W.M. John **2000**, Haploinsufficiency of the transcription factors FOXC1 and FOXC2 results in aberrant ocular development, *Hum. Mol. Genet.* 9, 1021–1032.

13 A. Blixt, M. Mahlapuu, M. Aitola, M. Pelto-Huikko, S. Enerbäck, P. Carlsson, P. **2000**; A forkhead gene, *FoxE3*, is essential for lens epithelial proliferation and closure of the lens vesicle, *Genes Dev.* 14, 245–254.

14 E. V. Semina, I. Brownell, H. A. Mintz-Hittner, J. C. Murray, M. Jamrich **2001**, Mutations in the human forkhead transcription factor *FOXE3* associated with anterior segment ocular dysgenesis and cataracts, *Hum. Mol. Genet.* 10, 231–236.

15 P. Hertwig **1942**, Neue Mutationen und Koppelungsgruppen bei der Hausmaus, *Z. ind. Abstammungs.Vererbungslehre* 80, 220–246.

16 E. Steingrímsson, K. J. Moore, M. L. Lamoreux, A. R. Ferré-D'Amaré, S. K. Burley, D. C.S. Zimring, L. C. Skow, C. A. Hodgkinson, H. Arnheiter, N. G. Copeland, N. A. Jenkins **1994**, Molecular basis of mouse *microiphthalmia* (*mi*) mutations helps explain their developmental and phenotypic consequences, *Nat. Genet.* 8, 256–263.

17 J. H. Hallsson, J. Favor, C. Hodgjinson, T. Glaser, M. L. Lamoreux, R. Magnusdottir, G. J. Gunnarsson, H. O. Sweet, N. G. Copeland, N. A. Jenkins, E. Steingrimsson **2000**, Genomic, transcriptional and mutational analysis of the mouse *microphthalmia* locus, *Genetics* 155, 291–300.

18 A. G. Hansdottir, K. Pálsdóttir, J. Favor, A. Neuhäuser-Klaus, H. Fuchs, M. Hrabé de Angelis, E. Steingrímsson **2004**, The novel mouse microphthalmia mutations $Mitf^{mi\text{-}enu5}$ and $Mitf^{mi\text{-}bcc2}$ produce dominant negative Mitf proteins, *Genomics* 83, 932–935.

19 I. Yajima, S. Sato, T. Rimura, K.-i. Yasumoto, S. Shibahara, C. R. Godine, H. Yamamoto **1999**, An L1 element intgronic insertion in the black-eyed white ($Mitf^{mi\text{-}bw}$) gene: the loss of a single Mitf isoform responsible for the pigmentary defect and inner ear deafness, *Hum. Mol. Genet.* 8, 1431–1441.

20 S. T. Santhiya, M. Shyam Manohar, D. Rawlley, P. Vijayalakshmi, P. Namperumalsamy, P. M. Gopinath, J. Löster, J. Graw **2002**, Molecular characterization of new alleles in the γ-crystallin genes demonstrating the genetic heterogeneity of autosomal dominant congenital cataracts, *J. Med. Genet.* 39, 352–358.

21 S. T. Santhiya, S. M. Manisastry, D. Rawlley, R. Malathi, S. Anishetty, P. M. Gopinath, P. Vijayalakshmi, P. Namperumalsamy, J. Adamski, J. Graw **2004**, Mutation analysis of congenital cataracts in Indian families: identification of SNPs and a new causative allele in *CRYBB2* gene, *Invest. Opthalmol. Vis. Sci.* 45, 3599–3607.

22 J. Graw, A. Neuhäuser-Klaus, N. Klopp, P. B. Selby, J. Löster, J. Favor **2004**, Genetic and alklelic heterogeneity of *Cryg* mutations in eight distinct forms of dominant cataract in the mouse, *Invest. Opthalmol. Vis. Sci.* 45, 1202–1213.

23 H. G. Krumpaszky, V. Klauß **1996**, Epidemiology of blindness and eye disease. *Ophthalmologica* 210, 1–84.

24 C. Rivolta, D. Sharon, M. M. DeAngelis, T. P. Dryja **2002**, Retinitis pigmentosa and allied diseases: numerous diseases, genes, and inheritance patterns, *Hum. Mol. Genet.* 11, 1219–1227.

24a C. Dalke, J. Graw **2005**, Mouse mutants as models for congenital retinal disorders. *Exp. Eye Res.* 81, 503–512.

25 M. Burmeister, J. Novak, M. Y. Liang, S. Basu, L. Ploder, N. L. Hawes, D. Vidgen, F. Hoover, D. Goldman, V. I. Kalnins, T. H. Roderick, B. A. Taylor, M. H. Hankin, R. R. McInnes **1996**, Ocular retardation mouse caused by *Chx10* homeobox null allele: impaired retinal progenitor proliferatrion and bipolar cell differentiation, *Nat. Genet.* 12, 376–384.

26 E. F. Percin et al. **2000**, Human microphthalmia associated with mutations in the retinal homeobox gene *CHX10, Nat. Genet.* 25, 397–401.

27 S. J. Pittler, W. Baehr **1991**, Identification of a nonsense mutation in the rod photoreceptor cGMP phosphodiesterase β-subunit gene of the rd mouse, *Proc. Natl. Acad. Sci. USA* 88, 8322–8326.

28 C. Dalke, J. Löster, H. Fuchs, V. Gailus-Durner, D. Soewarto, J. Favor, A. Neuhäuser-Klaus, W. Pretsch, F. Gekeler, K. Seinoda, E. Zrenner, T. Meitinger, M. Hrabé de Angelis, J. Graw **2004**, Electroretinography as a screening method for mutations causing retinal dysfunction in mice, *Invest. Opthalmol. Vis. Sci.* 45, 601–609.

29 D. WuDunn **2002**, Genetic basis of glaucoma, *Curr. Opin. Ophthalmol.* 13, 55–60.

30 B. S. Kim, O. V. Savinova, M. V. Reddy, J. Martin, Y. Lun, L. Gan, R. S. Smith, S. I. Tomarev, S. W.M. John, R. L. Johnson **2001**, Targeted disruption of the myocilin gene (*Myoc*) suggests that human glaucoma-causing mutations are gain of function, *Mol. Cell. Biol.* 21, 7707–7713.

31 M. Sarfarazi, I. Stoilov, J. B. Schenkman **2003**, Genetics and biochemistry of primary congenital glaucoma, *Ophthalmol. Clin. N. Am.* 16, 543–554.

32 H. B. Chase, E. B. Chase **1941**, Studies on an anophthalmic strain of mice, *J. Morphol.* 68, 279–301.

33 P. Tucker, L. Laemmle, A. Munson, S. Kanekar, E. R. Oliver, N. Brown, H. Schlecht, M. Vetter, T. Glaser **2001**, The *eyeless* mouse mutation (*ey1*) removes an alternative start codon from the *Rx/rax* homeobox gene, *Genesis* 31, 53–53.

34 J. Kratochvilova, U. H. Ehling **1979**, Dominant cataract mutations induced by γ-irradiation of male mice, *Mutat. Res.* 63, 221–223.

35 U. H. Ehling, D. J. Charles, J. Favor, J. Graw, J. Kratochvilova, A. Neuhäuser-Klaus, W. Pretsch **1985**, Induction of gene mutations in mice: The multiple end-point approach, *Mutat. Res.* 150, 393–401.

36 J. Favor, A. Neuhäuser-Klaus **2000**, Saturation mutagenesis for diminant eye morpholigical defects in the mouse *Mus musculus*, *Mamm. Genome* 11, 520–525.

37 J. Auwerx et al. **2004**, The European dimension for the mouse genome mutagenesis program, *Nat. Genet.* 36, 925–927.

38 C. P. Austin et al. **2004**, The knockout mouse project, *Nat. Genet.* 36, 921–924.

39 M. F.Lyon, S. E. Jarvis, I. Sayers, R. S. Holmes **1981**, Lens opacity: a new gene for congenital cataract on chromosome 10 of the mouse, *Genet. Res. Camb.* 38, 337–341.

40 P. B. Selby **1990**, Experimental induction of dominant mutations in mammals by ionizing radiations and chemicals, *Iss. Rev. Teratol.* 5, 181–253.

41 J. Graw, J. Löster, O. Puk, D. Soewarto, H. Fuchs, B. Meyer, P. Nürnberg, W. Pretsch, P. Selby, J. Favor, E. Wolff, M. Hrabé de Angelis **2005**, Three novel *Pax6* alleles in the mouse leading to the same small-eye phenotype due to different consequences at target promoters. *Invest. Ophthalmol. Vis. Sci.* 46, 4671–4683.

42 B. Chang, N. L. Hawes, R. E. Hurd, M. T. Davisson, S. Nusinowitz, J. R. Heckenlively **2001**, Retinal degeneration mutants in the mouse, *Vision Res.* 42, 517–525.

43 H. Kolb **2003**, How the retina works, *Am. Sci.* 91, 28–35.

44 R. D. Penn, W. A. Hagins **1969**, Signal transmission along retinal rods and the origin of the electrophretinographic a-wave, *Nature* 202, 201–204.

45 A. L. Lyubarsky, B. Falsini, M. E. Pennesi, P. Valentini, E. N. Pugh Jr **1999**, UV- and midwave-sensitive cone-driven retinal responses of the mouse: a possible phenotype for coexpression of cone photopigments, *J. Neurosci.* 19, 442–445.

46 M. Biel, M. Seeliger, A. Pfeifer, K. Kohler, A. Gerstner, A. Ludwig, G. Jaissle, S. Fauser, E. Zrenner, F. Hofmann **1999**, Selective loss of cone function in mice lacking the cyclic nucleotide-gated channel CNG3, *Proc. Natl. Acad. Sci. USA.* 96, 7553–7557.

47 A. Möller, T. Eysteinsson, E. Steingímsson **2004**, Electroretinographic assessment of retinal function in microphthalmic mutant mice, *Exp. Eye Res.* 78, 837–848.

48 B. Thaung, K. Arnold, I. J. Jackson, P. J. Coffey **2002**, Presence of visual head tracking differentiates normal sighted from retinal degenerate mice, *Neurosci. Lett.* 325, 21–24.

49 C. Schmucker, M. Seeliger, P. Humphries, M. Biel, F. Schaeffel **2005**, Illuminance-dependent grating acuity in wild-type mice, and in mic e lacking rod or cone function, or both, *Invest. Opthalmol. Vis. Sci.* 46, 398–407.

50 O. V. Savinova, F. Sugiyama, J. E. Martin, S. I. Tomarev, B. J. Paigen, R. S. Smith, S. W.W. John **2001**, Intraocular pressure in genetically distinct mice: an update and strain survey, *BMC Genetics* 2, 12 (online http://www.biomedcentral.com/1471–2156/2/12).

51 M. G. Anderson, R. S. Smith, N. L. Hawes, A. Zabaleta, B. Chang, J. L. Wiggs, S. W.M. John **2002**, Mutations in genes encoding melanosomal proteins cause pigmentary glaucoma in DBA/2J mice, *Nat. Genet.* 30, 81–85.

52 B. Chang, R. S. Smith, M. Peters, O. V. Savinova; N. L. Hawes, A. Zabaleta, S. Nusinowitz, J. E. Martin, M. L. Davisson, C. L. Cepko, B. L.M. Hogan, S. W.M. John **2001**, Haploisufficient Bmp4 ocular phenotypes include anterior segment dysgenesis with elevated intraocular pressure, *BMC Genetics* 2, 18 (online http://www.biomedcentral.com/1471–2156/2/128).

53 L. H. Pinto, C. Enroth-Cugell **2000**, Tests of the mouse visual system. *Mamm. Genome*, 11, 531–536.

54 A. Wegener **2004**, Light scattering measurements in mouse lenses. *Ophthal. Res.* 36 (S1), 140 (abstract).

55 C. Schmucker, F. Schaeffel **2004**, *In vivo* biometry in the mouse eye with low coherence interferometry, *Vision Res.* 44, 2445–2456.

56 A. I. Kontiola, D. Goldblum, T. Mittag, J. Danias **2001**, The induction/impact Tonometer: a new instrument to measure intraocular pressure in the rat, *Exp. Eye Res.* 73, 781–785.

57 J. Danias, A. I. Kontiola, T. Filippopoulos, T. Mittag **2003**, Method for the noninvasive measurement of intraocular pressure in mice, *Invest. Ophthalmol. Vis. Sci.* 44, 1138–1141.

58 H. A. Reitsamer, J. W. Kiel, J. M. Harrison, N. L. Ransom, S. J. McKinnon **2004**, Tonopen measurement of intraocular pressure in mice. *Exp. Eye Res.* 78, 799–804.

59 S. W. John, O. V. Savinova **2002**, Intraocular pressure measurement in mice: technical aspects. In *Systematic Evaluation of the Mouse Eye*, R. S. Smith (ed.), CRC, Boca Raton, USA, 313–320.

60 C. Dalke, U. Pleyer, J. Graw **2005**, On the use of Tono-Pen XL for the measurement of intraocular pressure in mice, *Exp. Eye Res.* 80, 295–296.

60a F. Schaeffel, E. Burkhardt **2005**, Pupillographic evaluation of the time course of atropine effects in the mouse eye. *Optom. Vis. Sci.* 82, 215–220.

60b M. W. Seeliger, S. C. Beck, N. Pereyra-Muñoz, S. Dangel, J.-Y. Tsai, U. F.O. Luhmann, S. A. van de Pavert, J. Wijnholds, M. Samardzija, A. Wenzel, E. Zrenner, K. Narfström, E. Fahl, N. Tanimoto, N. Acar, F. Tonagel **2005**, In vivo confocal imaging of the retina in animal models using scanning laser ophthalmoscopy. *Vision Res.* 45, 3512–3519.

61 N. L. Hawes, R. S. Smith, B. Chang, M. Davisson, J. R. Heckenlively, S. W.M. John **1999**, Mouse fundus photography and angiography: a catalogue of normal and mutant phenotypes, *Mol. Vis.* 5, 22.

62 E. Giménez, L. Montoliu **2001**, A simple polymerase chain reaction assay for genotyping the retinal degeneration mutation ($Pdeb^{rd1}$) in FVB/N-derived transgenic mice, *Lab. Anim.* 35, 153–156.

63 P. Artal, P. Herreros de Tejada, C. Munoz Teda, D. G. **1998**, Green retinal image quality in the rodent eye, *Vis. Neurosci.* 15, 597–605.

64 P. Massin, G. Duguid, A. Erginay, B. Haouchine, A. Gaudric **2003**, Optical coherence tomography for evaluating diabetic macular edema before and after vitrectomy, *Am. J. Ophthalmol.* 135, 169–177.

65 P. Carpineto, M. Ciancaglini, E. Zuppardi, G. Falconio, E. Doronzo, L. Mastropasqua **2003**, Reliability of nerve fiber layer thickness measurements using optical coherence tomography in normal and glaucomatous eyes, *Ophthalmology* 110, 190–195.

66 C. Strom, B. Sander, N. Larsen, M. Larsen, H. Lund-Andersen **2002**, Diabetic macular edema assessed with optical coherence tomography and stereo fundus photography, *Invest. Ophthalmol. Vis. Sci.* 43, 241–245.

13
EUMORPHIA and the European Mouse Phenotyping Resource for Standardized Screens (EMPReSS)

Steve Brown, Heena Lad, Eain Green, Georgios Gkoutos, Hilary Gates, Martin Hrabé de Angelis, and the members of the EUMORPHIA consortium (www.eumorphia.org)

13.1
Introduction

With the completion of the human genome sequence (International Human Genome Sequencing Consortium, 2001) [1], mammalian genetics is set to enter a new era of discovery. For the first time we will be able to interpret the relationship between gene and phenotype as well as gene and disease in the context of comprehensive knowledge of gene content of a mammalian genome. The mouse occupies a pivotal position amongst model organisms particularly with its annotated genome sequence now released into the public domain, (Mouse Genome Sequencing Consortium, 2002) [2]. The unique insight that the mouse can bring to many physiological and biochemical pathways underlying human disease, makes it an unparalleled vehicle for functional genomics studies.

13.2
The EUMORPHIA Project

EUMORPHIA (European Union Mouse Research for Public Health and Industrial Applications) is a large project funded by the European Commission comprising 18 research centers in eight European countries (Box 13.1). The project was founded with the intention of advancing research into functional mouse genetics to a level that will significantly enrich both the academic and commercial sectors of our medical research community. Incorporating the expertise and resources of many European mouse genetics centers, the main focus of the project has been on the development of novel approaches in phenotyping, mutagenesis and informatics to improve the characterization of mouse models for understanding human molecular physiology and pathology through integrated functional genomics.

Standards of Mouse Model Phenotyping. Edited by Martin Hrabé de Angelis, Pierre Chambon, and Steve Brown

ISBN: 3-527-31031-2

Box 13.1 Details of the EUMORPHIA project.

EUMORPHIA (European Union Mouse Research for Public Health and Industrial Applications) is an integrated research program within Framework Programne 5 – "Quality of Life" funded by the European Commission, and involves the collaboration of 18 research centers across Europe. The main objectives of the project are primarily concerned with:

- Development and standardization of phenotyping methods that fully characterize mutant mouse models of human disease.
- Improving reproducibility and comparability of phenotyping tests by validation and refinement across laboratories.
- Contributing to the mouse mutant resource.
- Development of informatics tools for the appropriate query and acquisition of phenome data.

UK	MRC Mammalian Genetics Unit
	MRC Human Genetics Unit, Edinburgh
	MRC Functional Genetics Unit, Oxford
	The Wellcome Trust Sanger Institute, Cambridge
	University of Manchester
France	GIE-CERBM: IGBMC, Institut Clinique de la Souris, Genopole Alsace-Lorraine, Strasbourg and University Louis Pasteur, Strasbourg
	CREATIS: ANIMAGE, ENS/Genopole, Lyon, Institut Pasteur, Orleans
	CNG, Evry
	CNRS, Institut de Transgenose-Orleans
Germany	GSF, Munich
	GBF, Braunschweig
Netherlands	Netherlands Cancer Institute, Amsterdam
Italy	EMBL Mouse Biology Program, Monterotondo
	CNR Institute of Cell Biology, Monterotondo
Sweden	Karolinska Institute, Stockholm
Switzerland	University of Lausanne, Center for Integrative Genomics
	University of Geneva
Spain	CNIO, Madrid

Further details and information about the project can be found on the EUMORPHIA website: www.eumorphia.org

13.2.1
Project Structure

Phenotyping Protocols The project is divided into work packages and is principally designed to encompass many of the disciplines necessary to phenotype the body systems of a mouse, including the following:

- Clinical chemistry and hematology
- Chemical and functional exploration of hormonal and metabolic systems
- Cardiovascular exploration
- Allergy and infectious diseases
- Renal function
- Sensory systems
- Central/peripheral nervous and skeletal muscle systems
- Behavior and cognition

- Cancer
- Bone, cartilage, arthritis and osteoporosis
- Gene expression
- Necropsy examination, pathology and histology
- Respiratory function

Whilst each work package is represented by scientists with expertise in the subject area, the fact that some work packages overlap has facilitated the integration of ideas and complementary activities of phenotyping. There are also several work packages that underpin these areas, namely:

- Standardization of animal handling
- Imaging
- Mutagenesis
- Informatics

Standardization of Animal Handling Addressing factors that may influence the outcome of results is critical in generating robust phenotyping data and extends to the standardization of animal handling. In defining common methods of animal husbandry and housing standards, including setting comparable standards for animal diets and animal welfare, we are minimizing the possibility of obtaining confounding results.

Imaging The implementation and development of new imaging techniques are valuable for analyzing gene expression *in vivo* and aid in determining and understanding the structure and function of organs and tissues, for example bone, heart, renal and sensory systems, and for general pathology.

Mutagenesis Developing methods for novel gene-driven approaches in mouse mutagenesis within EUMORPHIA, using ES cells and parallel archives of mutant mouse DNA and sperm, is contributing to the continual expansion of the breadth and depth of mouse mutant resources.

Informatics Adding to the value and proficiency of ongoing research developments through the dissemination of new tools, resources and skills to academia and industry, informatics is encouraging and building links with other mouse, human and clinical genetics groups, as well as biotechnology/biopharmaceutical industries, across Europe. The construction of appropriate data portals that link phenotype databases with an ontological framework will provide the consortium and the wider scientific community with access to phenotype information and mutant stocks.

13.3
Using Mouse Models

While considerable progress has been made in understanding monogenic disorders, there are many gaps in the identification of genes underlying multi-factorial disease.

Uncovering the function of genes in such diseases is proving to be a challenge. The use of mouse models of human diseases serves to improve our understanding of the complexities involved in gene function and expression. A key issue in mouse functional genomics is the development and standardization of phenotyping platforms. Approaches in mouse phenotyping to date, which fully characterize mutant models, have been modest despite the increasing need for defining common working standards to minimize variability and generate meaningful results. The integration of the work packages in the project has been used as a foundation for a comprehensive primary screening strategy to characterize a mouse, and to inform its passage to appropriate second-line screens. With particular emphasis on standardization for the efficient and reliable analysis of mouse mutant models, EUMORPHIA aims to increase reproducibility and comparability of phenotyping within and across different laboratories. Access to robust and improved phenotyping platforms will reduce the need for duplication and in turn the number of animals used in research.

13.4 European Mouse Phenotyping Resource for Standardized Screens (EMPReSS)

A major achievement in EUMORPHIA is the establishment of the European Mouse Phenotyping Resource for Standardized Screens (EMPReSS). EMPReSS constitutes a precise, evaluative platform of Standard Operating Procedures (SOPs) that have been incorporated into a searchable database for the systematic and standardized primary characterization of mouse mutants.

Over 100 scientists, representing the majority of the work packages in EUMORPHIA, have contributed to the development of phenotyping protocols to be included in EMPReSS. Working groups evaluated proposed phenotyping assays defining common standards. Complementary phenotyping protocols were derived where significant synergies emerged between work packages. The refinement and validation of the protocols, using chosen inbred and mutant lines, is being completed in collaboration with centers in EUMORPHIA and in consultation with clinicians, human geneticists and physiologists.

13.4.1 Development of the SOPs

The SOPs for EMPReSS have been developed to form a coherent battery of tests that can be performed to fully characterize a mutant mouse. While the majority of SOPs may be performed in isolation, some form part of a test battery where a specific order of testing is indicated, for example, in the behavioral work package, the order of the tests may influence the response of the mouse.

The SOPs are categorized according to the experience and amount of specialized equipment required to carry them out which indicates the depth of the phenotyping assay:

- Primary first-line protocols are simple to apply and require little specialist equipment.

Box 13.2 Format for the SOPs.

Purpose: details of the main objectives and the basic principles underlying the protocol.
Scope: who the procedure is intended for.
Safety requirements: details of the necessary precautions to be taken when carrying out the procedure.
Associated documents: lists SOPs within EMPReSS which are specifically associated with the protocol, including those that are referred to within the procedure and any other relevant documentation.
Notes: any important additional factors that must be taken into account before carrying out this procedure, for example, age of testing and time of testing that may have an affect on the test.
Quality control: any methods/means of calibrating equipment used for experimentation and how often it is undertaken. Details on how to validate the quality of a particular assay may also be contained within this section.
Equipment: any one-off purchases of equipment required for the procedure.
Supplies: any consumables that will be required for the procedure, including solutions.
Procedure: a step-by-step, detailed description of how the procedure must be carried out.
Supporting information: lists any available data or literature that supports the procedure.
History review: details of any other versions of the SOP that have since been modified.

- Primary extended protocols give further information on the phenotype but require more specialized skills or equipment.
- Secondary protocols are complex requiring expertise and specialized equipment and uncover more subtle aspects of the phenotype.

Each assay is designed to provide descriptive observations for given phenotypes. The aim is to link screens to defined phenotype outcomes within an ontological framework. This will enable users to undertake both comprehensive and subtle searches in EMPReSS to identify an assay that tests for clear phenotypes of interest. In view of the objectives for the ontology database, a comprehensive SOP format was devised to allow the systematic integration of the assay with the phenotype and also incorporates all aspects necessary for accurate phenotyping (Box 13.2). Some of SOPs are accompanied by appendices which provide more detailed descriptions of part of the procedure or the analysis of the procedure. There are also annexes. These contain other specific information which may apply to one or more SOPs, for example, lists of reagents or antibodies.

13.4.2 Review Status

Clarity and accuracy of the information contained within the SOP is of prime importance to ensure that the results obtained are reproducible. The reviewing process is crucial in the development of each SOP before its admission to EMPReSS. Once each SOP is written in the template, the SOP is reviewed by a number of routes. The status of review is indicated by status levels at the top of each SOP (Fig. 13.1) as follows:

eumorphia Standard Operating Procedure	Title: **Differential blood count**	
	Doc. Number: 3_001 Rev No: 0 Status: 3 Validation: ★★★	Date Issued: 01/06/04 Type: 1F

Fig. 13.1 Extract from EMPReSS showing the information at the top of each SOP.

1. Initial development, discussion and review within expert working groups leading to a SOP.
2. Review by the EMPReSS Resource team.
3. Review and sign-off of the SOP by a EUMORPHIA scientist, outside the working group, with expertise in the subject area.

13.4.3
Validation of SOPs

EUMORPHIA recognizes that validation of each SOP is fundamental in the progression towards reproducibility and comparability of phenotyping data. Selected inbred mouse strains and/or mutants have been used to acquire baseline data and validate each SOP in EMPReSS. The status of validation is indicated in the form of a star rating:

- ★ Bronze indicates validation within *one* EUMORPHIA laboratory.
- ★★ Bronze and silver indicate validation at *two* different EUMORPHIA laboratories.
- ★★★ Bronze, silver and gold indicate validation at *more than two* EUMORPHIA laboratories.

Many of the SOPs have already been validated in multiple laboratories. A primary example of the value of this process has been observed within the neurological work packages where there was significant collaboration in devising and validating the SOPs (see also Chapter 6). A high through-put battery of nine primary tests to be included in the first-line screen was identified: open field; modified SHIRPA immediately followed by grip strength assessment; accelerating rota-rod; Y-maze spontaneous alternation; acoustic startle and pre-pulse inhibition; tail flick; tail suspension; and swim ability. Validation of the nine tests was completed in the order they are listed with 10–15 male mice of the same four inbred strains (C57BL/6J, C3HeB/FeJ, BALB/cByJ, 129S2/SvPas), aged 8–10 weeks. Despite some differences in the equipment used, the results generated in these tests at four centers were largely analogous, proving the SOPs to be robust. An exception however was the accelerating rota-rod test, where results revealed some differences within and across laboratories. It was clear, when results for the accelerating rota-rod test were analyzed, that the SOP was lacking explicit detail in the method, and that the manner in which the equipment was used was significant because of the apparent wide-ranging variability in the test outcome. As a result, the accelerating rota-rod has been taken out of the behavioral battery of first-line screens for further development and refinement. The outcome highlighted the importance of standardization and the need for validation within phenotyping platforms. If research centers worldwide are to continue producing pioneering discoveries for the wider medical research community, it is essential that each laboratory adopts common standards and working practices.

13.5 Ontologies and Structure of the Empress Resource

To realize fully the potential of model organisms to bridge the gap between phenotype and genotype it is essential to provide structured descriptions of phenotypes that can be interpreted in a consistent fashion. We have recently suggested ways of using combinations of ontologies to describe mouse phenotypes [3, 4] and provided tools that allow the storage, active updating and visualization of multiple ontologies [5]. Central to this approach is the establishment of standardized methods (SOPs) for the measurement of phenotypic attributes.

Phenotypes need to be defined and described within a formal ontological structure [3]. Clearly, the screen or assay of the mouse governs the phenotype that will be detected and therefore plays a central role in our proposed schema for phenotype representation. We are in the process of constructing the controlled vocabularies of assays required for implementation of this schema. EMPReSS is a valuable resource where these assays have been defined.

To facilitate the storage and processing of the SOPs we have created SOPML, an XML language that allows the description of SOPs. The SOPs generated by different institutes are automatically annotated using a SOP template and stored in SOPdb, the underlying database. The documents produced by this process are automatically validated against the XML schema and manually curated by a domain expert before being committed to the database. The combination of XML-based markup, RDF document metadata and XSLT transforms facilitates filtering, advanced indexing, searching and rendering of the information [6]. The SOPs will eventually be authenticated and validated against digital signatures, in accordance with the idea of a Semantic Web of Trust [6] (see Box 13.3 for further details).

Box 13.3 Details of the implementation of EMPReSS.

The system is composed of two tiers: the database backend SOPdb and a web client (EMPReSS). SOPdb was implemented in eXist (http://exist.sourceforge.net/), an XML database.

The EMPReSS Web client is made up of two types of JAVA servlet:

- Query servlets – connect to SOPdb, execute client queries and retrieve results via the SOPdb XML:DB API.
- Browsing servlets – connect to SOPdb and retrieve requested documents via the SOPdb REST-style HTTP interface.

XML documents are transformed to HTML presentations via XSLT (http://www.w3.org/TR/xslt) for browser visualization, whilst FOP (http://xml.apache.org/fop/) is invoked with XSLT for PDF format transformation.

13.6 The EMPReSS Browser

The browser is composed of four frames (Fig. 13.2): the Navigation, Search, Main, and Links frames (panel A of Fig. 13.2). Panel B illustrates the result summary page generated as a result of carrying out a search from the Search frame (indicated by the arrow). In the example, the page shows all SOPs found containing the term "behaviour". The page consists of the document name, number of hits and a link to view these hits. The number of hits displayed refers, depending on the type

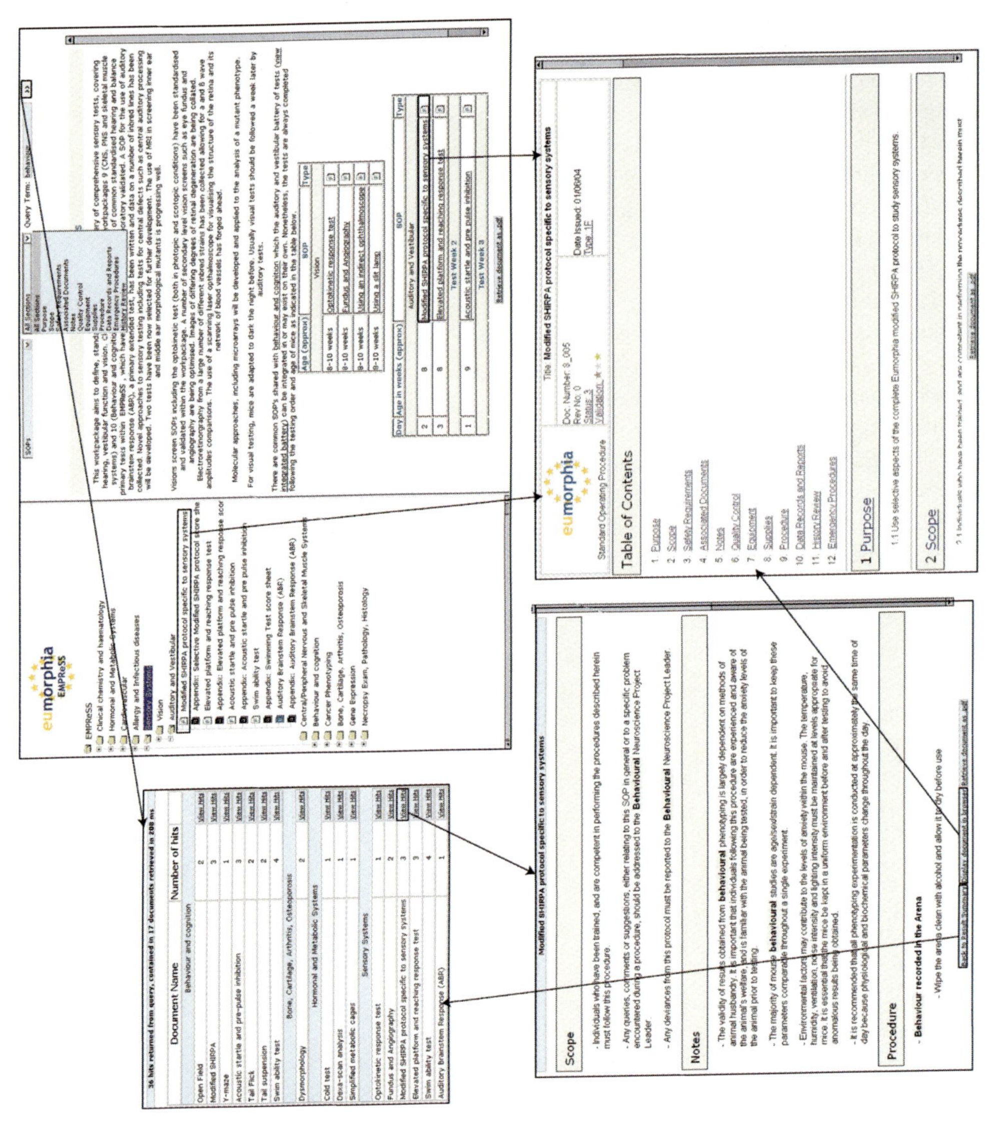

Fig. 13.2 EMPReSS browser.

of search, to either the number of sections within the document or the number of points within a named section. The query servlets group the returned documents according to phenotypic area and color-code them according to their type by accessing the appropriate metadata categories. Panel C, accessed via a link from the result summary page, shows the location of the search hits within the selected document. The relevant SOP can be displayed in HTML form (panel D). Individual SOPs can also be accessed from the Navigation frame, which contains a JavaScript-encoded hierarchical tree, with the documents categorized into phenotypic areas. This tree representation can be automatically generated from the assay vocabularies by custom parsers. Browsing a phenotypic area will result in its description being presented as well as its expansion to reveal its substructure. The description gives general and background information about the selected phenotypic area and denotes the usage of SOPs in phenotyping batteries in a table, for example age of mouse to be tested, order that the SOPs should be carried out, etc. These phenotyping battery tables, as well as any associated information, are automatically generated from the metadata carried in the XML documents.

EMPReSS is live on the EUMORPHIA website (www.eumorphia.org), available to both EUMORPHIA and mouse geneticists worldwide.

13.7 Future Work

EMPReSS represents the first stage in the development of a comprehensive set of phenotyping protocols by the EUMORPHIA consortium. Baseline data from the selected EUMORPHIA inbred strains acquired during SOP development and validation will be entered into the EMPReSS database. In addition, mutant phenome data acquired as part of protocol development and validation will also be entered into EMPReSS. The SOPs may be revised in the light of these results and they will be updated to improve their robustness and will include new techniques as they are developed. Subsequent updates of the system will present more specialized (secondary and tertiary) screens.

We are currently in the process of developing links between the EMPReSS resource, phenotype ontologies [4], assay-controlled vocabularies [4] and databases that contain phenotypic data (for published examples see [7, 8]). This will facilitate the automatic generation of phenotype descriptions and advanced mining, indexing and retrieval mechanisms. We are also considering ways of linking this dataset to that held by the mouse Phenome Project [9]. These efforts, coupled with tools such as CRAVE (Concept Relation Assay Value Explorer) which enables browsing of the schema for describing phenotypes [5], will facilitate the realization of a Semantic Web implementation [6] that is phenotype-aware.

Acknowledgments

The EUMORPHIA project is funded by the European Commission under contract number QLG2-CT-2002-00930.

References

1 International Human Genome Sequencing Consortium **2001**, Initial sequencing and analysis of the human genome, *Nature* 408, 860–921.

2 Mouse Genome Sequencing Consortium **2002**, Initial sequencing and comparative analysis of the mouse genome, *Nature* 420: 520–562.

3 Gkoutos, G. V., Green, E. C.J., Mallon, A. M., Hancock, J. M. and Davidson, D. **2004**, Building mouse phenotype ontologies, *Pac. Symp. Biocomput.* 9, 178–189.

4 Gkoutos, G. V., Green, E. C.J., Mallon, A. M., Hancock, J. M. and Davidson, D. **2004**, Using ontologies to describe mouse phenotypes, *Genome Biology*, 6, R8. The electronic version of this article can be found online at: http://genomebiology.com/2004/6/1/R8.

5 Gkoutos, G. V., Green, E. C.J., Greenaway, S., Blake, A, Mallon, A. M. and Hancock, J. M. **2005**, CRAVE: a database, middleware and visualisation system for phenotype ontologies, *Bioinformatics* 21, 1257–1262, http://bioinformatics.oupjournals.org/cgi/reprint/bti147v1.

6 Gkoutos, G. V., Murray-Rust P., Rzepa H. S., Wright M. **2001**, Chemical markup, XML, and the World-Wide Web. 3. Towards a signed semantic chemical web of trust, *J. Chem. Inf. Comp. Sci.* 41, 1124–1130

7 Strivens, M. A., Selley, R. L., Greenaway, S. J., Hewitt, M., Liu, X., Battershill, K., McCormack, S. L., Pickford, K. A., Vizor, L., Nolan, P. M., Hunter, A. J., Peters, J. Brown, S. D. **2000**, Informatics for mutagenesis: the design of mutabase-a distributed data recording system for animal husbandry, mutagenesis, and phenotypic analysis, *Mamm. Genome* 11, 577–583.

8 Blake, J. A., Eppig, J. T., Richardson, J. E., Davisson, M. T. and the Mouse Genome Database Group **2000**, The Mouse Genome Database (MGD): expanding genetic and genomic resources for the laboratory mouse, *Nucleic Acids Res.* 28, 108–111.

9 Bogue, M. **2003**, Mouse Phenome Project: understanding human biology through mouse genetics and genomics, *J. Appl. Physiol.* 95, 1335–1337.

Index

Standards of Mouse Model Phenotyping. Edited by Martin Hrabé de Angelis, Pierre Chambon, and Steve Brown

ISBN: 3-527-31031-2

d